Acta Zool. Fennica 204: 1–61 ISBN 951-9481-49-4 ISSN 0001-7299
Helsinki 3 December 1996

European Miocene Amphicyonidae — taxonomy, systematics and ecology

Suvi Viranta

Viranta, S., Division of Geology and Palaeontology, Department of Geology, P. O. Box 11, FIN-00014 University of Helsinki, Finland

Received 13 November 1996

Taxonomy, systematics and paleoecology of the Amphicyonidae (Mammalia, Carnivora) in the European Miocene was studied. The positions of *Pseudarctos* sp. and *Ictiocyon* sp. within the Amphicyonidae are questioned. Taxonomies of genera *Amphicyon* and *Pseudocyon* are revised, and a new genus, *Euroamphicyon*, is erected. Body-mass estimates and feeding ecologies of all the species are reconstructed. The family became extinct in Europe by about 8 Ma, although the main phase of the extinction took place in the Mid-Vallesian crisis, about 10.5 Ma. Changes in the structure of large mammalian community are revealed as the probable reasons for the amphicyonid extinction. It is suggested that reduced diversity of ungulates caused intensified competition in the carnivore community, and that amphicyonids were outcompeted by more specialized members of the community.

Contents

1. History of the study of amphicyonids

1.1. Introduction

The Amphicyonidae is an archaic carnivore family that originated in the Late Eocene and became extinct in the Late Miocene. During the Oligocene it was a diverse family in both the New and Old Worlds. Its species diversity was reduced in the Miocene, but it still formed a vital part of the carnivore guilds of that period. Amphicyonid ecologies varied from omnivory to hypercarnivory, and body masses ranged from few kilograms up to hundreds of kilograms.

Amphicyonids have been given the popular name "bear dogs" in reference to their combination of bear- and dog-like characters. However, although they evidently belong in the Caniformia there is no consensus regarding their relationship to either of these families (see below). Some general characteristics of the group are short limbs, a long tail, a primitive dentition with retained third molars, an only slightly inflated auditory bulla which does not extend posterior to the mastoid process, and absence of a suprameatal fossa.

1.2. Systematic position

Trouessart identified the amphicyonids as a taxonomic unit in 1885. Ever since, their systematic position has been a matter of debate. Trouessart's opinion was that they were a subfamily within the family Canidae.

Schlosser (1888) considered postcranial characters of some European amphicyonids, and concluded that the amphicyonids should be placed in a family of their own, Amphicyonidae. In his paper on "the bears and bear-like forms of the European Miocene" Schlosser (1899) suggested a possible common ancestor for the canids, ursids, and amphicyonids in the Eocene, from which the ursid line later split. His main argument was based on the upper molars. Ursids have an enlarged metaconule and reduced paraconule in the upper molars, and this results in broad talons. Neither canids nor amphicyonids have a posterior extension in the talons of their upper molars. The canid and amphicyonid lines were separated from their common ancestor in the Eocene–Oligocene. Schlosser's views were not fully accepted, however, and both American and European authors continued to include amphicyonids in either the Canidae or the Ursidae. Viret (1929a) placed the amphicyonids of La Grive in the family Amphicyonidae (famille des Amphicyonidés) with the exception of *Agnotherium grivense*, which he placed in a separate family Cynodontidae (famille des Cynodontidés). In his second publication on the La Grive Carnivora (Viret 1951) he moved all the amphicyonids to the family Canidae. He also placed Hemicyoninae (commonly considered a subfamily of Ursidae (Kretzoi 1937)) in the Canidae. In his study of the Carnivora of the upper Stampian (lower Oligocene) Helbing (1928) included the amphicyonids *Haplocyon, Cephalogale* and *Amphicyon* in the Canidae basing his decision on a study by Matthew (1924).

In this study of the Snake Creek Fauna in Nebraska, North America, Matthew (1924) presented a phylogeny that placed the amphicyonids in the family Canidae. He included the Oligocene genus *Daphoenus*, and the Miocene genera *Daphoenodon*, *Amphicyon* and *Pliocyon* in the "amphicyonine dogs" based on a series of shared features. They share a progressive enlargement of the molars and reduction of the premolars and carnassial, and the lower carnassial has a well-developed metaconid. Their skulls are elongate, with a high sagittal crest. Their tympanic bullae are small and limited to the anterior part of the otic region, and the space between them and the paroccipital is wide with a prominent lateral protrusion of the mastoid. Their cervical vertebrae are relatively short, the dorsals small with high spines, the lumbars large and moderately long. The ribs are short. The limbs are machairodontine-like; short and massive. In Matthew's (1924) view, canid phylogeny includes three major groups: Amphicyoninae, Simocyoninae, and Caninae. The Amphicyoninae separated from the other two already in the Early Oligocene. Matthew also discussed the possible relationships between ursids and amphicyonids.

The hypothesis that amphicyonids are most closely related to ursids was first proposed by Hough (1948a), and later supported in studies by Ginsburg (1966, 1977), and Olsen (1960). Hough based her argument on the bulla structure of amphicyonids, whereas Ginsburg and Olsen mainly emphasized postcrania. Hough (1948a) reported five ursid characters in amphicyonids: (1) the bullae are small; (2) the position of the promontorium is similar to that of ursids; (3) the form and position of the mastoid process, paroccipital process, and postglenoid process are similar to those in ursids; (4) the basioccipital is broad as in ursids; (5) the carotid circulation seems to have been similar to that of ursids. Ginsburg (1966) argued that the auditory region, skeletal characters, and size dimorphism observed in amphicyonids lend support to their close relationship with ursids. He also argued that the similarities seen in the dentitions of canids and amphicyonids are primitive retentions. In 1977, Ginsburg based his arguments on cranial blood circulations as well as limb morphology. He stated that the amphicyonid limbs are ursid-like with felid-like adaptations and that also the auditory region, carotid circulation, and intracranial venous system are like those in ursids. Olsen indicated that the general robustness of the limb elements and, especially, the morphologies of their distal parts (metapodials and phalanges) in amphicyonids resemble those of ursids, and separated these two groups from the other Carnivora on this basis. Heizmann (1973) placed the Steinheim amphicyonids in the Ursidae, but he didn't entirely reject the possibility of the existence of a distinct amphicyonid family.

Hough (1948b), apparently unaware of Schlosser's (1888) suggestion of a family Amphicyonidae, proposed family rank for the amphicyonids and named them Daphoenidae after the first Oligocene amphicyonids. Her main argument was with regard to the amount of variation: "the variation within the group is certainly more analogous to that of a family than a subfamily" and "the most pronounced difference between Daphoenidae and extant families is the range of brain capacity". De Bonis (1966, 1969) reintroduced the concept of the family Amphicyonidae. He noted the similarities between canids and amphicyonids, but argued for two different evolutionary lines, and thus for two different families. During the Oligocene amphicyonids resembled canids in their dentition, but in the Miocene they gained more ursid-like characters, such as robust canines, which De Bonis considered to have no phylogenetic significance.

Hunt (1974a, 1977) brought in further craniological evidence for treating the amphicyonids as a separate family. Hunt studied bulla structures of extant and extinct carnivores, and found that the Oligocene ursids, for example *Cephalogale*, had a bulla that fully enclosed the auditory region and apparently was made up of several fused elements. The Oligocene amphicyonids, in contrast, had more primitive bullae consisting of only a single preserved bone. The Miocene amphicyonids had a more advanced bulla structure similar to that of some living taxa, for example the ursids. Hunt used this as evidence for a divergence of the amphicyonids and ursids in the Oligocene. He also suggested that the foreshortening of the postauditory region distinguishes the family Amphicyonidae from the Canidae and Ursidae. The anatomy of the basicranium in the earliest true canids is nearly identical to that of the living canids. The ossification of an entotympanic bulla and the development of a low septum in the caudal entotympanic separates the early canids from their miacid ancestors (Wang & Tedford 1994).

Hunt (1974b) clarified the position of daphoenids by placing them in a subfamily, Daphoeninae, of the family Amphicyonidae. The Daphoeninae differed from the Amphicyoninae in their M2 and M3, which were not enlarged as in the Amphicyoninae, and in their anterior premolars which were not reduced as in most Miocene Amphicyoninae.

Springhorn (1976), studying two European Oligocene amphicyonine skull fragments, noted their close similarity to daphoenine skulls. He also argued for the family Amphicyonidae, but questioned its classification within the Arctoidea. He found support in the structure of the amphicyonid tympanic bullae and venous circulation for a close relationship with the Viverridae. Both Hough (1948b) and De Beaumont (1964a) had noted the primitive, viverrid-like auditory region of the amphicyonids. They proposed that the Early Oligocene *Cynodictis* was a probable ancestor to the amphicyonids. The small and anteriorly placed bulla is a primitive character, and not very useful for phylogenetic studies.

Radinsky (1980) studied cranial endocasts of different amphicyonids. He concluded that the brains of amphicyonids show affinities with the canids and ursids. The amphicyonids share an ectolateral sulcus with the canids, while a carotid loop in the inferior petrosal sinus ("ursid loop") is shared with the ursids. The ectolateral sulcus occurs in some Recent mustelids, as well as in the canids, and is associated with the secondary visual field. The carotid loop is only known in the ursids. Radinsky noted that at least one of these two characters must have evolved at least twice.

Flynn *et al.* (1988) showed that the osteology of the amphicyonids is close to the primitive arctoid morphology. They pointed to characters such as the elongated skull, massive skeleton, long tail, and presence of the M3 as evidence of this. The "ursid loop" of the internal carotid and post-scapular (subscapularis) fossa are synapomorphies shared with the Ursidae. In the phylogenetic trees presented by Wyss and Flynn (1993), the Ursidae and Amphicyonidae form a superfamily Ursoidea. The Amphicyonidae differ from the Ursidae in having a reduced and anterior paracone in the upper carnassial, an M3 and an unossified entotympanic. They noted that the unossified entotympanic is true only for the daphoenine amphicyonids. The diagnostic characters of the Ursoidea are the "ursid" loop in the basioccipital and the presence of m3. The Ursoidea was placed in the Ursida together with the Pinnipedia on the basis of the equal width of the palatine anteriorly and posteriorly, absence of a muscular process on the malleus, absence of an enlarged talonid in m2, absence of a hypocone in M1, unenlarged metaconules in the upper molars, reduced tail (with a reversal in amphicyonids), and conglomeratic kidneys (unknown in the Amphicyonidae). Ursida together with the Procyonidae, *Ailurus*, and the Mustelidae form the Arctoidea, which is a sister group to the Canoidea.

On the other hand, Wolsan (1993) separated the Amphicyonidae from the Mustelidae and Ursidae (his Arctomorpha) on the basis of the presence of the M3 and the absence of the suprameatal fossa in amphicyonids. He excluded the "ursid loop", presence of m3 and subscapularis fossa from his discussion.

In the recent literature on North America the family status of the amphicyonids seems to be widely accepted (e.g. Martin 1989), while in Europe some authors still consider amphicyonids to be canids (e.g. Alcala *et al.* 1989–1990) or ursids (e.g. De Beaumont 1984).

In this study I accept the view that amphicyonids are most closely related to ursids. This has its strongest support in the specific cranial circulations shared by these two groups. Although characters such as the lack of anastomoses between the extracranial and intracranial circulations and the arrangement of the cephalic arteries are primitive in ursids (Flynn *et al.* 1988), the fold of the internal carotid artery ("ursid-loop") is a derived character and is shared only by the ursids and amphicyonids. Another derived character of importance is the subscapularis fossa, which is also found only in the amphicyonids and ursids. Dental characters such as the absence of a hypocone in the M1, which the amphicyonids share with the ursids and not with the canids, are more difficult to evaluate because these vary even within families (some canids lack a hypocone), and seem to have evolved independently several times. Most of the arguments for amphicyonid-canid affinities are based on dental characters, such as the P4 cusp patterns. Other proposed characters, such as a long rostrum are clearly primitive. The most striking similarity between canids and amphicyonids is the presence of the ectolateral sulcus. In addition to the amphicyonids and canids, this is also present in some mustelids. According to Radinsky (1980), it may reflect an expansion of the visual cortex. Although it is clearly a derived character, it has appeared independently at least in some mustelid genera and it is likely to have appeared independently in the amphicyonids as well. Work on the basicranial anatomy of early canids and miacids by Wang and Tedford (1994) has recently clarified the phylogeny of the late Eocene Caniformia. They showed that already from that period the Cynoidea can be distinguished from the Arctoidea (represented by *Daphoenus* spp. in their study) on the basis of the entotympanic bulla (ossified in the Canidae) and promontorium (flattened in the Canidae). In addition, the "ursid loop", an enlarged basioccipital embayment, is already present in the Arctoidea.

It is also the position taken here that the amphicyonids form a family of their own, the Amphicyonidae. The primitive auditory bulla, which was ossified only in the later forms shows that their evolutionary history has been separate from the canids and ursids. Also, the fact that the postcranial bones retain many characteristic conditions such as an entepicondylar foramen in the humerus and short metapodials argues for family distinction.

1.3. Systematic studies of European amphicyonids

Kuss (1965) published an extensive study of the late Oligocene and Miocene amphicyonids of Europe. His focus was mainly on biostratigraphy, and he erected many subspecies in order to have a specific amphicyonid for each stratigraphic unit. Many of his new groupings have turned out to be arbitrary. Variation within amphicyonid species is considerable, and many of the characters he used to separate populations taxonomically are only normal variation observed within populations. For example, based on enhanced cinguli and an enhanced m1/m2 length ratio he distinguished the Pontlevoy and Eibiswald amphicyonids from *A. giganteus* and *A. major,* and placed them in a different genus and species, *Arctamphicyon tolosanus eibiswaldensis.* This material was earlier attributed to *Amphicyon major* or *A. giganteus*, and later placed in *A. giganteus* by Ginsburg and Telles Antunes (1968).

On the other hand, some of his subspecies have subsequently been made into species. This is true, for example of his *Amphicyon major gutmanni,* which Kretzoi (1985) placed in a species of its own, *Amphicyon gutmanni*, as was originally the case in Kittl (1891). The taxonomic status of *A. gutmanni* is difficult to ascertain, because the species is represented by only a single specimen, a lower carnassial tooth. The peculiar shape of the talonid of this carnassial separates it from other known species. Kuss' *Pseudarctos bavaricus* subspecies, *P. b. socialis* and *P. b. dehmi,* were later placed in a different genus, *Ictiocyon*, by Ginsburg (1982) on the basis of the presence of the p1 (absent in *Pseudarctos*), the enlarged talonids of m1 and m2, and the higher cusps on the upper molars.

The Paleogene radiation of European amphicyonids was thoroughly revised by Springhorn (1977a). The European Paleogene amphicyonids were divided into two tribes, Amphicyonini and Haplocyonini (Ginsburg 1966). Springhorn (1976, 1977a) pointed out the similarity between the American Daphoeninae and the European Haplocyonini, and suggested that the European Oligocene Amphicyonidae should also be divided into the two subfamilies Amphicyoninae and Daphoeninae, with Haplocyonini forming a tribe within the Daphoeninae. The Haplocyonini is characterized by absent or reduced diastemata between the premolars and by a lower carnassial with a high trigonid and low talonid. The occlusal contour of the upper molars differs from that of the Amphicyoninae in being more symmetric. Springhorn (1977b) mentioned that the limb bones of the Daphoeninae were felid-like. Only a few postcrania of the European Daphoeninae have been described. Helbing (1928) attributed a distal humerus fragment, a radius, ulna fragments, a scapholunar, a second metacarpal and a phalanx to *Haplocyon dombrowskii* Helbing, 1928. The humerus, which lacks an entepicondylar foramen, is in many ways similar to those of Recent canids. The radius is short with a wide proximal articulation which extends over the shaft. The radial tuberosity is strong as in felids. The distal facet for the ulna is twisted. The radius differs from both canids and felids in having a styloid process directly underneath the shaft. In canids and felids the styloid process is situated more radially. The groove for the extensor pollicis brevis and extensor pollicis longus is deeply excavated. The ulnar side of the bone has a strongly roughened area for the insertions of the extensors. A well-developed insertion for the brachioradialis gives a peculiar outline to the radius. The ulna is felid-like with a massive olecranon. The scapholunar has a distinct facet for the trapezium, which indicates that this species retained a proper first metacarpal. The second metacarpal is relatively long, differing in this respect from the those of the Amphicyoninae. The phalanx is very much like in the Recent *Canis*. More postcrania are known for the American Daphoeninae. Hough (1948b) found that the American Daphoeninae are in many ways similar to the Recent lynx (*Lynx lynx*) in the appearance of their postcranial skeleton. For example, the entepicondylar foramen is present in all the known Northern American daphoenine humeri. She also found that the metacarpals are of the same relative length and proportions in the lynx and the daphoenine species. The metatarsals of the Daphoeninae are shorter.

Of the two Paleogene subfamilies, the Daphoeninae in Europe consisted of only four genera, whereas the Amphicyoninae consisted of six genera in the European fauna. The earliest known amphicyonid in Europe is a daphoenine, *Simamphicyon helveticus,* from the Upper Eocene (Bartonian). Amphicyonids became diverse during the Lower-Middle Oligocene (Sannoisian-Stampian), and their diversity stayed high throughout the Oligocene (Springhorn 1977a). A high amphicyonid diversity with forms like those in the Oligocene persisted into the Early Miocene (Ginsburg 1965, 1966, Rothausen 1988).

There seems to be an enhanced turnover rate among amphicyonids in the Early Miocene. All the haplocyonine species and almost all the amphicyonine genera (with the exception of *Cynelos)* disappeared in the Early Miocene (Springhorn 1977a). Even the new species of *Cynelos* show clear morphological changes in the Early Miocene.

The present study concentrates on the Middle Miocene Amphicyonidae. This was the amphicyonid fauna that originated during the large turnover in the Early Miocene and lasted until the Late Miocene. All the Daphoeninae went extinct prior to or during this turnover. The amphicyonid fauna included new species of Amphicyoninae and species of a new subfamily, the Thaumastocyoninae (Hürzeler 1940). The new amphicyonine species that appeared in the Early Miocene were in many respects similar in Europe and North America. Among other features they share enlarged (hypertrophic) posterior molars (Hunt 1972). The Thaumastocyoninae is a rather poorly known group, which appears to have had functional morphology similar to that of the Daphoeninae. They have been described as feloid canids, mainly because of the reduced talonid of the lower carnassial and the reduced molars.

Due to a lack of comparative studies, the relationships between American and European forms are poorly known, and, for example, Springhorn (1977a) mentioned that American *Amphicyon* spp. is not identical to European *Amphicyon* spp., but he didn't provide a basis for this argument. Of the earlier authors, Matthew (1924) stated that the European *Amphicyon* (e.g. *A. major)* and the American *A. sinapius* appear very similar, and despite their geographical distance must be closely related. He especially noted the similarity in the upper molars and the reduced premolars of these two species. Hunt (1977) studied the basicranium of European *Cynelos* and observed that the basicranial structure of this form is similar to that of the American forms. Besides these observations, very few attempts have been made to compare the American amphicyonids with European forms. The amphicyonid material of North America is much more abundant than that of Europe.

1.4. The simocyonid question

An additional factor in understanding the history of the amphicyonids is the simocyonids. These are small- to medium-sized carnivores that appear in the fossil record of Europe during the Early Miocene and survive until the Late Miocene. They had a shortened snout and a derived dentition with reduced premolars. They have variously been placed in the Hyaenidae, Canidae, Procyonidae, Mustelidae and Amphicyonidae.

Zdansky (1924) included amphicyonids and simocyonids in the Canidae. Matthew (1924) also considered *Simocyon* to belong in the Canidae, placing it in the group Simocyoninae with the living *Icticyon* (a junior synonym of *Speothos) Cyon* (correct spelling *Cuon*) and *Lycaon*. The Simocyoninae is characterized by unicuspid m1 and reduction of the tubercular parts of the dentition. In addition to the Simocyoninae Matthew's canid phylogeny consisted of the Amphicyoninae and Caninae. Viret (1951) placed the simocyonid from La Grive (*Alopecocyon goeriachensis*) in the Canidae, while Pilgrim (1931) indicated affinities with the Amphicyoninae. Pilgrim pointed out that the stoutly built m1, broad M2, low-cusped m2, small, depressed auditory bulla, prominent mastoid process, high sagittal crest, wide zygomatic arches and wide glenoid processes are common characteristics of the simocyonids and amphicyonids. Based on these features he argued for a common subfamily for these two groups.

De Beaumont (1964b) united the simocyonids (*Alopecocyon, Simocyon*) with *Broiliana* found at Wintershof-West (MN 3) to form a subfamily Broilianinae which he placed in the Mustelidae. He showed that these three genera are similar in the morphology of their carnassials and upper molars, the elongation of the m2, and the dental formula. He found affinities to the Mustelidae in the simocyonid skull. The alisphenoid canal, auditory bulla, basicranium, angular process of mandible and jugal arch are similar in Broilianinae and the other Mustelidae. Schmidt-Kittler (1981) accepted the view of De Beaumont (1964b) and placed *Alopecocyon* and *Simocyon* in his superfamily Musteloidea, in a group consisting of animals with a procyonid-like middle ear.

Wolsan (1993) has again argued for a position distinct from the amphicyonids for the simocyonids. His argument was mainly based on the absence of M3 in the simocyonids, along with the presence of a suprameatal fossa, which is absent in the amphicyonids (Wolsan, pers. comm.). Wang (1996) brought in further evidence for a distinction of the simocyonids from the amphicyonids. He studied Chinese simocyonid material and found affinities to the Recent red panda (*Ailurus*) in characters such as the long bony external auditory meatus and the ventrally ridged paroccipital process. These are characters not found in amphicyonids.

Based on the above mentioned studies a hypothesis of the simocyonids being amphicyonids can be rejected, and simocyonids are omitted from the present study.

1.5. Studies of the functional morphology of amphicyonids

The family Amphicyonidae was a diverse group with many different ecological adaptations. Many of the European species are known only from dental material, and for these species reconstruction of the mode of locomotion is almost impossible although in some cases reconstructions can be attempted with reference to closely related taxa. Two complete amphicyonine skeletons are known from the European Miocene; an *Amphicyon major* female from Sansan and an *Amphicyon* sp. from Cassots, Valles Pénedes. A report on the former was published by Bergounioux and Crouzel (1973) and the latter is currently under study by Lluis Rius in the Institut Paleontològic Dr. M. Crusafont in Sabadell. Ginsburg (1961) studied the abundant *Amphicyon major* material from Sansan, but not the complete skeleton, which was found later.

One of the first proper comparisons between living carnivore and amphicyonid skeletons was made by Olsen (1958, 1960) when he studied an American Miocene amphicyonid species, *Amphicyon longiramus*, from Thomas Farm, California. He noted that the skull is like that of a canid in shape and proportions. The high sagittal crest, with two compact layers of bone for the attachment of the temporalis muscle of *A. longiramus* is more strongly developed than in Recent canids or ursids. The dentition approaches that of canids in shape and proportions, and accordingly the morphology of the upper and lower jaw regions is very similar to that of *Canis*. Olsen claimed that the braincase was relatively small, but Radinsky (1980) in his study of the endocasts of amphicyonids found that the encephalization quotients (indices which are calculated from the relationship between brain size and body mass) in earlier amphicyonids were in the range of living carnivores, and in the later amphicyonids were toward the high end of the range.

Olsen (1960) found many felid characters in the vertebral column of *A. longiramus*; the spines of the thoracics indicate strong extensor muscles of the back, and the muscle attachments of the lumbars indicate both strong extensors and flexors. The sacrum is stout with fused neural spines as in felids or canids. The strong fusion of the sacral vertebrae probably served to support the muscles used in controlling a long tail composed of 28 vertebrae, which is as many as found in the longest carnivore tails today. The number of caudal vertebrae in living felids varies between 14 and 28 (Ewer 1973). Olsen suggested that the subscapularis fossa present in bears and amphicyonids may indicate a capacity for climbing; Davis (1964) has pointed out that the short muscles (the subscapularis minor, teres minor, and posterior head of the triceps longus) which insert in the subscapularis fossa are important in fixing the joint, and strong fixing may be needed in climbing. Whether any climbing was actually done by amphicyonids is difficult to prove from the skeleton alone. The shape of the ungual phalanges has been shown to correlate with climbing behaviour so that today's climbing carnivores have curved and dorsoventrally deep claws (Van Valkenburgh 1985, 1987). Van Valkenburgh (1987) included a daphoenine amphicyonid, *Daphoenus vetus* in her analyses. This animal showed pronounced curvature of the ungual phalanges, and thus probably climbed. Olsen (1960) noted that the ungual phalanges of *A. longiramus* are similar to those of bears. He used the American black bear (*Ursus americanus*) as his comparative bear. In the analyses of Van Valkenburgh (1987) this species has moderate ungual depth, but relatively curved claws (approaching those of felids). The black bear is an agile climber. In the other limb and foot bones Olsen (1960) found bear-like characters and no similarities to felids.

The morphology of *A. major* from Sansan described by Ginsburg (1961) and Bergounioux and Crouzel (1973) is very similar to that of *A. longiramus* (Olsen 1958, 1960). Ginsburg concluded that *A. major* was an accomplished predator of the size of the Recent American grizzly bear. He pointed to the well-developed molars and their horizontal abrasion, typical of bone-crushing carnivores like the Recent hyenas, and the long canines similar to those of the Recent large felids. The body was long, and the limbs short and stout, with the feet pentadactyl with non-retractile claws and very short phalanges. The stance was probably plantigrade. He suggested an ecological role similar to that of the Recent lion for *A. major* in the Miocene of the Pyrenées.

The work by Bergounioux and Crouzel (1973) on the *A. major* skeleton from Sansan agrees in most respects with that of Ginsburg. They concluded that the mixture of felid, canid and ursid characters seen in the skeleton indicated that *A. major* was an opportunistic carnivore.

Despite the great range of feeding adaptations found among the Amphicyoninae, the postcranial morphology of the known species seems to have been fairly conservative. The short distal limb segments described above were apparently common features of the Amphicyoninae. This has been taken to indicate non-cursorial, forest-dwelling habits. Bakker (1983) claimed that of the American Amphicyonidae the *Daphaenodon-Pliocyon* lineage produced running predators. He based this on the shape of the trochlea of the astragalus, which became deep in later *Pliocyon* spp. In her analysis of locomotor diversity in past and present carnivore guilds, Van Valkenburgh (1985) showed that the femoral/metatarsal ratio of an American Oligocene amphicyonid, *Daphoenus vetus*, was below all the Recent terrestrial carnivores (except *Ursus arctos*) and closer to arboreal or fossorial ones. This, together with the ungual phalanx shape, olecranon length and metacarpal/phalangeal index indicated that this species had scansorial habits. Olsen (1960) mentioned the possibility that the presence of an entepicondylar foramen in the distal humerus in both felids and amphicyonids indicated a crouching habit. The carnivorous amphicyonids probably used crouching and pouncing when hunting.

Only a few postcranial finds have been attributed to the Thaumastocyoninae. They include a fifth metatarsal and the proximal end of a radius from Bled Douarah (Tunisia) described by Kurtén (1976). The massive radius shows an orientation of the proximal articular surface similar to that of canids, and different from amphicyonids. In amphicyonids the articular surface is oblique to the long axis of the bone, whereas in this specimen (and in canids) they lie at right angles. The fifth metatarsal was only tentatively attributed to the Thaumastocyoninae. It is very long, and again shows more dog-like and cursorial conditions than found in the Amphicyoninae. It has a large articular surface for the cuboid and a poorly developed lateral protrusion at the proximal end. Kurtén concluded that this animal was very large, heavy and powerful, and preyed on large, slow-footed game like rhinoceroses, or even mastodonts. Crusafont and Kurtén (1976) suggested that a calcaneus found at Can Ponsich might belong to a thaumastocyonine. This specimen has an unusually great distance between the astragalar and cuboid facets, and a well developed groove for the flexor digitorum. Crusafont and Kurtén interpreted those characters to indicate a digitigrade stance. The fifth metatarsal from Bled Douarah also clearly indicates a digitigrade stance.

A rare piece of information on amphicyonid behaviour was provided by Hunt *et al.* (1983), when they described excavated dens with partial skeletons of four *Daphoenodon superbus* in a Miocene floodplain. According to Hunt *et al.* (1983) dens were used for protection of the young and as shelter for the adults. Remains of a small canid, mustelid and a temnocyonine amphicyonid were also found in the dens. It was concluded that these carnivores were probably occasional occupants of the dens. They were not likely to have been carried in by the *Daphoenodon superbus* as prey, because no remains of any ungulate or other suitable prey species were found, and thus *D. superbus* probably didn't bring carcasses to the dens.

1.6. Studies on amphicyonid extinction

Amphicyonids became extinct in North America before the Early Hemphillian (Hunt 1996) in Africa a little later than this (6.2–5.5 Ma, Werdelin, pers. comm.), and in Asia no later than 7 Ma (Siwaliks formation, Barry & Flynn 1989). The last European amphicyonids are about 9 to 7 million years old (MN 10 and 11). The last Thaumastocyoninae are around 10 million years old (MN 9).

Early explanations for amphicyonid extinctions have been based on their archaic appearance, and, for example, Schlosser (1899) believed that the amphicyonid adaptations to omnivory led to inflexibility and that this was the reason for their extinction. Additional fossil finds and further studies have shown a diversity of feeding adaptations among Miocene amphicyonids, so they cannot be all considered omnivores. On the contrary, many of the latest amphicyonids show hypercarnivorous adaptations (e.g. Gabunia 1973, Kurtén 1976). Werdelin (1996b) suggested that the decline in the amphicyonid diversity in the Middle and Late Miocene in Europe and Western Asia reflected the environmental changes of the time, and they were replaced by omnivorous forms on the one hand, and cursorial predators on the other.

The present study is an attempt to approach the extinction of amphicyonids by clarifying their phylogeny and studying the ecology of the Miocene amphicyonids of Europe.

2. Taxonomy

2.1. Material and methods

The fossil material for this study was studied in the following institutes:

— Institut Paleontologic "M.Crusafont", Sabadell (Barcelona)
— Museo Nacional de Ciencias Naturales, Madrid
— Muséum National d'Histoire Naturelle, Paris
— Naturhistorisches Museum Basel
— Forschungsinstitut und Naturmuseum Senckenberg, Frankfurt am Main
— Staatliches Museum für Naturkunde, Stuttgart
— Universitäts-Institut und Staatssammlung für Paläontologie und historische Geologie, München
— Naturhistorisches Museum Wien, Geologisch-Paleonto-

logische Abteilung
— Institut für Paläontologie der Universität Wien
— Magyar Állami Földtani Intézet, Budapest.

These collections do not cover all the Middle and Late Miocene amphicyonid remains from Europe, but they represent the known amphicyonid fauna well. Relying on earlier publications, some additional specimens from other collections have also been included in the study. These are indicated in the text, and the relevant literature is cited.

The study was restricted to the Middle and Late Miocene species. For the species that appeared in the Lower Miocene and survived to the Middle Miocene the early specimens were also included and studied, when possible. The taxonomic decisions and the systematic analysis were mainly based on dental, and in few cases also on cranial material. Because of their scarcity, and also because of the uncertain associations of post- cranial with dental material, postcrania were omitted from the taxonomic study. Characters of the skeleton are, however, discussed in the sections dealing with amphicyonid paleobiology.

In general, only material from localities that have been assigned to an MN unit was included in the study. The MN (Mammals, Neogene) zonation is a common system in Europe, and are based on entries and exits of specific mammalian taxa (Mein 1975, De Bruijn *et al.* 1992). It is a compromise between a true biozonation and a biochronology based on the stage of evolution (De Bruijn *et al.* 1992). The MN zonation is widely used, and currently it is the most useful correlation system for mammalian fossil localities in Europe, where there are many localities without any stratigraphic context (e.g. fissure fillings). The main range of the present work is MN 4 to MN 10, which equals the late Astaracian to Vallesian in European faunal units, and 18–8.9 Ma on the geochronometric scale (Steininger *et al.* 1990). Some later, thus Turolian, amphicyonid specimens exist and are included in the present work. In cases where an amphicyonid species is only known from localities without an assignment to an MN unit the material was included with comments about the apparent stratigraphic position of the locality. The localities are assigned to MN units as in De Bruijn *et al.* (1992) unless otherwise indicated.

In the phylogenetic analyses, parsimony analysis was used (see Farris 1983, also Werdelin & Solounias 1991). I used four different programs. More information on the analyses are provided in Section 3.

The comparative material for body mass estimates was measured in the following museums: the American Museum of Natural History (AMNH), Museum of Comparative Zoology, Harvard, Boston (MCZ). the Finnish Museum of Natural History (FMNH), the Zoological Museum in Copenhagen (ZMUK); the Zoological Museum at University of Oulu, and the University of Maine, Department of Anthropology (UM). The following literature was used for body masses and head to tail lengths: Nowak (1991), Pulliainen (1973) and Van Valkenburgh (1990).

Most of the dental measurements of the amphicyonid material were taken in the museums listed above, except for the following taxa which were taken from the literature: *A. caucasicus* (Gabunia 1973), *Gobicyon macrognathus* (Colbert 1939, Pavlović & Thenius 1959) and *Hubacyon pannonicus* (Kretzoi 1985).

2.2. Taxa

Amphicyonidae incertae sedis

Genus *Pseudarctos* Schlosser, 1899

Type and only species. — *P. bavaricus* Schlosser, 1899.

Pseudarctos bavaricus Schlosser, 1899

Type specimen. — 1898I35, Staatssammlung für Paläontologie und historische Geologie, München, Germany.

Type locality. — Flinz de Tutzing, Germany.

Synonymy. — *Pseudarctos bavaricus beaucensis* nov. ssp., *P. b. pontignensis* nov. ssp., *P. b. bavaricus* Schlosser, 1899 in Ginsburg (1992) and references therein.

Occurrence. — *Pseudarctos bavaricus* is known from Austria (Göriach), Czechia (Neudorf-Sandberg), France (Baigneaux-en-Beauce, Castelnau-d'Arbieu, Esvres, La Grive St. Alban, La Romieu (Pellechaus), Sansan, Simorre), Germany (Erkertshofen 2, Flinz de Tutzing, Gau Weinheim, Häder), and Switzerland (Anwil).

Stratigraphic range. — MN 5–MN 9.

Referred material. — Göriach (housed in the Naturhistorisches Museum Basel): T.O.965, mandibular fragment with m1–m3 sin; T.O.965, maxillary fragment with M1–M3 sin; T.O.905, M1 dex. Anwil (housed in the Naturhistorisches Museum Basel): AL173, M2 sin; AL173, m3 dex. Häder (housed in the Naturhistorisches Museum Basel): No number, M1 sin.

Other data. — Ginsburg 1992.

Description of the type. — The holotype is a right mandibular fragment with p4 to m3 preserved. The p4 is a unicuspid oval with a conspicuous cingulum. The m1 has a large protoconid, a small and low paraconid and a posteriorly placed, high metaconid. The talonid is broad with three low cusps. The buccal border of the talonid is high. The m2 has a short trigonid with a high paraconid. The m3 is round and single rooted. None of the lower molars has a cingulum.

Extended description. — The dentition of the mandible from Göriach is very similar to that of the type mandible. The m1 has well-developed trigonid cusps and a very flat talonid with low cusps. Even the hypoconid is relatively low. The margins of the talonid are slightly serrated. In the maxilla of *P. bavaricus* from Göriach the M1 and M2 are quadratic in outline because the talon is as wide as the trigon. The paracone is as large as but a little higher than the metacone. The talon is lower than the trigon. The M3 is a small tooth with a single cusp. The M1 from Häder is slightly smaller than that of Göriach, but similar in morphology.

Genus *Ictiocyon* Crusafont, Villalta & Truyols, 1955

Type and only species. — *Ictiocyon socialis* (Schlosser, 1904).

Ictiocyon socialis (Schlosser, 1904)

Type specimen. — 1904V1, Universitäts-Institut und Staatssammlung für Paläontologie und historische Geologie, München, Germany.

Type locality. — Solnhofen-Spalte, Germany.

Synonymy. — *Ictiocyon socialis* (Schlosser, 1904) in Ginsburg (1992) and references therein.

Known localities. — *I. socialis* is known from France (Artenay), Germany (Wintershof-West and Solnhofen-Spalte) and Spain (El Canyet). Solnhofen-Spalte was assigned to MN 4 by Ginsburg (1992).

Stratigraphic range. — MN 3–MN 4.

Referred material. — Wintershof-West (housed in the Universitäts-Institut und Staatssammlung für Paläontologie und historische Geologie, München): 1937II12301, mandibular fragment with p4–m1 sin.

Other data. — Ginsburg 1992.

Description of the type. — The holotype is a right mandible with m1 to m2 (Schlosser provides figures of a mandible with a complete dentition from p1–m3, but the figured premolars and the m3 were found isolated). The m1widens greatly posteriorly. The paraconid is very small, the protoconid large, and the metaconid well developed and posteriorly situated. The talonid lacks a hypoconulid. The hypoconid, which is lingually shifted, and the entoconid are of about the same height. The talonid is surrounded by a cingulum. The m2 is a large tooth surrounded by a cingulum. It is as wide as the talonid of the m1 and about the length from the anterior edge of the m1 paraconid to the posterior end of that tooth. The m2 is large in relation to the m1 as compared with the situation in *P. bavaricus.*

Extended description. — The Wintershof-West material consists of a left mandible with p4 and m1 preserved. The mandible is broken at the p1 alveolus, and there are alveoli for all the premolars. There is no spacing between the teeth and the mandible is relatively short. The p4 is unicuspid. The alveoli show that the p3 and p2 had two roots and that the p1 was one-rooted. Ginsburg (1992) described some isolated teeth of *I. socialis* from Artenay. This collection also includes upper teeth. The P4 is a stout tooth with a relatively short metastyle, large paracone, and small protocone. The anterior margin of the protocone extends beyond that of the paracone. There is no parastyle blade. All the upper molars, as well as the P4, are surrounded by heavy and serrated cinguli. The M1 and M2 have three cusps (paracone, metacone and protocone). The M1 is quadratic in occlusal outline, whereas M2 is anteriorly arched. The metacone of the M2 is caudally shifted and the cingulum above it is enhanced. The M3 is small, with a single cusp and a thick, serrated cingulum.

Amphicyonidae

Genus *Agnotherium* Kaup, 1832

Type species. — *Agnotherium antiquum* Kaup, 1832.

Referred species. — *A. grivense* (Viret, 1929).

Agnotherium antiquum Kaup, 1832

Type specimen. — EP65, Hessisches Landesmuseum, Darmstadt, Germany.

Type locality. — Eppelsheim, Germany.

Synonymy. — *Agnotherium antiquum* Kaup 1832 in Kuss (1962) and references therein; *Agnotherium antiquum* Kaup, 1833 in Kurtén (1976); *Agnotherium antiquus* in Alcala *et al.* (1989–1990).

Known localities. — *Agnotherium antiquum* is known from Spain (Pedregueras), Germany (Eppelsheim) and Hungary (Rudabánya). Outside of Europe it is recorded at Bled Douarah, Tunisia.

Stratigraphic range. — MN 9.

Referred material. — Bled Douarah (housed in the Finnish Museum of Natural History): P-6188, maxillary fragment with P4 dex (cast of T-370). Rudabánya housed in Magyar Állami Földtani Intézet, Budapest: V12566 m1 dex, m1 sin, p4 sin.

Other data. — Alcala *et al.* (1989–1990).

Description of the type. — The holotype is a right lower carnassial. The protoconid is high and large. The talonid is short with the hypoconid as the dominant cusp. There is a small, unworn entoconid. The rest of the cusps, the paraconid, protoconid, and hypoconid, have heavy vertical wear. The enamel on the buccal side of the paraconid is worn all the way to the base of the tooth. Two thirds of the enamel from the tip of the protoconid is worn off, and almost all the enamel on the buccal side of the hypoconid as well. From this vertical wear it follows that the occlusal edges of the cusps are very sharp.

Extended description. — The material from Pedregueras is only mentioned as unpublished in Alcala *et al.* (1989–1990), and I have not examined it. A maxillary fragment with P4 from Tunisia (Bled Douarah) was attributed to the species by Kurtén (1976). The specimen also preserves alveoli for the M1 and P3. In relation to the estimated lengths of these teeth, the P4 is long in comparison with that of *A. major*. The parastyle blade of the P4 is well-developed, but the protocone is reduced to a relatively small bump that is posteriorly situated.

This poorly known species is similar to the better-known *Agnotherium grivense*, but separated from it by its smaller size and the more reduced talonid of the m1.

Agnotherium grivense (Viret, 1929)

Type specimen. — 1120, Museum Lyon, France.

Type locality. — La Grive St. Alban, France.

Synonymy. — *Agnotherium grivense* (Viret) 1929 in Kuss (1962) and references therein.

Known localities. — *A. grivense* is known from France (Baigneaux-en-Beauce, Esvres and in La Grive St. Alban).

Stratigraphic range. — MN 4–MN 7/8.

Referred specimen. — La Grive St. Alban (housed in the Naturhistorisches Museum Basel): G.A.5731, m1 sin (cast of 1128); G.A.5943, m2 dex (cast of unnumbered); G.A.5730, M2 sin. Baigneaux-en-Beauce (housed in the Naturhistorisches Museum Basel): S.O.4459, m1 dex; S.O.874, P4 sin; S.O.4460, P4 sin; S.O.1559, M2 sin; S.O.1046, c sin.

Other data. — Viret 1951.

Description of the type. — The type specimen is a single lower carnassial. The tooth is very narrow. There is no metaconid, but the posterolingual side of the protoconid has a strong ridge. The tooth has no clear cingulum, but a faint trace of one is present on the lingual side. It has heavy tooth wear: the buccal side of the tooth has all the enamel worn off.

Extended description. — The *A. grivense* material from Baigneaux-en-Beauce includes an m1, two P4s, and an M2. The m1 is unworn. It is similar to the type in morphology. It shows a character apparently typical of this species: the three main cusps (paraconid, protoconid and ectoconid) are heavily crested. The tooth tapers towards the anterior and posterior margins of the cusps, which makes it look as if its occlusal surfaces are crested. The P4s are small compared to the m1 (another typical character). The paracone is massive, the metastyle blade short, and the parastyle blade and protocone faint. The teeth are surrounded by a thick cingulum. The M2 is small and almost oval in outline. The paracone is lingually shifted and much smaller than the metacone.

Genus *Amphicyon* Lartet, 1836

Type species. — *Amphicyon major* Blainville, 1841.

Referred European species. — *A. giganteus* Schintz, 1825, *A. gutmanni* Kittl, 1891, *A. prior*, Springhorn, 1977 and *A. hubacyon* (Kretzoi, 1985).

Definition of the genus. — Based on the phylogenetic analysis presented here, only the four species listed above are included in the genus. They all share the presence of spacings between their premolars, the unirooted p2, and enlarged M2. They also preserve the entoconid on the m1.

Fig. 1. Holotype of *A. major* (Sa 38 housed in Muséum National d'Histoire Naturelle, Paris).

Amphicyon major Blainville, 1841

Type specimen. — Sa 38, Muséum National d'Histoire Naturelle, Paris, France.

Type locality. — Sansan, France.

Synonymy. — *Amphicyon major major* Blainville, 1841; *Amphicyon major eppelsheimensis* Weitzel, 1930 in Kuss (1965), and references therein.

Known localities. — *A. major* is known in Czechia (Neudorf Spalte), France (La Grive St. Alban, Sansan), Germany (Eppelsheim, Oberkirchberg, Sandelzhausen, Stätzling), Spain (Can Llobateres, Hidroeléctrica, Paracuellos 3, Puente de Vallecas, Tarazona, Torrijos, Los Valles du Fuentiduena), and Turkey (Candir).

Stratigraphic range. — MN 5–MN 9.

Referred material. — Eppelsheim (housed in the Universitäts-Institut und Staatssammlung für Paläontologie und historische Geologie, München): 1903I5, M1 dex (cast); 1959XIII18, m1 dex (cast); 1905I 502, m1 dex (cast). Neudorf-Spalte (housed in the Naturhistorisches Museum Wien, Geologisch-Paleontologische Abteilung): No number, C sin; No number, c sin. Oberkirchberg (housed in the Staatliches Museum für Naturkunde, Stuttgart): No number, C dex (cast); No number, P4 sin (cast). Paracuellos III (housed in the Museo Nacional Ciencias Naturales, Madrid): (no7426) Pa1628, I3 dex; (no7426)Pa1689, m3 sin. Sansan (housed in the Muséum National d'Histoire Naturelle, Paris): Sa9709, premaxillary fragment with I3; Sa3, premaxillary and maxillary fragment with I3–C, P2–P4, M2 dex and I2-C, P2–M2 sin; Sa5, maxillary fragment with P4, M2 sin; Sa4, max fragment with P1–M1 dex; Sa2, maxillary fragment with P4, M2 sin and M1–M2 dex; Sa34, mandibular fragment with c–p2, m1–m2 and dp4 sin; Sa38, mandibular fragment sin; Sa207, mandibular fragment with c, p4–m2 dex; Sa37, mandibular fragment with p3–m2 sin (same individual as Sa3); Sa35, mandibular fragment with c, p2–m1; Sa33, mandibular fragment dex; Sa208 c sin; Sa30, M2 dex; Sa31 M3; Sa47, m3. Stätzling bei Augsberg (housed in the Naturhistorisches Museum Basel): TD1137, M1 dex (cast); TD1134, c dex (cast); TD1135, c dex (cast); TD1140, m1 dex (cast); TD1141, m2 sin (cast); TD1142, m3 sin (cast). Torrijon-Toledo (housed in the Museo Nacional de Ciencias Naturales, Madrid): no15280, m1; no15279 m2. Los Valles de Fuentiduena (housed in the Museo Nacional de Ciencias Naturales, Madrid): VF-9, p4 dex; VF-5, m1 dex.

Other data. — Alcala *et al.* (1989–1990), Gürbüz (1974) and Viret (1951).

Description of the type specimen. — The holotype is a left fragment of maxillary bone with a canine, P1, P4 and the molars preserved (Fig. 1). It is difficult to determine the original shape and structure of the skull as the bone is crushed. The canine is transversely compressed and the medial part has three strong vertical enamel ridges, of which the two lateral ones are conspicuous (the double cutting edge of Matthew, see Olsen 1958), while the one between them is rather faint. The P1 has one root and one cusp, and is relatively reduced. In the P4, the anterior part of the paracone has a keel which terminates in a very small parastyle blade. The protocone makes a low, lingual bump in the cingulum. It is situated posterior to the parastyle blade and anterior to the paracone. The cingulum is continuous on the lingual side. The metastyle blade is short and simple. The M1 has a well-developed paracone, metacone, paraconule, and protoconule, and a small metaconule. The buccal cusps are slightly worn, whereas the lingual ones show no wear. The crown is surrounded by a cingulum which rises to make a triangular expansion for the P4 and M2 facettes. The M1 is almost triangular in lingual outline. The

Fig. 2. Upper dentition of a complete *A. major* skeleton (housed in Muséum National d'Histoire Naturelle, Paris).

M2, on the other hand, is rectangular in outline. This difference is produced by the lingual part, which in the M1 narrows almost to a point, but in the M2 remains almost as wide as the buccal part. The paracone is the highest cusp in the M2. The metacone is only about half of the width of the paracone, and shifted somewhat lingually. There is an alveolus for a single-rooted M3. The angle between the premolars and molars is 155 degrees.

Extended description. — Due to the abundant material from Sansan *A. major* is the best known Middle and Late Miocene amphicyonid species in Europe. The Sansan collection was fully described by Ginsburg (1961a), but excavations have produced more amphicyonids, including a complete skeleton, which was briefly described by Bergounioux and Crouzel in 1973.

A. major is a good representative of the amphicyonids of the period. In the lower jaw the premolars are widely spaced. The three anterior ones consist of only the main cusp and a cingulum. The p4 usually (in four cases out of five) has a posterior accessory cusp in addition to the main cusp, and is also surrounded by a cingulum. The lower carnassial (m1) has a small paraconid, a stout protoconid with a metaconid merging with the protoconid, and a high and massive hypoconid. There is no hypoconulid, and the entoconid is small. The talonid is broad as is typical in the Middle and Late Miocene *Amphicyon*. It is also relatively long. The talonid makes up about one third of the total length of the tooth. The m2 has two high anterior cusps, the protoconid and metaconid. The paraconid is present only as a small rudiment. The hypoconid and entoconid are well-developed, but not as massive as the anterior cusps. The m3 has a well-developed metaconid and protoconid, and a flat talonid which shows only rudiments of the cusps. In occlusal outline, the m1 is a longish rectangle, with little buccal arching. The m2 is a shorter rectangle of about the same width as m1, and shows slight lingual arching. The m3 is oval and widens anteriorly.

The maxillary dentition also has a pattern typical of the Middle and Late Miocene amphicyonids: the premolars are reduced and the molars are enlarged. The three anterior premolars are small unicuspid teeth. They remain unworn even when the molars are moderately worn (e.g. the skeleton (Fig. 2) from Sansan and Sa3). The P4 is short and stout. The metastyle blade and paracone are of about the same length. The development of the parastyle blade varies. It may be distinct (e.g. Sa 5), represented by a small protuberance (e.g. Sa 2) or totally absent (e.g. Sa 3). A specimen from Oberkirchberg attributed to *A. major* even shows two small cusps in place of the parastyle blade. The M1 and M2 consist of a paracone and a metacone of about same size and a well-developed, anteriorly positioned protocone. The cingulum is enhanced anterior to the protocone. Due to a tapering talon, the M1 has a triangular occlusal contour. The M2 is quadratic in outline. There are two small additional cusps in the M1 and M2. These are the metaconule and paraconule. Although their size varies, they are never very prominent. There are no hypocones in any of the Middle or Late Miocene amphicyonid M1s or M2s. The enhanced lingual cinguli increase the total occlusal area of these teeth. The M3 is a small tooth with an almost round contour. Only one cusp, the metacone, is well developed.

The incisors gradually increase in size from i1/I1 to i3/I3. They are conical teeth with two lingual cusplets. The canines are relatively long and massive. The double cutting edges in the upper canines are characteristic not only of *A. major*, but apparently of all the Miocene amphicyonids.

The only known complete skull of *A. major* is that belonging to the complete skeleton from Sansan (Fig. 2). The skull is crushed, but preserves all the bones and most of the dentition. The snout is elongated, and has widely spaced premolars and a primitive dental formula (four premolars and three molars). There are spaces between the canine and third incisor, and between the canine and first premolar. The anterior part of the palate is narrow, expanding sharply posterior to P4. The suture between the maxilla and palatine is difficult to see because of deformation of the skull. It may be relatively anterior. The ventral foramina of the premaxilla are deeply excavated and extend anteriorly almost to the alveolus of the I1. The palatine foramina are impossible to see due to deformation. The nasal opening seems large. The nasals extend anteriorly only as far as immediately behind the P1. The skull is dominated by a long parietal with a strong sagittal crest. The occipital region seems narrow (partly due to deformation). The auditory region of this specimen is relatively well preserved and is described in Bergounioux and Crouzel (1973).

Both halves of the mandible are preserved in this specimen. The mandible is long and slender. The premolars are widely spaced, and there are long spacings, especially between the canine and p1. Two mental foramina are placed under the p1 and p3. The mandibular symphysis is shallow, but extends far posteriorly. The large and deep masseteric fossa covers the posterior part of the mandible. The angular process is short and blunt. The condyloid process does not stick out remarkably in posterior aspect, but it makes a marked lateral extension out of the deep masseteric fossa.

Amphicyon giganteus Schinz, 1825

Type specimen. — M7753, Muséum d'Orléans, France.

Type locality. — Avaray, France.

Synonymy. — See Ginsburg and Telles Antunes (1968).

Known localities. — *A. giganteus* has been found at localities in Austria (Eibiswald, Riedern), France (Artenay, Avaray, Baigneaux-en-Beauce, Bézian, Captieux, Chevilly, Chilleurs-aux-Bois, Esvres, Hommes, Neuville-aux-Bois, Pontlevoy, Pontigné, La Romieu, Savigné-sur-Lathan), Germany (Wintershof-West), Portugal (Lisboa), Spain (Henares 1), and Switzerland (Benken).

Stratigraphic range. — MN 3–MN 5.

Referred material. — Artenay-Autroche (housed in the Naturhistorisches Museum Basel): S.O.3524, m1 dex; S.O.3325, m2 dex; S.O.3551, m2 sin. Baigneaux-en-Beauce (housed in the Naturhistorisches Museum Basel): S.O. 873, M1–M3 dex; S.O.1275, M1 sin; S.O.6587, M1 dex; S.O.4906, p4 sin; S.O.4905, m1 sin; S.O.6589, m2 sin; S.O.1863, m2 dex; (housed in the Staatliches Museum für Naturkunde, Stuttgart): Nr. 43351, m1 sin (cast of S.O.742); Nr. 43350, m1 sin (cast of S.O.3574). Chilleurs-aux-Bois (housed in the Muséum National d'Histoire Naturelle Paris): No number (Ch1), mandible with c, p4–m2. Henares 1 (housed in the Museo Nacional Ciencias Naturales, Madrid): No. 6954 (H1.134), M3 dex. Pontlevoy-Thenay (housed in the Naturhistorisches Museum Basel): Pont. 709, I3 dex (cast of Fa54); Pont. 70, M1 sin (cast of Fa7); Pont. 89, M2 sin (cast of Fa89); Pont. 693, mandible with m1–m2 dex (cast of Fa5). Neuville aux Bois (housed in the Naturhistorisches Museum Basel): S.O.360, P4 dex; S.O.2458, I3 dex; S.O. 4377, mandible with m1 dex and sin; S.O.6531, mandible with p4–m2 sin; S.O.282, m1 dex; S.0.6667, C dex. La Romieu (housed in the Muséum National d'Histoire Naturelle; Paris): M868, mandible with p4–m1 dex (cast of an original housed in Lyon). Savigné-sur-Lathan (housed in the Staatliches Museum für Naturkunde Stuttgart): Nr.41676, M1 sin.

Other data. — Ginsburg and Telles Antunes 1968; Ginsburg and Antunes 1995.

Description of the type. — The type specimen of *Amphicyon giganteus* is an almost complete left M1. The metacone and paracone are both well developed, with the former larger and higher than the latter. The paraconule and metaconule are confluent through a prominent ridge. The protocone is also well developed. The crown is surrounded by a cingulum, which rises to create a triangle-shaped space for the P4 and M2 facettes.

Extended description. — The dentition of *A. giganteus* is generally very similar to that of *A. major*. No complete upper dentitions or crania are known for *A. giganteus*. A right half of a mandible with p4–m1 and alveoli for c–p3 is known from La Romieu, a mandibular fragment with m1–m2 is known from Pontlevoy and a left hemimandible with m1 is known from Neuville-aux-Bois (Fig. 3). A good sample of isolated teeth from many localities is also available. The canines are robust, with two strong vertical enamel ridges. The p1 was probably very reduced in this species and the p2 and p3 are simple, unicuspid teeth. The p4 has a posterior accessory cusp, a main cusp and in many cases an anterior accessory cusp. The lower carnassial of *A. giganteus* gives a very stout impression. This is due to the lack of a small depres-

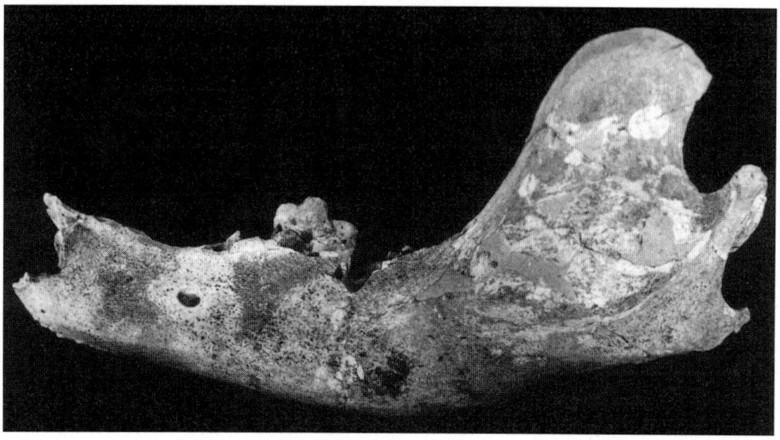

Fig. 3. Mandible of *A. giganteus* (S.O.4377 housed in Naturhistorisches Museum Basel).

sion on the lingual side in front of the anterior margin of the metaconid that is seen in the m1 of *A. major*. The m1 has all the trigonid cusps, although the metaconid is often reduced. The talonid is massive with the hypoconid as the most prominent cusp. The tooth is surrounded by a cingulum, which is stronger around the talonid. The m2 is usually as high as the talonid of the m1. It has a well developed talonid and trigonid. The m3 is small but double rooted. The P4 has a protocone but no trace of a parastyle blade. The M1 has a heavy trigon and smaller talon, as described for the type. The M2 is relatively large in comparison with the M1. It has a narrower trigon. The M3 is smaller than the M1 or M2, but has a similar morphology with a clear talon and trigon. The M3 of *A. giganteus* is a well-developed, large tooth, whereas in *A. major* it is very small.

The Chilleurs-aux-Bois material represents one of the earliest known occurrences *A. giganteus*. Accordingly, it shows many primitive characters. Telles Antunes and Ginsburg (1968) placed this material in a distinct subspecies, *A. g. carnetense*. In the m1 of *A. g. carnetense* the talonid is relatively small, and the entoconid large. In the carnassial of later *A. giganteus* the hypoconid makes up most of the talonid.

Amphicyon gutmanni Kittl, 1891

Type specimen. — Unnumbered m1, Naturhistorisches Museum Wien Geologisch-Palaeontologische Abteilung, Austria.

Type locality. — Mannersdorf, Germany.

Synonymy. — *Amphicyon major gutmanni* Kittl, 1891 in Kuss (1965) and references therein, *"Amphicyon" gutmanni* Kittl, 1891 in Kretzoi (1985), (cf. *Amphicyon giganteus, sensu* Ginsburg and Telles Antunes, 1968).

Known localities. — *A. gutmanni* is known from Mannersdorf, Germany, and tentatively from Kohfidisch, Germany.

Stratigraphic range. — Uncertain, probably up to Pannonian (MN 9–MN 13). See Discussion.

Referred material. — Kohfidisch, Austria (MN 11) (housed in the Naturhistorisches Museum Wien, Geologisch-Paleontologische Abteilung): 1956/i, I3 sin, C dex, M1 dex, M2 dex (cf. *A. gutmanni*).

Description of the type. — The type is a single right lower carnassial. This tooth is robust and high-crowned. It has a low and short paraconid, a triangular, high protoconid, and a well-developed metaconid lying slightly posterior to the protoconid. In the talonid there is a very high hypoconid rising up to the level of the metaconid. There is no visible hypoconulid, but a very low though not very distinct entoconid. Instead of forming a real cusp the lingual margin of the talonid is serrated. The tooth tapers towards the trigonid. The hypoconid is placed posterolingually. At the base of the crown there is thickening of the cingulum in the posterior-buccal corner, which gives the tooth a peculiar contour in lingual view. Apart from this there is no clear cingulum.

Discussion. — This species was described on the basis of a single tooth, a lower carnassial, by Kittl (1891). The m1 is many ways similar to that of *A. major* and *A. giganteus*. Based on similarities such as the wide talonid and extensive hypoconid it has been included in *A. major* (Kuss 1965). Kretzoi (1985) pointed out the unmistakable difference in the contour of the talonid of *A. gutmanni*, and argued for a specific distinction. In both *A. major* and *A. giganteus* a line drawn from the posterolingual corner to the posterobuccal corner is approximately perpendicular to the sagittal line of the tooth, whereas in *A. gutmanni*, because of the extension of the posterobuccal corner, the angle is greater on the buccal side. Another clear difference between these species is in the position of the hypoconid. Its highest point is more posterior in *A. gutmanni* than in the two other species. In these two characters, *A. gutmanni* is very similar to *Thaumastocyon* spp. and *Agnotherium* spp. The type locality is of uncertain age but based on the presence of hipparionine horses (*"Hipparion gracilis"* in Kittl 1891) it must be not older than MN 9.

Four amphicyonid teeth (I3, C, M1, M2) from Kohfidisch (Austria), which is placed in MN 11, were attributed to cf. *Amphicyon giganteus* by De Beaumont (1984). These teeth are large and robust and fit well into the variation of those of *A. giganteus*. Besides these specimens, the youngest *A. giganteus* is known from MN 5. I suggest that the specimens from Kohfidisch may belong to *A. gutmanni* based on a correspondence in size and stratigraphy. This hypothesis remains to be verified.

Amphicyon pannonicus (Kretzoi, 1985)

Type specimen. — V.84.209, Musei Nationalis Hungaric a Magyar Nemzeti Muzeum, Budapest, Hungary.

Type locality. — Pecs, Hungary.

Synonymy. — *Hubacyon (Kanicyon) pannonicus* n.sp. in Kretzoi (1985).

Known localities. — This species is known only from its type locality, Pécs in southern Hungary.

Stratigraphic range. — Lower Pannonian (MN 11–MN 12).

Description of the type. — I have not seen this tooth so the description is based on Kretzoi, 1985. The species is known only from the holotype, which is a left lower carnassial. Only the crown of the tooth is preserved. It is relatively low-crowned, and belongs to an amphicyonid of large size. The paraconid is very low, whereas the protoconid is massive and upright. The posterior margin of the protoconid is almost vertical, and forms a distinct ridge. The posterior margin of the metaconid lies well anterior to this ridge. In the talonid the hypoconid is a high and well-developed cusp. There are other distinct cusps, and the occlusal surface of the lingual margin has a roughening (serration). In occlusal view, the outline of the talonid is more less triangular: the postero-lingual corner is truncated.

The lower carnassial of *A. pannonicus* is in many ways similar to that of other *Amphicyon* spp. It differs from them in the contour of the talonid, and in the lingual side of the talonid, which is never so strongly serrated in *Amphicyon* spp., except in *A. gutmanni*. It differs from *A. gutmanni* in its lower crown and more anteriorly placed metaconid.

Genus *Pseudocyon* Lartet, 1851

Type species. — *Pseudocyon sansaniensis* Lartet, 1851.

Referred European species. — *P. steinheimensis* (Fraas, 1885), *P. styriacus* (Heizmann, 1973) and *P. caucasicus* (Gabunia, 1973).

Definition of the genus. — According to the present phylogenetic study *P. steinheimensis*, *P. styriacus* and *P. caucasicus* are closely related and they are placed here in a genus *Pseudocyon*. This clade is diagnosed by apomorphic triangle-shaped M2.

Pseudocyon sansaniensis Lartet, 1851

Type specimen. — Sa 207, Museum National d'Histoire Naturelle, Paris, France.

Type locality. — Sansan, France.

Synonymy. — *Pseudocyon sansaniensis sansaniensis* Lartet, 1851 in Kuss (1965) and references therein.

Known localities. — *P. sansaniensis* is known from several localities in France (Baigneaux-en-Beauce, Bézian, Esvres, La Romieu, Sansan).

Stratigraphic range. — MN 4–MN 6.

Referred material. — Baigneaux-en-Beauce (housed in the Naturhistorisches Museum Basel): S.O.2793, P4 dex; S.O.1700, P4 dex; S.O.3338, M1 sin; S.O.2461a, M2 dex; S.O.3389, p4 sin; S.O.2574, m1 sin; S.O.1273, m2 sin. Sansan (housed in the Museum National d'Histoire Naturelle, Paris): Sa208, c sin; Sa9709, maxillary fragment with I3; Sa45, maxillary fragment with P2–M1 dex and P2–M1, M3 sin; Sa210 M2 sin; Sa207 mandible with c, p4–m2 dex.

Other data. — Ginsburg and Antunes 1995.

Description of the type. — The type specimen, a right hemimandible, preserves the dentition from p4 to m2, and a broken canine (Fig. 4) and was first figured by Filhol (1891). The alveoli of two incisors are also preserved. All the premolars have two roots. The preserved teeth are worn and show postmortem damage. The p4 con-

Fig. 4. Holotype of *P. sansaniensis* (Sa 207 housed in Muséum National d'Histoire Naturelle, Paris).

sists of a main cusp, a posterior accessory cusp and a strong cingulum. The tooth is broken anteriorly. The cingulum is thick posteriorly and almost forms a second posterior accessory cusp. In the m1 the protoconid is the most prominent cusp. The metaconid is well-developed, but lies in close proximity to the protoconid. The hypoconid is the most distinct and highest of the cusps of the talonid. The entoconid and hypoconulid can not be seen, which partly could be due to obscuring root marks in this area. The metaconid is more anteriorly placed than in *Amphicyon* spp., and the paraconid is relatively longer. The m2 has a well developed metaconid, protoconid, and entoconid. The talonid is long relative to the trigonid. In occlusal outline the tooth is a narrow rectangle with a bump for the metaconid. The molars are relatively low-crowned.

Extended description. — From Sansan there is also a crushed skull with most of the dentition that has been attributed to *P. sansaniensis*. The incisors are not preserved. P2 and P3 are preserved on both sides of the maxilla. They are unicuspid teeth with spacings separating them. The P4 has a relatively short metastyle blade, a cone-shaped paracone, a low but wide protocone, and a faint extension in front of the paracone indicating a small parastyle blade. The M1 is almost triangular. It is asymmetrical with a well-developed protocone in the center, while the lingual cingulum makes a posteriorly directed extension. The M2 is missing from both sides. The left M3 is a small, round tooth. There is also a left premaxillary fragment, with a broken C, I3 and I2. The bone is broken medially to I2. I3 is wide and massive.

The Baigneaux-en-Beauce collection includes isolated *P. sansaniensis* teeth. The two upper molars are similar to the type specimen in form and size. A lower carnassial, which preserves only the talonid, is larger than in the type material but shows a similarly reduced entoconid and narrow talonid. An m2 in this collection is also attributed to *P. sansaniensis,* although it is more quadratic in occlusal contour, lacking the posterior tapering displayed by the m2 of the type specimen. However, it shares a reduced talonid and well developed protoconid with the type specimen.

Discussion. — The type mandible preserves two incisors which are broken at the level of the bone, so that only the roots are preserved in the alveoli. Ginsburg (1961a) noted the robustness of the second incisor in relation to the lateral most one. He also found a small cavity in the mandible between these two teeth. He interpreted the cavity as a rudiment of the i2 alveolus, and concluded that *P. sansaniensis* had reduced the i2, preserving only two lower incisors, the i3 and i1. I prefer, however, to be more cautious about the occurrence of this peculiar condition in *P. sansaniensis*. Of the Recent Carnivora all but the sea otter (*Enhydra lutris*) preserve all the six incisors in the lower jaw. In my opinion it is more likely that the teeth represented in the type specimen are actually i3 and i2, and that i1 (and its alveolus) is missing. Another possibility is that the type specimen is pathological.

Pseudocyon steinheimensis (Fraas, 1885)

Type specimen. — 6225, Staatliches Museum für Naturkunde Stuttgart, Germany.

Type locality. — Steinheim am Albuch, Germany.

Synonymy. — *Amphicyon steinheimensis* O. Fraas, 1885 in Heizmann (1973) and references therein.

Known localities. — *P. steinheimensis* is known from localities in Czechia (Neudorf Spalte), Germany (Steinheim, Stätzling), and France (Castelnau-d'Arbieu).

Stratigraphic range. — MN 5–MN 7/8.

Referred material. — Neudorf-Spalte (housed in the Naturhistorisches Museum Wien, Geologisch-Palaeontologische Abteilung): No numbers, I3, C, P4, m3. Steinheim

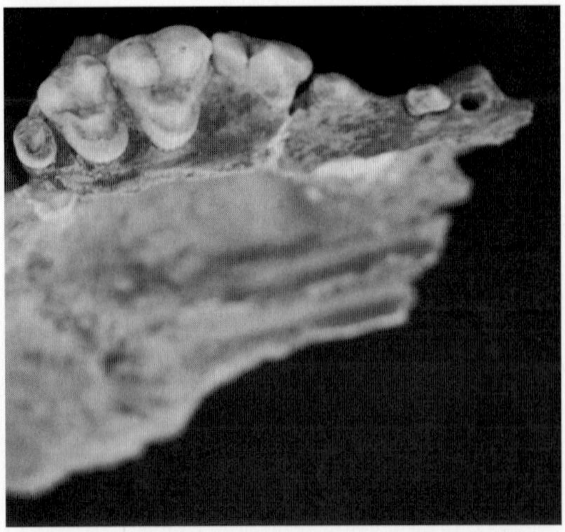

Fig. 5. Holotype of *P. steinheimensis* (6225 housed in Staatliches Museum für Naturkunde, Stuttgart), upper dentition.

am Albuch (housed in the Staatliches Museum für Naturkunde, Stuttgart): Nr. 5810, P4 sin; Nr. 4843, c sin; Nr. 4778, mandible with m1–m2 sin; no number, dI3 dex.

Description of the type. The holotype of *P. steinheimensis* consists of left mandibular and right maxillary fragments of one individual (Figs. 5 and 6). The skull is crushed, and no palatine is preserved. In the right maxilla the anterior part of the temporalis is preserved. The right dentition is partly retained; for the canine and P1 there are only alveoli, but P2–M3 are preserved. The P2 is long and slender and has one pointed cusp, which is keeled on both anterior and posterior sides. The tooth is surrounded by a weak cingulum. The P3 is similar to P2, although the cusp is more anteriorly situated. The P4 has a prominent paracone and a short metastyle blade. The protocone and the parastyle blade are very low, really only enhancements of the cingulum. The parastyle blade is connected to the paracone by a ridge. There are spacings between the premolars, except between P3 and P4, which are situated very close together but do not have contact facettes. The first molar has a very prominent paracone, which is much higher and larger than the metacone. The lingual side consists of two folds of which the middle fold forms a cusp. The folds unite to form a cingulum with two tuberosities: one for the facette for the P4, the other for the facette for the M2. In occlusal view the overall shape of M1 is an arched triangle with one corner (the lingual one) rounded. In the M2 as well, the paracone is better developed than the protocone, but the difference is not as pronounced as in the M1. The M2 has two folds; the buccal fold forms a cusp, the lingual one is very thick and has a bulbous surface. The M3 is rudimentary. It is very small and low-crowned in comparison with the two anterior molars.

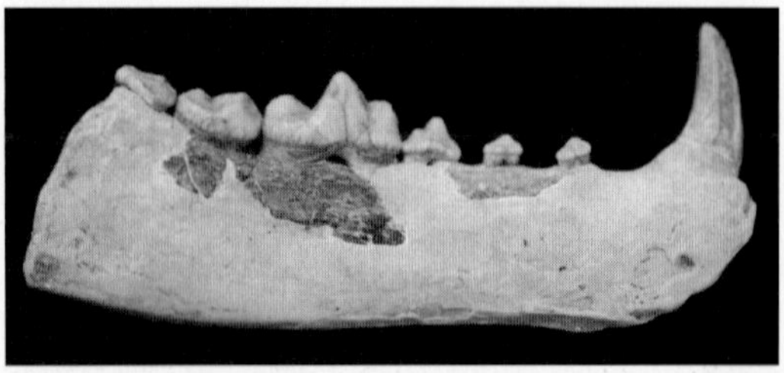

Fig. 6. Holotype of *P. steinheimensis* (6225 housed in Staatliches Museum für Naturkunde, Stuttgart), mandible.

Only the upper parts of the type mandibles are preserved. The lower canine has two strong, slightly serrated ridges, one situated on the posterior side, and the other anteromedially. These are the double cutting edges of Matthew. The p2 is single-rooted, long, low and slender. The p3 is similar to the p2. The p4 has a main cusp and a posterior accessory cusp. There are spacings between the premolars. The m1 is slightly worn on the buccal side, the posterior and anterior sides of the protoconid, the occlusal and lingual sides of the hypoconid, and the occlusal side of the entoconid. The paraconid is low and short. The protoconid is very prominent, the hypoconid high, the hypoconulid absent, and the entoconid very low but distinct. The metaconid has a common posterior border with the protoconid, and is very close to it and seems to blend into it. In the m2, as in the m1, the buccal side is much higher than the lingual side. The m2 has three cusps, the protoconid, metaconid, and hypoconid, and a cingulum. The m3 is very flat, and has only one cusp, the protoconid.

Extended description. — The type collection from the Steinheim Basin includes a deciduous tooth, dI3, and some isolated permanent teeth which are similar to the type and were attributed to *P. steinheimensis* by Heizmann (1973). His referral of the dI3 to this species was based on the size (that of a small amphicyonid), and its similarities with incisors of *C. helbingi* . He notes that the unlikely possibility remains that this is a tooth of an adult of some smaller amphicyonid species. There is also a canine fragment, m1 trigonid and m2, which Heizmann (1973) placed in *Amphicyon* spec. indet. The m1 has a stronger metaconid than the type and the m2 has a wider talonid without the anterolingual extension found in the type. These three teeth fit well within the range of variation of *P. steinheimensis* (Heizmann, pers. comm.), and in the present study they are included in this species.

The material from Neudorf-Spalte attributed to *P. steinheimensis* consists only of isolated teeth. There is a P4 with a very low paracone and distinct parastyle blade. An m3 in this material is small and round as in the type specimen of the species.

Pseudocyon caucasicus (Gabunia, 1973)

Type specimen. — No. 77 in Institute of Paleobiology, Academic Science of Georgia, Tbilisi.

Type locality. — Belomechetskaja, Georgia.

Synonymy. — *Amphicyon caucasicus* nov. sp. in Gabunia, 1973.

Known localities. — *P. caucasicus* is known only from Belomechetskaja.

Stratigraphic range. — MN 6.

Referred material. — See Gabunia (1973).

Description of the type material. — I have not examined the material, so the following description is based on Gabunia (1973). The holotype is a right hemimandible with c to m2 of a large amphicyonid. The p1 and p2 are lost, leaving a long diastema between the c and p3. The p3 is very small. The carnassial is elongated and has a rudimentary metaconid. The m2 is small and has a tapering talonid. In this specimen there is an alveolus for m3, which itself is lost.

Gabunia also mentions a second mandible with incomplete dentition that he attributes to this species.

Pseudocyon styriacus (Heizmann, 1973)

Type specimen. — Nr. 1421, Joanneum Graz, Austria.

Type locality. — Göriach, Austria.

Synonymy. — *Amphicyon steinheimensis* Fraas, 1885 in Thenius (1949); *Amphicyon styriacus* in Heizmann (1973).

Known localities. — *P. styriacus* is only known from the type locality of Göriach (Steiermark).

Stratigraphic range. — MN 6.

Referred material. — *P. styricaus* is known only from its type.

Description of the type. — The type specimen of this species is a mandible with p3–m3. The lower premolar row is widely spaced. The p3 is a unicuspid tooth. The p4 preserves a distinct posterior accessory cusp, which is worn, although the main cusp does not show any wear. The height of the accessory cusp is about half that of the main cusp. These two cusps are connected by a distinct ridge. The lower carnassial is worn on the paraconid, protoconid, metaconid and hypoconid. The paraconid is relatively low. The protoconid has an anterior ridge and is posteriorly inclined. The metaconid is well-developed, and its posterior margin is slightly posterior to the posterior margin of the protoconid. The talonid has a visible entoconid and a tall hypoconid. The entire talonid remains much lower than the trigonid, however. The talonid is almost round in occlusal outline. The m2 is a robust tooth with high trigonid cusps. Because of the high protoconid and metaconid the anterior part of this tooth is higher than the talonid of m1. In the talonid of m2 the hypoconid is the only distinct cusp. It is transversely wider than the m1. The m3 is almost round in occlusal outline, but much better developed and relatively wider than the m3 of *P. steinheimensis*. It has a low metaconid and paraconid in a generally low crown.

Comments. — The amphicyonid mandible from Göriach was attributed to *P. steinheimensis* by Thenius in 1949. Kuss (1965) recognized the specimen as a primitive subspecies of *P. steinheimensis* (*A. s. bohemicus*). Heizmann (1973) found these interpretations unlikely since the Göriach amphicyonid shows similarities not only to *P. steinheimensis,* but also to *A. major*. The characters shared with *A. major* but not with *P. steinheimensis* are a strong metaconid and broad entoconid of the m1, and a high crown of the m2. He concluded that the combination of characters displayed in the Göriach material suggests a separate species, for which he proposed the name *styriacus*.

Genus *Cynelos* Jourdan, 1862

Type species. — *Cynelos lemanensis* (Pomel 1846).

Referred European species. — *C. crassidens* (Filhol, 1874), *C. rugosidens* (Schlosser, 1899), *C. helbingi* (Dehm,

1950), *C. schlosseri* (Dehm, 1950) and *C. piveteaui* (Ginsburg, 1965).

Cynelos helbingi (Dehm, 1950)

Type specimen. — 1937II12293, Staatssammlung für Paläontologie und historische Geologie, München, Germany.

Type locality. — Wintershof-West, Germany.

Synonymy. — *Amphicyon helbingi* nov. sp. in Dehm (1950); *Amphicyon acutidens* nov. sp. in Dehm (1950); *Cynelos lemanensis helbingi* (Dehm) 1950 in Kuss (1965); *Cynelos* cf. *helbingi* (Dehm 1950) in Roth (1989); *Cynelos helbingi* (Dehm 1950) in Belinchón and Morales (1989).

Known localities. — *C. helbingi,* like *C. schlosseri,* is known from Wintershof-West as well as from other sites in Germany (Erkertshofen 2, Petersbuch) and Spain (Buñol).

Stratigraphic range. — MN 3–MN 4.

Referred material. — Buñol (housed in the Museo Nacional de Ciencias Naturales, Madrid): N-33863, P4 dex (cast of Bu-6), N-23864, m2 sin (cast of Bu-7). Wintershof-West (housed in the Universitäts-Institut und Staatssammlung für Paläontologie und historische Geologie, München): 1937II12395, P4 sin; 1937II12397, P4 dex; 1937II12388, P4 sin; 1937II12390, P4 sin; 1937II12401, P4 sin; 1937II12400, P4 dex; 1937II12396, P4 dex; 1937II12386, P4 sin; 1937II12393, P4 sin; 1937II12389, P4 sin; 1937II12394, P4 sin; 1937II12402, P4 dex; 1937II12387, P4 sin; 1937II12851, M1 sin; 1937II12346, M1 dex; 1937II12347, M1 dex; 1937II12362, M1 dex; 1937II12352, M1 dex; 1937II12349, M1 dex; 1937II12360, M1 dex; 1937II12348, M1 dex; 1937II12350, M1 dex; 1937II12359, M1 sin; 1937II12363, M1 dex; 1937II12351, M1 dex; 1937II12357, M1 sin; 1937II12356, M1 sin; 1937II12364, M1 dex; 1937II12354, M1 sin; 1937II12355, M1 sin; 1937II12361, M1 dex; 1937II12380, M2 sin; 1937II12382, M2 dex; 1937II12381, M2 sin; 1937II12373, M2 sin; 1937II12383, M2 sin; 1937II12378, M2 dex; 1937II12375, M2 dex; 1937II12384, M2 sin; 1937II12385, M2 sin; 1937II12377, M2 dex; 1937II12379, M2 dex; 1937II12376, M2 dex; 1937II12850, M3 sin; 1937II12848, M3 sin; 1937II12849, M3 sin; 1937II12373, M3 sin; 1937II12880, M3 sin; 1937II12879, M3 sin; 1937II12847, M3 dex; 1937II12293, mandibular fragment with m1–m3 dex; 1937II12292, mandibular fragment with p4–m1; 1937II12872, p4 dex; 1937II12871, p4 dex; 1937II12870, p4 dex; 1937II12878, p4 sin; 1937II12875, p4 sin; 1937II12874, p4 dex; 1937II12873, p4 dex; 1937II12869, p4 dex; 1937II12876, p4 sin; 1937II12877, p4 sin; 1937II12322, m1 sin; 1937II14325, m1 dex; 1937II12334, m1 sin; 1937II12317, m1 sin; 1937II12323, m1 sin; 1937II12321, m1dex; 1937II12316, m1 dex; 1937II12327, m1 sin; 1937II12291, m1 sin; 1937II12315, m1 dex; 1937II12328, m1 sin; 1937II12331, m1 dex; 1937II12332, m1 sin; 1937II12319, m1 sin; 1937II12318, m1 dex; 1937II12320, m1 dex; 1937II12324, m1 dex; 1937II12333, m1 sin; 1937II12329, m1 sin; 1937II12335, m1 dex; 1937II12338, m1 sin; 1937II12829, m2 sin; 1937II12828, m2 sin; 1937II12828, m2 dex; 1937II12833, m2 sin; 1937II12837, m2 dex; 1937II12839, m2 dex; 1937II12831, m2 sin; 1937II12835, m2 dex; 1937II12850, m2 dex; 1937II12843, m2 dex; 1937II12841, m2 dex; 1937II12830, m2 sin; 1937II12844, m2 dex; 1937II12832, m2 dex;1937II12838, m2 dex; 1937II12836, m2 dex; 1937II12842, m2 dex; 1937II12834, m2 sin;1937II12857, m3 dex; 1937II12858, m3 sin; 1937II12861, m3 sin; 1937II12855, m3 dex; 1937II12856, m3 sin; 1937II12860, m3 sin; 1937II12856, m3 dex; 1937II12836, m3 sin;1937II12863, m3 sin; 1938II12860, m3 sin; 1937II12864, m3 sin.

Other data. — Roth 1989.

Description of the type. — The holotype is the posterior part of a right mandible preserving the molars. The m1 has a broad, large talonid. The paraconid is high, the metaconid distinct and caudally positioned, and the protoconid small. The hypoconid is high and pointed, the entoconid much lower but still forming a proper cusp. The hypoconulid is not present. The m2 has very high and pointed anterior cusps. The protoconid is as high as the hypoconid. The metaconid is posteriorly shifted, leaving a plateau in front of it. It is a little higher than the entoconid. Posterior to the protoconid there is a well-developed hypoconid. The m3 has a high protoconid at the antero-buccal corner. The anterior part of the masseteric fossa, which is well excavated, is preserved.

Extended description. — From the type locality there is also a large number of upper teeth. Unfortunately none of them are associated. The metastyle blade of P4 is short. The protocone is usually situated near the anterior margin of the paracone, and in some specimens it forms a small cusp. The upper molars have a strong, continuous cingulum. The M1 is rectangular in occlusal outline and has three well-defined cusps. The M2 is something between a rectangle and triangle in shape, and is curved. It also has three cusps. The M3 is small and has either one cusp at the anterior end, or two cusps, one situated at the anterior end and one at the posterior end.

As also noted by Dehm (1950), the Wintershof-West material illustrates the great range of variation in the dentition of *C. helbingi*. This probably was typical of amphicyonid species in general. The development of the cusps in p4 varies from one main cusp only to one main cusp with a posterior accessory cusp. Eight out of the ten p4s have an accessory cusp. Specimen 1937II12872 has only a faint band at the posterior end and

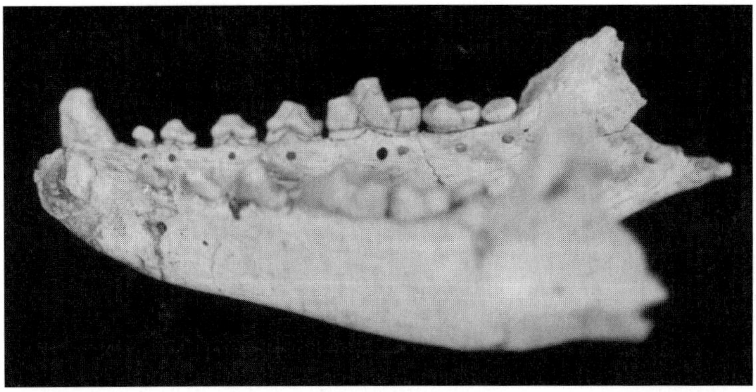

Fig. 7. Holotype of *C. schlosseri* (1937II13562 housed in Staatssammlung für Paläontologie und historische Geologie, München).

1237II12876 has a very low bump. The m1 may have a cingulum around the trigonid, and the entoconid is in some cases low, whereas in others it may be high. The anterior cusps of m2 and m3 are usually elevated, but not in all the specimens (e.g. 1937II12829). In 1937II12831 the paraconid consists of two separate, pointed cusps. In the other m2s there is only one cusp. The protocone of P4 is often near the paracone. Specimen 1937II12402 has a paracone that extends far lingually and forms a distinct cusp. There is always a cingulum around the P4, but its thickness varies. It is extremely thick in 1937II12397. In some M1 the talonid narrows lingually to a point making the tooth triangular in outline. Usually, the occlusal outline is inclined caudally. There are M3s that are diminutive and with a simple morphology (1937II12847) as well as fair sized M3s (1937II12880) that maintain the main cusps.

The other *C. helbingi* collections include isolated teeth that are only tentatively attributed to this species. The only *Cynelos*-like tooth from Erkertshofen 2 is a trigon of M2, which matches that of *C. helbingi*. There is an m3 from Petersbuch 2 which is similar in cusp morphology but narrower than those of *C. helbingi* from Wintershof-West. It is also tentatively attributed to this species. The *Cynelos* material from Buñol consists of an m2 and P4, both of which match *C. helbingi* in morphology, but are shorter than the teeth in the type collection. The m2 is also relatively wider.

*Cynelos schlosseri (*Dehm, 1950)

Type specimen. — 1937II13562, Staatssammlung für Paläontologie und historische Geologie, München, Germany.

Type locality. — Wintershof-West, Germany.

Synonymy. — *Amphicyon schlosseri* nov. sp. in Dehm (1950); *Cynelos rugosidens schlosseri* (Dehm) 1950 in Kuss (1965).

Known localities. — *Cynelos schlosseri* is known from abundant material from Wintershof-West, Germany. It is also known from France (Bézian, Artenay, Savigné-sur-Lathan).

Stratigraphic range. — MN 3–MN 5.

Referred material. — Artenay-Autroche (housed in the Naturhistorisches Museum Basel): S.O.536, M2 sin; S.O.34, P4 sin. Savigné-sur-Lathan (housed in the Staatliches Museum für Naturkunde, Stuttgart): Nr. 41975, m1 sin (cast of Coll. Ford 1980). Wintershof-West (housed in the Universitäts-Institut und Staatssammlung für Paläontologie und historische Geologie, München): 1937II12345, maxillary fragment with P4–M2 dex; 1937II12343, P4 sin; 1937II12344, P4 sin; 1937II12403, P4 sin; 1937II12398, P4 sin; 1937II12392, P4 dex; 1937II12336, M1 dex; 1937II12368, M1 sin; 1937II12366, M1 dex; 1937II12365, M1 dex; 1937II12372, M2 sin; 1937II12371, M2 sin; 1937II12370, M2 sin; 1937II12343, M2 sin; 1937II12344, M2 sin; 1937II12307, mandibular fragment with p1–p4 sin; 1937II12306, mandibular fragment with p4–m1 sin; 1937II12311, mandibular fragment with p4–m1 dex; 1937II12303, p3 dex; 1927II12304, mandibular fragment with p4–m1 sin; 1937II12302, mandibular fragment with m1–m2 sin; 1937II12312, mandibular fragment with m1–m2 dex; 1937II12305, mandibular fragment with m1–m2 sin; 1937II12336, m1 dex; 1937II12337, m1 dex; 1937II12310, m2 dex.

Description of the type. — The holotype consists of a mandible with two almost complete tooth rows (Fig. 7) and two maxillary fragments with some of the dentition. Both halves of the mandible are broken at the masseteric fossa, so that no coronoid, condylar or angular process is preserved. The mandible is long and slender. There are two foramina mentalia; one 9.8 mm below the posterior edge of p1, and another larger one 6.4 mm below the posterior edge of p2. The alveoli for

Fig. 8. Holotype of *T. bourgeoisi* (cast housed in Muséum National d'Histoire Naturelle, Paris).

the incisors are small. The i2 is more lingually placed than the other incisors, so they do not form an even row. The canines are very worn, as are the lower premolars. The p1 is diminutive. All premolars have only one cusp. The premolar row is rather spread out and there is no contact between the teeth. The m1 is very worn on the buccal side, and the metaconid is slightly worn. The m2 has two distinct anterior cusps (protoconid and metaconid). These are of about the same height as the talonid of the m1. The two posterior cusps (hypoconid and entoconid) are much lower. On the lingual side there is a cingulum. The buccal side of m2 is worn. The m3 is small, with only one antero-buccally situated cusp. The tooth is not worn.

The maxillary teeth have about the same amount of wear as the mandibular ones, and are of same size. Moreover, the color of the maxillary bone and the upper teeth is similar to that of the type specimen. Therefore Dehm (1950) believed that the maxilla and mandible belong to a single individual. The left maxilla fragment preserves the P4 to M2, while the fragment from of the right side preserves only M2 and M3. The P4 is worn on the lingual side and on the paracone and metastyle blade. It is surrounded by a continuous cingulum. The protocone is very small and there is no parastyle blade. The M1 is rectangular in occlusal outline. It has a paracone and metacone, and a well-defined cingulum. The M2 is similar in shape to M1, but is smaller. The M3 is very small and rounded.

Extended description. — The abundant material from the fissure filling of Wintershof-West includes a number of specimens of *C. schlosseri*. In addition to the permanent teeth there are deciduous teeth attributed to this species (Dehm 1950: figs. 16 and 17). The single tooth (M2) of a small amphicyonid from Artenay is smaller than *C. schlosseri* but closer to this species in morphology than to any other species. The youngest specimen attributed to *C. schlosseri* is from Savigné-sur-Lathan (MN 5). It is a lower carnassial, which is similar in morphology to, but a little larger than m1s from Wintershof-West.

Genus *Thaumastocyon* Stehlin & Helbing, 1925

Type species. — *Thaumastocyon bourgeoisi* Stehlin & Helbing, 1925.

Referred species. — *T. dirus* Ginsburg, Morales & Soria, 1981.

Thaumastocyon bourgeoisi Stehlin & Helbing, 1925

Type specimen. — The type is a mandible that was destroyed during the Second World War, but casts of it are preserved. A cast (Pontlevoy-Thenay 69), housed in the Muséum National d'Histoire Naturelle, Paris, was used in this study.

Type locality. — Pontlevoy-Thenay, France.

Synonymy. — *Thaumastocyon bourgeoisi* Stehlin & Helbing 1925, in Kuss (1962) and references therein.

Known localities. — *T. bourgeoisi* is known from France (Esvres, Pontlevoy).

Stratigraphic range. — MN 5.

Referred material. — Pontlevoy-Thenay (housed in the Muséum National d'Histoire Naturelle, Paris): Pont. 705, M1 sin (cast); Pont. 5, m1 sin (cast).

Description of the type. — The holotype is a right hemimandible with c and m1–m2 (Fig. 8).

The specimen was destroyed during the Second World War, but a few casts have been preserved in museums. There are no alveoli for the p2 and p3 visible in the cast, but based on the description of Stehlin and Helbing (1925) there were small, round alveoli for p1 to p3 and one for a double rooted p4. All the premolars appear to have been small. There is no alveolus for an m3. The two molars and the canine are moderately worn. There are three conspicuous cusps in the m1, the paraconid, protoconid, and hypoconid. The anterior end of the paraconid is turned lingually, as is the posterior end of the protoconid. The postero-buccal corner of the talonid has a buccal extension. In occlusal view the tooth forms the letter Z. The m2 is simple and short.

Extended description. — There are casts of the other *T. bourgeoisi* material from Pontlevoy in Paris. This material includes M1 and m1. The second m1 is like that of the type. The M1 is triangular in occlusal outline. The metacone is lingually shifted and small compared with the large paracone. The whole talon is reduced and short.

Thaumastocyon dirus Ginsburg, Morales & Soria, 1981

Type specimen. — LVF-6, Museo Nacional de Ciencias Naturales, Madrid, Spain.

Type locality. — Los Valles de Fuentiduena, Spain.

Synonymy. — *Thaumastocyon dirus* nova sp. in Ginsburg *et al.* (1981), *Thaumastocyon dirus* Ginsburg, Morales & Soria, 1981 in Alcalá *et al.* (1989–1990).

Known localities. — *T. dirus* is known only from its type locality, Los Valles de Fuentiduena, Spain.

Stratigraphic range. — MN 9.

Referred material. — Los Valles de Fuentiduena (housed in the Museo Nacional de Ciencias Naturales, Madrid): LVF6, M1 dex; LVF2303, m2 sin.

Description of the type. — The type specimen is a right upper M1. It has a very large and high paracone, making up more than one third of the crown. The talon is very reduced. It is broken at the posterior corner, which enhances the impression of a very narrow talon. The tooth is surrounded by a cingulum.

Other specimen. — Another tooth, an m2, from Los Valles de Fuentiduena has been attributed to this species. It has postmortem breakage on the crown. Based on what is left it seems to have had a low talonid basin, without any high cusps. The protoconid and metaconid are well developed and lie posteriorly (at about the midpoint of the crown). A paraconid is visible at the anterior edge of the tooth.

Genus *Gobicyon* Colbert, 1939

Type species in Europe. — *Gobicyon macrognathus* Colbert, 1939.

Referred species. — *G. zhegalloi* Gabunia, 1981.

Gobicyon macrognathus Colbert, 1939

Type specimen. — AMNH 26597, American Museum of Natural History, New York, USA.

Type locality. — Tung Gur (Wolf Camp Quarry), Mongolian autonomous region, China.

Synonymy. — *Gobicyon macrognathus* new genus and species in Colbert (1939); *Gobicyon macrognathus* Colbert, 1939 in Simpson (1945); *Gobicyon macrognathus* Colbert in Pavlović and Thenius (1959).

Known localities. — In Europe, *G. macrognathus* is known only from Prebreza (Serbia).

Stratigraphic range (in Europe). — MN 6.

Description of type specimen. — The following description is based on Colbert (1939). The holotype is a mandible which is broken just behind the m2s. The mandible is large and heavy and has a very strong symphysis. It has only one foramen mentale which is very large and situated directly below the p2. The jaw retains most of the dentition although it is damaged. The incisors are all broken, but seem to have been large and widely spaced. Because of the heavy symphysis of the mandible they form a transverse row between the canines. These are large and lack the double cutting edge. The p1 is a diminutive, single-rooted tooth. The p2 is broken at the base but it was evidently similar to the p3 and p4. The p3 and p4 have main and posterior accessory cusps, as well as a faint anterior accessory cusp. In addition, the cingulum forms an extra posterior cusp in the p4. The m1 preserves only the talonid. It consists of a well developed hypoconid and entoconid that are of equal size. Based on the alveolus, the talonid was small in relation to the trigonid. The m2 is low-crowned. The posterior end of the m2 has a facet for m3.

Extended description. — I have not examined the material and the following descriptions are based on Colbert (1939) and Pavlović and Thenius (1959). In addition to the type specimen, an associated right mandible with i3, dc, c, and p1–p4 and a maxilla with I2–I3, dC, and P1–P3 from Tung Gur are assigned to this species. The I2 and I3 are large, the latter being the larger. They are long and have lateral keels. The P1 is small and single rooted, while the P2 and P3 are large with high main and posterior cusps. The anterior accessory cusps of P2 and P3 are reduced to keels. They are surrounded by cinguli.

Only one European specimen has been attributed to this species (Pavlović & Thenius 1959). It is a mandible with the incisors, canine, premolars, and fragments of the molars of the left side preserved. The mandible is heavy, with a massive symphysis and one large foramen mentale. The incisors increase in size from the median to the lateral, and are not widely spaced as suggested for the type. The premolars are similar to those of the type. The m1 differs from the type in having a large, lingually shifted hypoconid, and no entoconid. The m2 is relatively larger than in the type.

Comments. — Colbert (1939) placed *G. macrognathus* in the Canidae. He noted its similarities to hyaenognathids (borophagines), but suggested a parallel evolution of these two lineages from a common ancestor (*Tomarctus*) in the lower Miocene. Simpson (1945) placed *G. macrognathus* in the borophagine canids. Pavlović and Thenius (1959) pointed out that the most likely Eurasian affinity for *G. macrognathus* is to the amphicyonids, but without a basicranium these relationships are impossible to establish. Recently cranial material of this species preserving the auditory bullae has been uncovered, and this has shown that it clearly is an amphicyonid (Tedford, pers. comm.).

Although I believe that the specimen from Prebreza is closely related to *G. macrognathus*, its taxonomic postition remains unclear. It may represent a new species. The taxonomy of the *Gobicyon*-group is clearly in a need of reconsideration.

Genus *Amphicyonopsis* Viret, 1951

Type species. — No type species.

Referred species. — *A. serus* (Kuss, 1965).

Amphicyonopsis serus (Kuss, 1965)

Type specimen. — Sth672, Naturhistorisches Museum, Basel, Switzerland.

Type locality. — Steinheim am Albuch, Germany.

Synonymy. — *Amphicyonopsis ? serus* (Kuss, 1965) in Heizmann 1973 and references therein.

Known localities. — *Amphicyonopsis serus* is known from Steinheim (Germany) and tentatively from La Grive St. Alban (France).

Stratigraphic range. — MN 6?–MN 7/8.

Referred material. — Steinheim (housed in the Staatliches Museum für Naturkunde Stuttgart): 6226, M2 dex. La Grive St. Alban (housed in the Naturhistorisches Museum Basel): L.Gr. 1136, m1 sin (cf. *A. serus*) (cast of Nr. 1136 in Lyon).

Description of the type. — The holotype of *A. serus* is a mandibular fragment with the m1 talonid and m2 from Steinheim. The m1 talonid is very short, and in contrast to the other amphicyonids with truncated talonids, it is wide and has a quadratic shape. The hypoconid is a high, triangular cusp. The entoconid is present and lies in a relatively posterior position. There is no hypoconulid. The m2 has a high trigonid, with prominent protoconid and metaconid. The talonid is very low-crowned and flat.

Extended description. — This species is only known from very poor material. In addition to the type, an M2 from Steinheim is attributed to this species. Kuss (1965) claimed that the similarity in the reduction and overall size of this tooth and those of *P. sansaniensis* (he considered this species as a subspecies of *P. sansaniensis*) show that they come from similar animals. Heizmann (1973) agreed with this conclusion, but indicated the possibility that this is actually an unusual M2 of *P. steinheimensis*. It is a narrow tooth which has only two roots. The paracone is the highest and most prominent cusp. It lies buccally to the metacone, and this gives the tooth a peculiar occlusal outline. Instead of being triangular with the paracone and metacone forming two corners, the occlusal outline is round. The lingual margin is surrounded by a cingulum which blends into the buccal cusps.

A lower carnassial from La Grive St. Alban shows affinities with this species. It is a complete tooth with all the trigonid cusps and a short talonid. Heizmann (1973), however, remarked that the talonid is not as reduced as and the entoconid is blunter than in the Steinheim specimen. This specimen is tentatively attributed to *A. serus*.

There is an M2 from Wintershof-West, which Dehm (1950) attributed to aff. *P. sansaniensis*. Heizmann (1973) proposed that this tooth, as well as three slender canines from Wintershof-West, could represent *A. serus*. The M2 is similar to that from Steinheim: it has one prominent cusp, the paracone on the antero-buccal corner. The metacone is shifted lingually, but not as much as in the Steinheim specimen. It is reduced, but it is not as small as in the Steinheim specimen. Wintershof-West (MN 3) is probably stratigraphically too old for a derived Miocene amphicyonid, and Heizmann did not include this material in *A. serus*.

Discussion. — The genus *Amphicyonopsis* was introduced by Viret (1929a) based on the m1 from La Grive St. Alban. The type species for the genus was not defined. Because the specimen where the genus was based on, is only tentatively attributed to *A. serus,* the only described species, no type species exists for *Amphicyonopsis*.

With their tendency toward a short and blunt lower carnassial this specimen and the *A. serus* material from Steinheim clearly represent another peculiar trend in the amphicyonid radiations of the Middle Miocene. Unfortunately the scarceness of the material makes more detailed studies on this group difficult.

Genus *Ysengrinia* Ginsburg, 1965

Type species. — *Y. gerandiana* (Viret, 1929).

Referred species. — *Y. tolosana* (Noulet, 1876), and *Y. depereti* (Mayet, 1908) and *Y. valentiana* Belinchón & Morales, 1989.

Ysengrinia valentiana Belinchón & Morales, 1989

Type specimen. — PaN359, Museo Nacional de Ciencias Naturales, Madrid, Spain.

Type locality. — Buñol, Spain.

Synonymy. — *Ysengrinia valentiana* nov. sp. in Belinchón and Morales (1989); *Ysengrinia valentiana* in Alcalá *et al.* (1989–1990).

Known localities. — *Y. valentiana* is only known from its type locality, Buñol, Spain.

Stratigraphic range. — MN 4.

Referred material. — Buñol (housed in the Museo Nacional de Ciencias Naturales, Madrid): PaN359, m2 sin; PaN319, M1 sin.

Description of the type. — The holotype of this species is a left M2. The paracone and metacone are of about the same size and symmetrical. The metacone is slightly lingually shifted. The tooth tapers lingually and is inclined caudally. The protocone is well-developed and lies between the paracone and metacone. The cingulum forms an extra enhancement in the postero-lingual corner.

Extended description. — There are also an m2 and M1 from Buñol attributed to *Y. valentiana*. The m2 has a trigonid that is much larger than the talonid and a very prominent metaconid. The M1 is triangular and has a prominent protocone. The cingulum has a strong enhancement on the postero-lingual side of the tooth. The paracone is larger than the metacone, but both of these cusps are well developed.

Affinities with other *Ysengrinia* spp. and the approximately corresponding sizes are the main arguments for assigning these teeth the same species.

Discussion. — *Ysengrinia* is characterized by a pear-shaped M1, a small and pointed p4, and an elliptic m2 with a strong hypoconid. In addition to the Miocene *Y. valentiana* the genus includes three Oligocene forms, *Y. gerandiana, Y. tolosana* and *Y. depereti* (Ginsburg 1965), which have previously been placed in *Amphicyon, Arctamphicyon*, and P*seudocyonopsis* (Kuss 1965).

Genus *Euroamphicyon* nov. gen.

Euroamphicyon olisiponensis (Telles Antunes & Ginsburg, 1977)

Type specimen. — No number in the collection of Prof. Dr. M. Telles Antunes, Monte de Caparica, Portugal.

Type locality. — Quinta do Narigao IV-b Lisboa, Portugal.

Synonymy. — *Pseudocyon sansaniensis* aff. *sansaniensis* Lartet in Kuss (1965); *Amphicyon olisiponensis* nova sp. in Telles Antunes and Ginsburg (1977); *Amphicyon olisiponensis* Telles Antunes & Ginsburg, 1977 in Belinchón and Morales (1989).

Known localities. — *E. olisiponensis* is known from Lisboa (Portugal) and Buñol (Spain). Morales *et al.* (1993) also list *E. olisiponensis* from La Retama but don't describe the material.

Stratigraphic range. — MN 3–MN 4.

Referred material. — Buñol (housed in the Museo Nacional de Ciencias Naturales, Madrid): N-23859, P4 dex

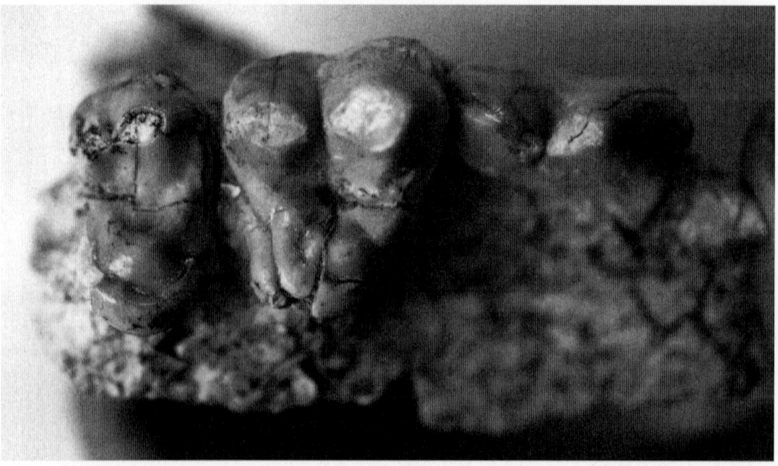

Fig. 9. Holotype of *"A." castellanus* (LVF-C2/1-31 housed in Museo Nacional de Ciencias Naturales, Madrid).

(cast of Bu-2, PaN 978); N-23858, p4 dex (Bu-1, PaN 352); N-23860, m2 dex (cast of Bu-3, PaN 352); N-23867 m3 dex (cast of Bu-10, BSC 1).

Other data. — Ginsburg and Antunes 1995.

Description of the type. — A cast of the type housed in the Institut Paleontologia "M. Crusafont", Sabadell (2378) was studied. The type specimen of *E. olisiponensis* is a part of a left hemimandible with p4 to m2 and the alveolus for the m3 preserved. The p4 has two cusps: a main cusp and an anterior accessory one. The m1 has a prominent protoconid and a smaller and lower paraconid. It is relatively wide in comparison to the m1 of *A. major* and *A. giganteus*. The metaconid is distinct, but shows a common border with the protoconid. The hypoconid is the prominent cusp of the talonid. The tooth is surrounded by a cingulum which is more developed on the buccal side. The cingulum is widest at the anterior-buccal corner where it juts out at the base of the crown, forming a terrace. The m2 has all the three cusps of the trigonid, and in the talonid the hypoconid is present. At the base of the crown it has an extension in the anterobuccal corner, which reaches the level of the m1 extension.

Extended description. — Of the upper teeth only the carnassial, third molar and canine found at Buñol are known for this species. The P4 is represented by a badly deformed tooth, which has a short metastyle blade, a low, long parastyle blade and a protoconid. I have not seen the M3, but based on the description in Belinchón and Morales (1989) it is a small tooth which is as wide as it is long. The Buñol collection also includes an m3, which is a very low-crowned, small tooth with posteriorly placed trigonid cusps.

Amphicyonidae genus indet.

"Amphicyon" castellanus Ginsburg, Morales & Soria, 1981

Type specimen. — LVF-C 2/1-31, Museo Nacional de Ciencias Naturales, Madrid, Spain.

Type locality. — Los Valles de Fuentiduena, Spain.

Stratigraphic range. — MN 9.

Synonymy. — *Amphicyon castellanus* nova sp. in Ginsburg *et al.* (1981), *Amphicyon castellanus* Ginsburg, Morales & Soria, 1981 in Alcalá *et al.* (1992).

Known localities. — *"A." castellanus* is only known from its type locality, Los Valles de Fuentiduena, Spain.

Referred material. — Los Valles de Fuentiduena (housed in the Museo Nacionales de Ciencias Naturales Madrid): No. 3839: 2080, M1 dex; 6211, M1 dex; 1730, C dex; No. 3836, mandibular fragment with p4–m2 dex.

Description of the type. — The holotype of *"A." castellanus* is a right maxillary fragment with P3 to M2, and an alveolus for M3 (Fig. 9). The P3 is very small and broken at the tip. Due to a posterolingual extension it seems almost triangular in occlusal outline. In the P4 the paracone is very prominent and the metastyle blade very short. The parastyle blade is evident only as a faint ridge in front of the paraconid. The protoconid is flat, but extends far lingually. It is situated posterior to the parastyle blade ridge but is well anterior to the paraconid. The M1 has a very large paracone and slightly smaller metacone and is surrounded by a thick cingulum. The outline of the tooth is almost triangular, and it narrows sharply towards the lingual edge. The M2 is anteroposteriorly very narrow and transversely wide. The paracone is larger than the metacone, but the latter is not lingually shifted as in many other amphicyonid taxa. The alveolus for the M3 is round. The alveolus for the canine is preserved, but the bone between it and P3 is deformed, so there is no indication of the presence or absence of P2 or P1. The snout, in any case, seems to have been very short.

Extended description. — There is a right C in the Los Valles de Fuentiduena collection, that is at a similar wear stage as the teeth of the maxilla, and thus could belong to same individual as the type. It is flattened posteriorly and has two strong serrated ridges. There is also a right mandible fragment with p4 to m2 of "*A.*" *castellanus* from Los Valles de Fuentiduena ("paratipo"). The bone is broken in front of the p4. The p4 is a relatively small, unicuspid tooth. The m1 and m2 are very worn, so some of the detailed morphology is lost. The m1 is narrow with a metaconid that seems to merge well into the paraconid. The hypoconid is the prominent cusp of the talonid, and in this specimen it has worn almost to the level of the entoconid. There is no alveolus for the m3. The m3 of this individual was either never present or very small and lost in life. Although the Amphicyonidae characteristically have three lower molars, some individuals may lack the third molar. The hypercarnivorous Miocene amphicyonids, at least *Thaumastocyon bourgeoisi*, have only two lower molars. As this is the only mandible known for this species, it is impossible to say if this was an unique character for the species or just for this individual.

3. Systematic paleontology

3.1. Characters

Here I provide a list of characters used in the systematic analysis. The characters are identified by a sequential numbering system, and these numbers are used throughout this chapter. Table 1 shows the coding for each species. Owing to a dearth of skeletal material, mainly dental characters are used.

Character 0. Robust c in relation to m1
Coding: 0 (not robust); 1 (robust)

This character can be coded for only a few species. Associated lower canines and carnassials are known for *A. major, A. giganteus, P. steinheimensis, P. sansaniensis* and *T. bourgeoisi.* Only *P. sansaniensis* has coding 1 (Fig. 10). "Robustness" of the lower canine is measured as a transverse width and is compared to the length of the lower carnassial.

Character 1. Presence or absence of p1
Coding: 0 (present); 1 (absent)

The presence of p1 is a common, primitive character for the Arctoidea that is only lost in Mephitinae and Lutrinae (Mustelidae) (Wyss & Flynn 1993). Among the European Miocene amphicyonids, p1 is absent in *C. helbingi* and *P. caucasicus.*

Character 2. Spacing between lower premolars
Coding: 0 (present); 1 (absent)

In the lower tooth rows of the Miocene amphicyonids there are typically gaps between premolars. This is not the case in the Oligocene forms and in *G. macrognathus, I. socialis* and *P. bavaricus*. The tooth row is also more crowded in *Cynelos* spp.

Character 3. Presence of p4 posterior accessory cusp
Coding: 0 (present); 1 (absent)

This character shows intraspecific variation in *Cynelos* spp. The posterior accessory cusp is absent in *P. bavaricus* and *I. socialis*, but present in the rest of the amphicyonid species (including the Oligocene ones) in which p4 is known. Unfortunately no p4 is known for *Agnotherium* spp. or *Thaumastocyon* spp. Normally hypercarnivorous species of other families, for example all the felids, have well-developed cusps in their posterior premolars.

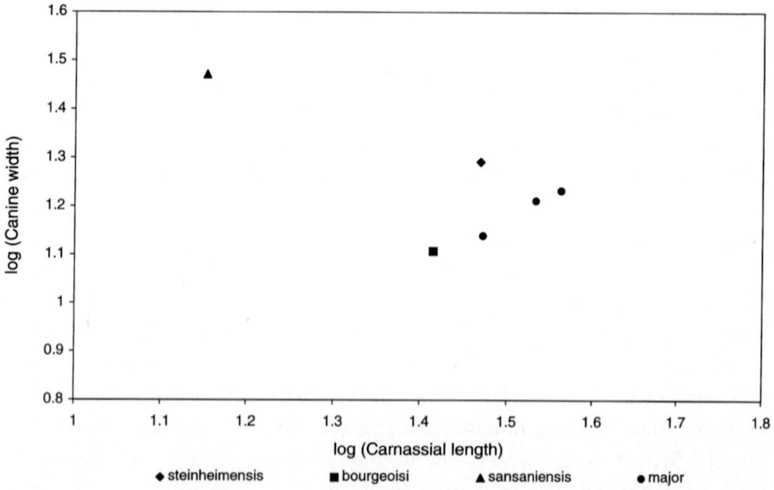

Fig. 10. Transversal widths of the lower canines as compared with lengths of the lower carnassials.

Table 1. List of the character coding for the studied taxa (* means that this taxon was not included in the analysis).

Taxa	Character	1	2
	0123456789	0123456789	01234
Miocene			
major	0010110000	0110000110	00001
giganteus	0010110000	0110000110	00001
**styriacus*	??10111000	0?????????	?00??
steinheimensis	0?1?111000	0210001110	0001?
**castellanus*	???1111001	011010?1?0	1000?
olisiponensis	?0?0110000	000???????	?10?1
**gutmanni*	?????1100?	??????????	?110?
**pannonicus*	?????1100?	??????????	?11??
caucasicus	?110111001	0?????????	?00?1
sansaniensis	1010111001	0100001100	00011
**serus*	?????101??	?????1????	?0???
**macrognathus*	?000011010	0?????????	?00??
schlosseri	?000110000	011000?0?0	00000
helbingi	?100110000	011010?000	00001
bavaricus	?001000000	1??000?1?0	00001
**valentiana*	??????????	???01????1	???0?
**antiquum*	?????1111?	?11??????0	?100?
grivense	?????1111?	?111?0???0	0100?
**dirus*	??????????	???11????0	???0?
**bourgeoisi*	00???11111	0???1??0?0	?100?
Oligocene			
Gouplictis minor	?000?00?01	000?00?0?0	0???0
Pseudocyonopsis landesquei	?000000001	00000010?1	10100
Pseudocyonopsis ambiquus	0000010001	0000000001	10000
Brachycyon gaudryi	0010011001	0??????0??	?0000
Cynelos crassidens	?010000001	0??00??0??	?00?0
Recent			
Ailurus fulgens	0100000000	1000000110	00000
Outgroup			
Canis lupus	0000010000	0200002000	00000

Character 4. The size of p4 in comparison to m1
Coding: 0 (enlarged); 1 (not enlarged)

Most of the Miocene amphicyonids have small premolars. *G. macrognathus* and *P. bavaricus* have relatively large p4s. This is also typical of the Oligocene forms.

Character 5. Presence of hypoconulid on m1
Coding: 0 (present); 1 (absent)

I. socialis and *P. bavaricus* have all three talonid cusps (hypoconid, hypoconulid and entoconid) in their m1. The remaining Miocene amphicyonids lack hypoconulids, as do a few of the Oligocene ones.

Character 6. Presence of entoconid on m1
Coding: 0 (well developed); 1 (reduced)

Coding of this character varies among the Oligocene amphicyonid species. *I. socialis* and *P. bavaricus* again have an entoconid on m1, but *A. major, A. giganteus, E. olisiponensis, A. serus, C. schlosseri and C. helbingi* also retain a distinct entoconid. *P. steinheimensis, P. styriacus, "A." castellanus, A. gutmanni, P. caucasicus, P. sansaniensis, A. antiquum, A. grivense,* and *T. bourgeoisi* all have reduced entoconids and appear to have only one cusp in their m1 talonid.

Character 7. Talonid in m1
Coding: 0 (not reduced); 1 (reduced)

The talonid of the lower carnassial is shortened in *A. serus, Agnotherium* spp. and *Thaumastocyon* spp. among the Miocene forms and in *Amphicyanis felinoides* and *Pseudamphicyon lupinus* among the Oligocene forms. This condition is associated with a hypercarnivorous ecology.

Character 8. Metaconid of m1
Coding: 0 (present); 1 (absent)

In *Agnotherium* spp., *Thaumastocyon* spp. and *G. macrognathus* the metaconid of the lower carnassial is lost. In the rest of the amphicyonids studied here, it forms a cusp which may be rudimentary or well-developed. The metaconid is present in the lower carnassials of all the Oligocene forms.

Character 9. Size of m2 (in comparison to m1)
Coding: 0 (normal); 1 (very small)

A reduced m2 is probably linked to a more hypercarnivorous diet. *T. bourgeoisi, P. sansaniensis, G. macrognathus, P. caucasicus* and "*A.*" *castellanus* all share a very short m2. *P. steinheimensis* shows moderate shortening of m2, which is not, however, great enough to be coded 1. Unfortunately, associated m1 and m2 are not known for the other hypercarnivorous species, and thus it is impossible to code these characters for them.

Character 10. Relative size of lower molars
Coding: 0 (size increases from back to front); 1 (subequal in size, only slightly increasing from back to front)

A marked anterior to posterior decrease in size of the lower molars is a synapomorphy of the Carnivora. In *P. bavaricus* and *I. socialis* the length of the m2 almost equals the length of the m1.

Character 11. Parastyle blade of P4
Coding: 0 (present); 1 (variable); 2 (absent)

This character is difficult to use, because it varies intraspecifically, and for species with only a few known specimens this variation may not show. This is particularly the case with *P. steinheimensis* and *I. socialis* which are the only two species that are coded 2. *P. steinheimensis* is represented by three specimens of P4 and *I. socialis* by only one. The parastyle blade of the upper carnassial is never very enhanced in amphicyonids, but presence of a small cusp is coded 0 here.

Character 12. Protocone of P4
Coding: 0 (present); 1 (varies); 2 (absent)

This character has the same problem as Character 11. It seems to correlate with Character 11 as well, so that when the paracone is present the protocone is also present. *I. socialis* is the only exception, having a paracone but lacking the protocone.

Character 13. Protocone of M1
Coding: 0 (normal); 1 (reduced)

In *Thaumastocyon* spp. and *Y. valentiana* the protocone of the M1 is small and the cingulum where this cusp is situated is reduced forming only a small extension in the postero-lingual corner. In the other species it covers the entire lingual side of the tooth.

Character 14. Size ratio of paracone and metacone of M1
Coding: 0 (of equal size); 1 (paracone is larger)

In the upper M1 of most of the amphicyonids, the buccal cusps are about of equal size. In some forms the anterior cusp, the protocone, takes up a greater area of the crown and is higher than the metacone. Of the Miocene species the paracone is enlarged in *Thaumastocyon* spp, *C. helbingi* and "*A.*" *castellanus*.

Character 15. Relative size of M2 metacone
Coding: 0 (equal to paracone); 1 (small and lingually shifted)

As in the M1, the metacone and paracone of M2 are usually of equal size. They are also placed in a line parallel to plane. *A. serus* differs from all other taxa in having shifted the small metacone on the M2 lingually. This is typical of some Miocene hemicyonine Ursidae. However, the M2 assigned to *A. serus* differs from ursid M2s in other characters, such as the smoothness of the cingulum and the well-developed metaconulus, and it is clearly an amphicyonid tooth.

Character 16. Condition of M3
Coding: 0 (normal); 1 (reduced); 2 (lost)

This potentially important character can be coded only for a few species due to the lack of associated M3s. *P. sansaniensis* and *P. steinheimensis* both have a very small M3, whereas *A. major* has a large one. The three known M3s of *A. giganteus* are all larger than known M3s of *A. major* although M1s of the two species are about the same size. Based on this observation, *A. giganteus* can be coded 0. In the European Oligocene forms *Pseudamphicyon* spp. M3 is lost. This might be true for the Miocene species "*A.*" *castellanus* as well.

Character 17. Transverse width of M2
Coding: 0 (smaller than M1); 1 (about the width of M1)

An enlarged M2 is typical of Miocene amphicyonids. It is not displayed, however, in *C. schlosseri* or *C. helbingi*. This trait is unfortunately unknown in *Agnotherium* spp. and *Thaumastocyon* spp. None of the Oligocene species have an enlarged M2.

Character 18. Number of roots in P2
Coding: 0 (2 roots); 1 (1 root)

Of the species in which P2 is known, *A. major, P. steinheimensis* and *P. sansaniensis* share a single-rooted P2. *C. helbingi* has two roots in P2. Single-rooted P2s are obviously associated with a general reduction of the premolars, which is apparent in all the Miocene forms except *Cynelos* spp.

Character 19. Position of protocone on M1
Coding: 0 (parallel to metaconule); 1 (lingual)

In *Y. valentiana*, the lingual cingulum stretches inwards and places the protocone far from the metaconule. This gives the tooth a peculiar occlusal contour. Amphicyonid M1s are typically triangular in shape, but in *Y. valentiana* the shape is caudally arched.

Character 20. Shape of M2
Coding: 0 (normal); 1 (slim)

"*A.*" *castellanus* has a reduced M2 that is anteroposteriorly very narrow. In other forms the M1 and M2 are of almost equal size, and the shape of the M2 is similar to that of M1. An M2 is also known for *A. grivense*. It is very similar to the "normal" amphicyonid ones in shape.

Character 21. Shape of talonid of m1
Coding: 0 (posterior contour square); 1 (postero-buccal corner makes an extension)

In most of the species the talonid is square. *A. gutmanni, A. pannonicus* and *E. olisiponensis* all have a postero-buccal talonid extension. A similar extension is displayed in the antero-buccal corner of the m2 of *E. olisiponensis*. Unfortunately, no m2s are known for the two other species.

Character 22. Orientations of protoconid and paraconid of m1
Coding: 0 (angle between protoconid and paraconid); 1 (no angle)

In the m1 of *A. gutmanni* the paraconid and protoconid form a straight line, whereas in other forms the paraconid is lingually shifted making an arch between paraconid and protoconid.

Character 23. Triangular shape of M1
Coding: 0 (stretched out); 1 (even sides)

P. steinheimensis and *P. sansaniensis* share a unique, apparently derived condition. In their M1s the ridge forming the paraconule and metaconule is situated closer to the buccal cusps (paracone and metacone) than in other amphicyonids, and this makes the tooth shorten in the buccolingual width. In the other amphicyonid taxa (at least in the ones considered here) there is a wider space between the buccal cusps and the inner cusps.

Character 24. Place of posterior mental foramen
Coding: 0 (below p3–p4); 1 (below p2)

C. schlosseri is the only Miocene species having the posterior mental foramen situated below p3–p4. This is the condition in all the Oligocene species.

3.2. Cladistic analysis

Although commonly regarded as a nearly automatic process nowadays, the search for the most parsimonious trees is a far from trivial undertaking. Recently highlighted problems include outgroup questions (Kluge & Farris 1969, Farris 1972, Nixon & Carpenter 1993), polymorphic and missing character states (Nixon & Davis 1991, Platnick *et al.* 1991) as well as zero-length branches and ambiguous character optimizations (Goloboff 1993a, Coddington & Scharff 1994, Wilkinson 1995, Carpenter 1996, Nixon & Carpenter 1996).

In order to find the best supported, most parsimonious solutions four different programs were used: NoNa and Pee-Wee (Goloboff 1993ab) Hennig96 (Farris 1988) and Clados (Nixon 1994). NoNa was used to obtain strictly supported trees with no zero-length branches and with nodes supported under all optimizations. The length of the minimum length trees were checked with Hennig86 and Clados. The trees were printed with Clados. The steps taken were largely in agreement with those suggested by Carpenter (1996) and Nixon and Carpenter (1996), see also Muona (1996).

Several taxa were removed from the analysis because more than a half of their character states remained unknown. In the end the character matrix included 17 taxa and 25 characters. All characters were treated as unweighted and unordered. The wolf (*Canis lupus*) was used as the outgroup, as it is related to the Arctoidea but clearly is not a member of the studied ingroup, and *Ailurus fulgens* was added as a representative of living arctoids (e.g. Flynn *et al.* 1988, Wozencraft 1989, Wyss & Flynn 1993, Vrana *et al.* 1994).

NoNa produced five trees with the option "hold*; hold/30; mult*20" (Fig. 11), one of which was not minimum length after zero length branches were collapsed with the "ksv*" option and length checked with Clados (Fig. 12). The remaining four trees were 42 steps long, with a CI of 0.61 and RI of 0.72. The Hennig86 result with the option "ie*" confirmed that this was indeed the length of the most parsimonious reconstructions. The four most parsimonious, strictly sup-

Fig. 11. Four trees (A–D) of minimum length produced by NoNa and confirmed by Hennig86. All are 42 steps long. CI = 0.61 and RI = 0.72.

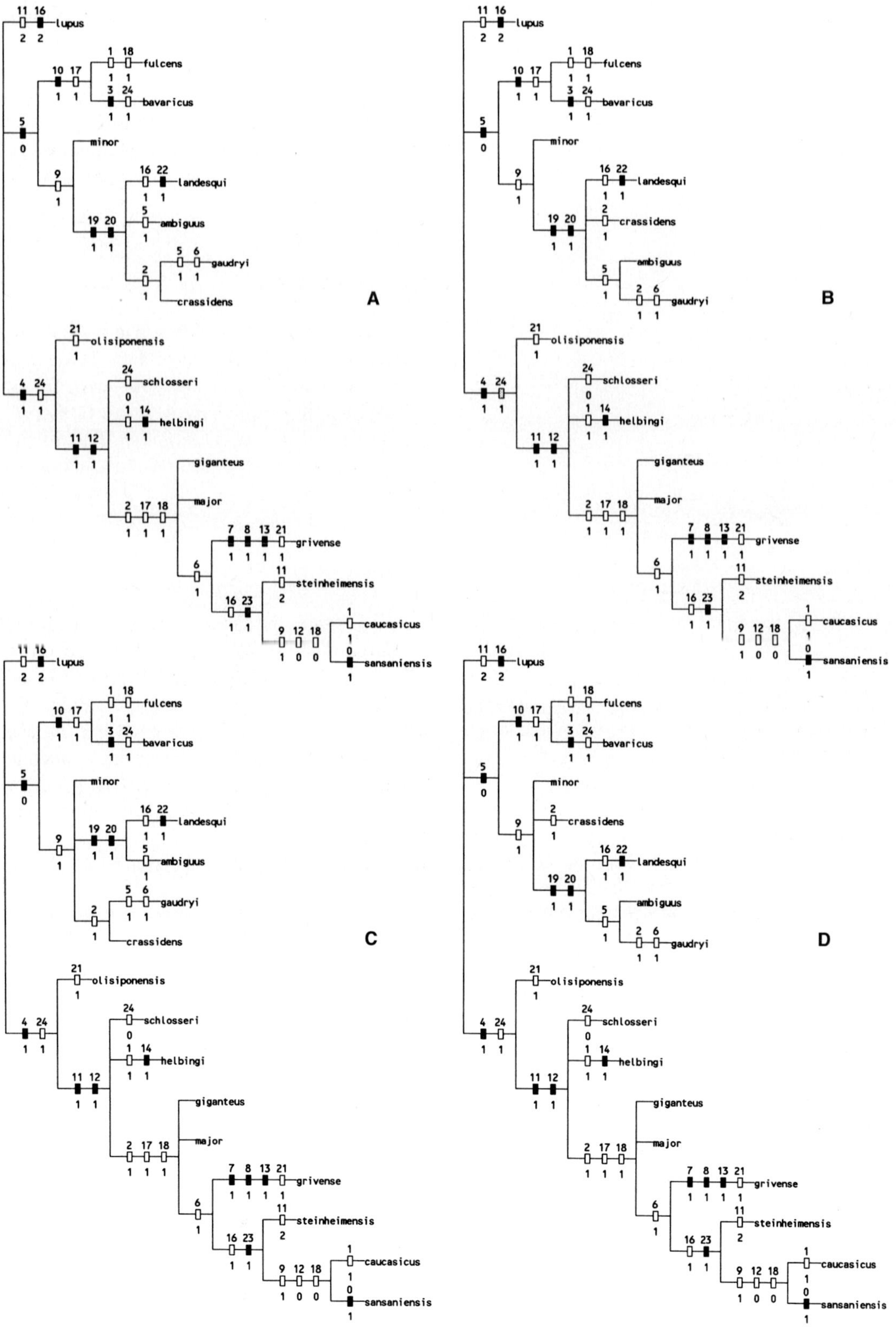
A
lupus
fulcens
bavaricus
minor
landesqui
ambiguus
gaudryi
crassidens
olisiponensis
schlosseri
helbingi
giganteus
major
grivense
steinheimensis
caucasicus
sansaniensis
B
lupus
fulcens
bavaricus
minor
landesqui
crassidens
ambiguus
gaudryi
olisiponensis
schlosseri
helbingi
giganteus
major
grivense
steinheimensis
caucasicus
sansaniensis
C
lupus
fulcens
bavaricus
minor
landesqui
ambiguus
gaudryi
crassidens
olisiponensis
schlosseri
helbingi
giganteus
major
grivense
steinheimensis
caucasicus
sansaniensis
D
lupus
fulcens
bavaricus
minor
crassidens
landesqui
ambiguus
gaudryi
olisiponensis
schlosseri
helbingi
giganteus
major
grivense
steinheimensis
caucasicus
sansaniensis

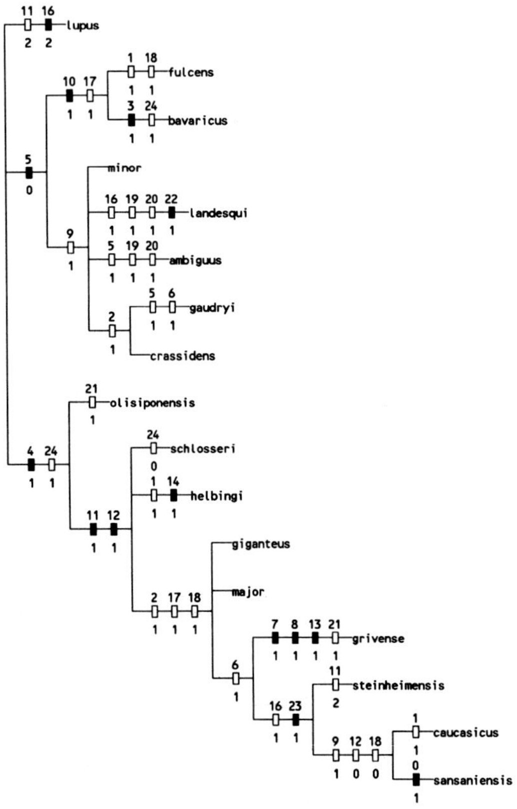

Fig. 12. A fifth minimum length tree produced by NoNa. This tree appeared longer (44 steps) when zero-length branches were collapsed.

ported trees differed from each other only in the resolution within one clade with the Oligocene species *Goupilictis minor, Pseudocyonopsis landesquei, Pseudocyonopsis ambiguus, Brachycyon gaudryi* and *Cynelos crassidens*. In order to check the stability of the topology, both Pee-Wee and the successive weighting option of Hennig86 were used. A fittest trees analysis with Pee-Wee produced five trees identical to the most parsimonious resolutions showing variation only in the same clade that remained unsolved in maximum parsimonious solutions as well. Similarly, the successive weighting option of Hennig86 produced five trees and these agreed completely with those obtained with Pee-Wee.

Clearly, with the exception of the one clade that was not resolved, the results were robust. They suggest that two separate clades exist. One of them included *P. bavaricus, A. fulgens* and the Oligocene forms. The other clade included the Miocene forms with the exception of *P. bavaricus*. The interesting implications of the Oligocene clade, and relations of *A. fulgens* and *P. bavaricus* will be explored in another connection, the following discussion will concentrate on the relationships of the Miocene clade.

3.3. Missing taxa

Several taxa could not be included in the cladistic analysis because of incomplete data. These taxa were: *I. socialis, P. styriacus, "A." castellanus, A. gutmanni, A. pannonicus, A. serus, G. macrognathus, Y. valentiana, A. antiquum, T. dirus* and *T. bourgeoisi*. Their apparent positions in the tree are discussed here.

I. socialis

I. socialis shares the apomorphic subequal size of the lower molars (Char. 10) and the hypertrophic M2 (Char. 17, homoplastic) with *P. bavaricus* and *A. fulgens*. Further, it shares the apomorphic absence of the posterior cusp in p4 (Char. 3), and homoplastic condition of the posterior mental foramen (Char. 24) with *P. bavaricus*. It can be expected to belong to that clade. Since the current data question the position of *P. bavaricus* within the Amphicyonidae, the same is true for *I. socialis*.

"A." castellanus

The reduced talonid of m1 (Char. 7) and reduced protocone of M1 (Char. 13) would group *"A." castellanus* with *A. grivense*. A small metaconid (Char. 8) is present in the m1 of *"A." castellanus*.

P. styriacus

The overall similarity between *P. styriacus* and *P. steinheimensis*, and the shape of M1 (Char. 23) suggests a placement of *P. styriacus* in the *Pseudocyon* group.

A. gutmanni and *A. pannonicus*

A. gutmanni and *A. pannonicus* are probably near the genus *Amphicyon* based on the sparse material known.

A. serus and *G. macrognathus*

The position of *A. serus* remains unclear in the current study because of the many peculiar characters of the species. The same applies to *G. macrognathus*. More material of this taxon is available from Asia, and a thorough study of that material will probably allow a phylogenetic analysis on the intra- and inter-generic relations of *Gobicyon*.

Y. valentiana

Y. valentiana is a probable sister taxon of the Oligocene *Ysengrinia* (*Y. gerandiana, Y. tolosana,* and *Y. depereti*), and may not belong to the Miocene clade. Further analysis of the Oligocene forms including *Ysengrinia* spp. will probably contribute resolution to this question.

A. antiquum, T. dirus and *T. bourgeoisi*

The traditional division of the Miocene amphicyonids into two separate subfamilies is not supported by the present phylogeny. However, *A. antiquum, T. dirus* and *T. bourgeoisi* appear very similar to *A. grivense*, and all four "thaumastocyonines" share a reduced talonid in m1 (Char. 7), lack of a metaconid in m1 (Char. 8) and a reduced protocone in M1 (Char 13). These species are probably closely related. This hypothesis and the justification for recognizing two separate genera also will have to await for additional data.

3.4. European Miocene Amphicyonidae revisited

The two different subfamilies, the Amphicyoninae and Thaumastocyoninae, are not supported in the present phylogenetic analysis. Rather, the analysis suggests that the European Miocene amphicyonids seem a monophyletic group. This group is separated from the Oligocene forms, *A. fulgens* and *P. bavaricus* by the apomorphic reduction of p4 (Char. 4) and the absence of a hypoconulid on m1 (Char. 5). They are also distinct from the Oligocene forms in the enlarged M2 (Char. 17). Both *P. bavaricus* and *A. fulgens* have the enlarged M2. The enlarged M2 is a character also synapomorphic within the Procyonidae (Wang 1996), and may suggest a relationship between *P. bavaricus* and procyonids. In any case, this character seems to be homoplastic in the evolution of the Arctoidea.

3.5. *Pseudocyon* revisited

My interpretation of the genus *Pseudocyon* contradicts previous studies. Heizmann (1973) dismissed the possibility of including *P. steinheimensis* in *Pseudocyon* based on all three lower incisors and on the shape of the lower canine. I am sceptical about a reduced number of incisors in *P. sansaniensis*. Canine proportions were included in the present cladistic analysis (Char. 0) and do indeed diverge *P. sansaniensis*. Despite this, the analysis groups *P. sansaniensis* with the other species here referred to *Pseudocyon*.

3.6. New genus *Euroamphicyon* gen. nov.

Based on the topology of the obtained trees, a new generic name is required. This has already been used in the text, but is formally described here.

Etymology. — *Euro* designating European distribution of the genus. (*Amphicyon* from the name of the family Amphicyonidae, *amphi*, from Greek both or around, *cyon*, from Greek kyon, canid).

Type and only species. — *E. olisiponensis* (Telles Antunes & Ginsburg, 1977)

Diagnosis: Amphicyonids of large body size, reduced premolars and well-developed molars. The lower carnassial has a characteristic postero-buccal extension. Similar extension is found on the antero-buccal side of m2.

4. Biology

4.1. Body masses of the European Miocene amphicyonids

4.1.1. Body mass estimation, an introduction

Body mass is probably the single most descriptive character of an animals' size and ecology (Damuth & MacFadden 1990). Since only a few body mass estimates have been published for amphicyonids, I will here attempt to estimate the body masses of a suite of European Miocene amphicyonids.

Several studies determining regression relations between body mass and different dental, cranial and postcranial characters of carnivores have been published. For living carnivores Van Valkenburgh (1990) found that head-body length correlates better with body mass than any dental or cranial character. For comparison she used skull length, orbit to occiput length, and lower carnassial length, choosing the lower carnassial as a dental index because of its key role in the carnivoran dentition. Body mass correlated poorly with carnassial length when the regression was based on a sample of all carnivore families, but relatively well within families. The correlation was best within the Felidae. The felid lower carnassial has a very simplified morphology lacking talonid and metaconid, which may explain why the correlation between body mass and lower carnassial length in living felids is high. Results indicate isometry for the slopes of Felidae and the total sample, positive allometry for the slope of the Mustelidae, and negative allometry for the Canidae, meaning that the larger mustelids have shorter teeth and larger canids have longer teeth relative to their body mass than smaller representatives of these families. Using the regression equations, Van Valkenburgh (1990) also predicted body masses for some New World Oligocene carnivores, including the amphicyonid *Daphoenus vetus*. From her least squares regressions for body mass against head length and body mass against head-body length based on the total sample of Carnivora, she obtained weights of 33 kg and 36 kg, respectively, and she concluded that the relative head length of this species was probably similar to that of Recent species. Body mass estimates based on lower carnassial length for this amphicyonid was not included in her study.

Creighton (1980) studied the correlation between head-body length and m1 area in terrestrial mammals by means of reduced major axis regression. He found an isometric slope for Carnivora, but, interestingly, within family regression analyses deviated more from isometry. In the Mustelidae and Canidae m1 area scaled with positive allometry, which means that larger animals have relatively smaller teeth, while the felid slope showed significant negative allometry. In other words, contrary to canids, larger felids have larger teeth relative to head-body length, which probably means higher body mass too. Similarly, Legendre and Roth (1988) studied the correlation between lower carnassial crown area and body mass using principal axis and least squares regression. They also used their equations to estimate body masses of extinct carnivores. They predicted a body mass of 630 kg for an amphicyonid (*A. major*) that is also examined in the present study. This estimate will be commented on in the discussion. In general, dentally based estimates that include the widths of teeth have been suggested to be more dependent on diet than body mass (Damuth 1990, Fortelius 1990, Janis 1990).

Anyonge (1993) made a thorough study of relationships between cross-sectional properties, distal articular surface area, lengths and circumferences of proximal limb bones (femur and humerus) and body mass in Felidae, Canidae, Ursidae, and Hyaenidae. He found that even when based on the total sample of these four families all measures except limb lengths correlated well with body mass. He obtained equations to estimate body masses for extinct Plio-Pleistocene felids.

Viranta (1994) studied the morphology of the limb bones and the body mass of the extinct Pleistocene cave bear (*Ursus spelaeus*). In that study I compared the morphology of Recent bears with that of the cave bear, and found that the cave bear had wider shafts of the long bones than Recent bears, but that the articular surfaces scaled with bone length in a similar manner in all bears. As the cave bear long bones are not longer than those of the polar bear (and the articular surfaces accordingly not wider), body mass estimates within the range of variation of the polar bear were obtained for the cave bear.

4.1.2. Methods

During the Miocene there were in Europe ecologically very different groups of Amphicyonidae. Four genera consisted of animals ranging in dietary adaptations from omnivory to carnivory (see 4.2.). The genus *Agnotherium* and the closely related *Thaumastocyon* were hypercarnivorous predators with slicing carnassials and reduced postcarnassial teeth. For many of the amphicyonid species few or no cranial or postcranial bones are known. Species have often been recognized solely on the basis of their dentition or even on the basis of individual teeth. There is, however, a complete skeleton of *A. major* from Sansan, France, and a few amphicyonid bones from other localities. Most of the known postcrania support the idea that the Miocene amphicyonids maintained a primitive body plan with relatively graviportal limbs, lacking any special adaptations to cursoriality. Because this is not necessarily true of the hypercarnivorous clade, they are considered separately here. In addition the genus *Cynelos* shows a tendency towards more cursoriality and a set of metapodials attributed to *C. schlosseri* is studied in comparison with that of *A. major* in order to test this hypothesis.

The relative lengths of the limb bones and the dimensions of the proximal limb bones of the Sansan *A. major* skeleton were compared with living carnivores by means of ratio diagrams (Simpson 1941). Owing to the lack of known back lengths, and the absence of vertebrae in the comparative material, head-body lengths were not included in the ratio diagrams. Body length ranges for the reference species were taken from the literature and compared with the body length of *A. major* in a separate table.

Because of the above mentioned lack of skeletal material (both cranial and postcranial) for most of the species, I needed a method based on teeth in order to estimate body masses for all the known European Miocene Amphicyonidae. Having a complete amphicyonid skeleton from Sansan available and assuming that it is representative of the body plan of most of the Miocene amphicyonid species, I used the available methods from the literature as reviewed in the introduction (Anyonge 1993, Van Valkenburgh 1990, Viranta 1994) to estimate body mass for this specimen. To estimate body masses of other taxa I chose a method based on the teeth on the basis of how well its estimate was in accordance with the results from these other methods in estimating the body mass of the Sansan skeleton. I then used the chosen method based on teeth. However, I first investigated the presence of sexual dimorphism in species where the sample size was large enough, and when this was observed, I calculated the body masses using estimated midpoints of the range of each sex.

Very few if any postcranial remains are known of *Agnotherium* spp. or *Thaumastocyon* spp. The few bones attributed to the group, if they really belong here, suggest that the body plan of these species was very different from that of the rest the Amphicyonidae. Apparently, the postcranial skeleton and dentition both differ from those of the other amphicyonids, an observation that would be likely to affect the correlation of many characters with body mass. Therefore, I have omitted these two genera from the limb dimension and body mass study. Rough body mass estimates are given for each of these species.

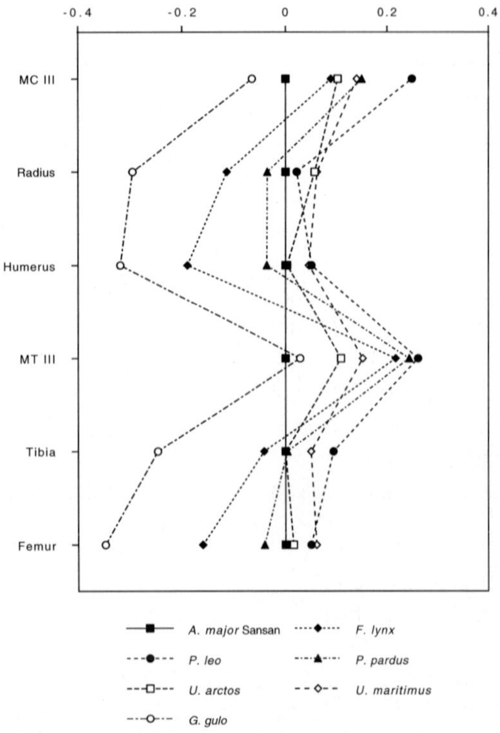

Fig. 13. A ratio diagram of limb lengths. Recent carnivores compared with *A. major* (zero).

4.1.3. Results

4.1.3.1. Limb bone proportions

A. major differs in relative limb bone length from all other species investigated in that the metapodials are shorter (Fig. 13). The felids, wolf, and wolverine have relatively longer tibiae relative to their femora than *A. major*. The length of the bear tibiae are about the same relative to the femora as in *A. major* or slightly longer. The radius is long relative to the humerus in the long-legged felid, the lynx, whereas in the other felids, the wolf and the wolverine this length relationship is similar to that in *A. major*. The humerus/femur length ratios in the felids, wolf, and wolverine are very similar to that of *A. major*. The bears have relatively slightly longer humeri than *A. major*.

The humeri of the felids, wolf and wolverine have relatively narrower epiphyses than that of the amphicyonid. This is true especially of *L. lynx*. In the lion, the distal epiphyses are only slightly narrower than in *A. major*. The bear humeri have relatively greater proximal widths, but notably narrower distal surfaces. The femora of the bears have wider distal epiphyses than the femur of *A. major,* with the exception of the American black bear, which has proportions similar to the amphicyonid. The lion femur is very similar to *A. major* in its proportions, while the rest of the felids and the wolf have relatively narrower epiphyses. The wolverine has a wider distal epiphysis than *A. major.*

4.1.3.2. Head–body length-based mass estimates for A. major

Head–body lengths and body masses of recent species are given in Table 2. The body mass predictions for this female *A. major* from Sansan vary between 47 and 92 kg, when calculated using the least squares regression equations for head–body lengths given by Van Valkenburgh (1990). The greatest value is returned by the equation for the Ursidae, and the smallest by the equation for the Canidae. When based on the total sample the predicted weight for this animal is 67 kg.

Body length might be a better indicator of body mass than head–body length, since head lengths vary a great deal without contributing significantly to the body mass of a carnivore. This is seen, for example, when the Recent felids are compared with the Recent canids (Van Valkenburgh 1990). Data on body lengths of recent species are not readily available, and the measurements given in the literature usually pertain to head–body length. Therefore, body lengths were not used in this study.

Table 2. Head–body length and body mass for extant species and head–body length for the female *A. major* skeleton from Sansan.

Species	Head–body length (mm)	Body mass (kg)
A. major	1 420	–
Ursus arctos female	1 580–2 840	82
U. a. male	1 580–2 840	115
U. maritimus female	2 000–2 500	150–300
U. m. male	2 000–2 500	300–800
U. americanus female	1 500–1 800	92–140
U. a. male	1 500–1 800	115–270
Gulo gulo female	820–980	11(–20)
G. g. male	820–980	16(–28)
Canis lupus female	1 200–1 700	32(–37)
C. l. male	1 200–1 800	40(–53)
Felis (Profelis) aurata	616–1016	5.3–16
F. (Leptailurus) servalis	670–100	8.7–18
F. (Lynx) lynx female	550–1150	16(–20)
F. l. male	550–1150	19(–24)
Panthera pardus male	910–1910	37–90
P. leo female	1 400–1 750	120–182
P. l. male	1 700–2 500	150–250

4.1.3.3. Head–body length vs. other indices

A. major has a greater skull length relative to head–body length than do canids, ursids, mustelids or felids. The difference is most clear when *A. major* is compared with the short-snouted mustelids and felids (Fig. 14a). In its relationship between occiput-to-orbit and head–body length *A. major* is similar to ursids, mustelids and felids. The amphicyonid seems to have a slightly longer occiputto-orbit length than canids (Fig. 14b). The carnassial of the amphicyonid is longer in comparison to head–body length than in felids, but within the ranges of variation of the other three families (Fig. 14c).

4.1.3.4. Comparison of different estimates

Most of the measurements of this *A. major* female are within the range of variation of the leopard (*Panthera pardus*) and approaching those of the female lion (*P. leo*) on the basis of original data and data from Nowak (1991). An exception is the lower carnassial, which is much shorter in leopards. The body mass range given for leopards is 28 to 90 kg, and for female lions 120–182 kg. Body mass estimates for the amphicyonid based on head–body length lie within the leopard range. All the estimates based on skull length are higher; 117 to 393 kg. Occiput to orbit length gives estimates from 78 to 133 kg. The estimates based on the lower carnassial vary between 29 and 218 kg, being highest when the felid equation is used and lowest when the canid equation is used (Table 3).

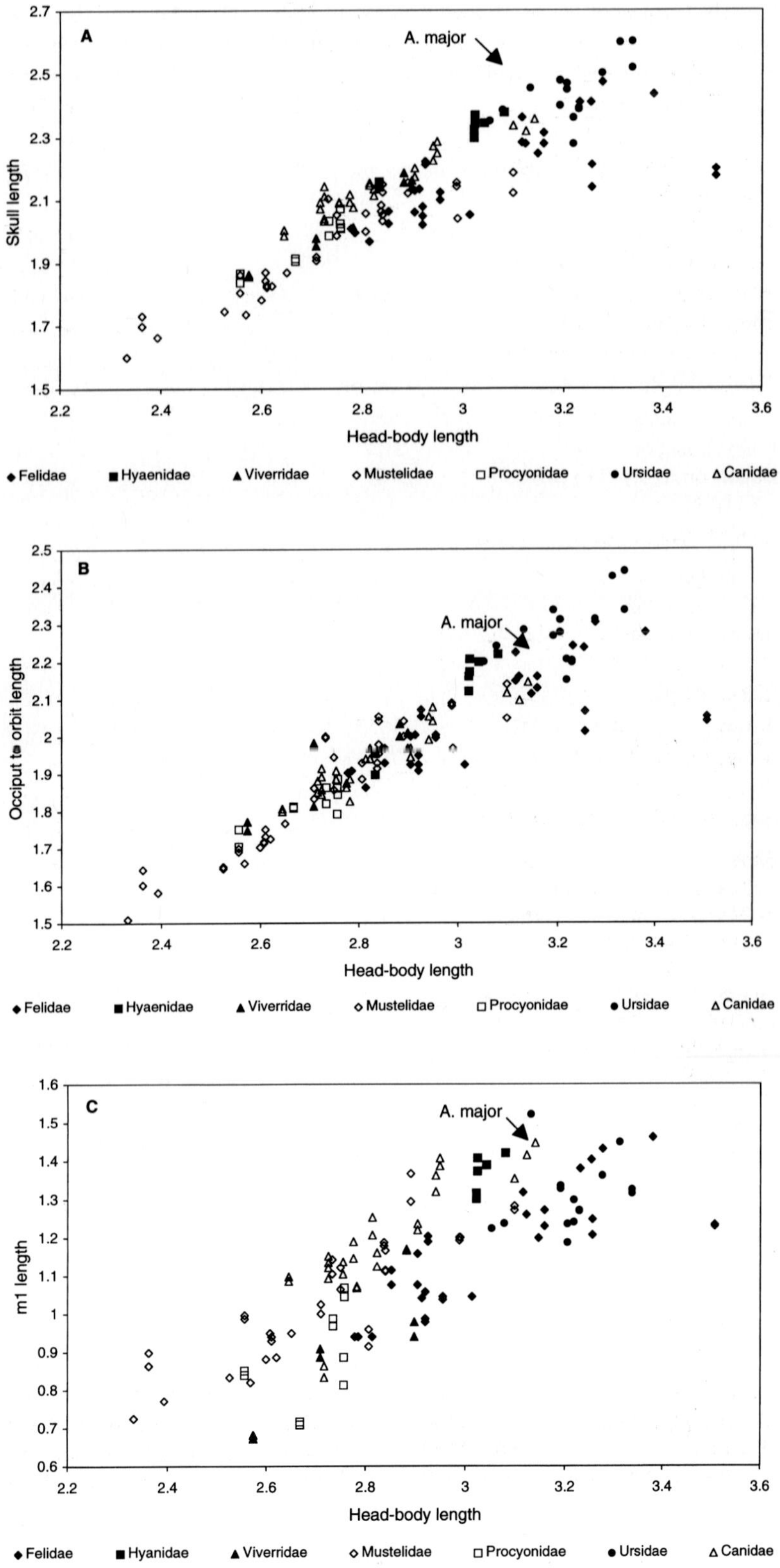

Fig. 14. Head–body length of recent carnivores (data from Van Valkenburgh 1990) and *A. major* plotted against: skull length (A), occiput to orbit length (B), and lower carnassial length (C).

Table 3. Body mass estimates for Sansan *A. major* female skeleton from published regressions.

Taxa included in the analysis	Body mass (kg) estimate returned for *A. major*
Regressions from Van Valkenburgh (1990).	
Based on head-body length	
Total (All Carnivoran families)	69
Canidae	47
Ursidae	92
Mustelidae	60
Felidae	55
Based on skull length	
Total (All Carnivoran families)	236
Canidae	116
Ursidae	218
Mustelidae	393
Felidae	340
Based on occiput to orbit	
Total	104
Canidae	82
Ursidae	122
Mustelidae	78
Felidae	133
Based on lower carnassial length	
Total	126
Canidae	29
Ursidae	x
Mustelidae	120
Felidae	217
Regressions from Anyonge (1993).	
Based on length of humerus	
Total	103
Felidae	120
Canidae	41
Ursidae	91
Based on length of femur	
Total	121
Felidae	144
Canidae	46
Ursidae	119

Bone	Body mass (kg) estimate returned for *A. major*
Regressions from Viranta (1994). All analysis were run on modern ursids only.	
Based on proximal articular surface widths	
Radius	146
Tibia	89
Based on distal articular surface widths	
Humerus	60
Femur	153
Tibia	84
Based on total lengths	
Humerus	22
Radius	67
Femur	87
Tibia	98

4.1.3.5. Body mass estimates for European Miocene amphicyonids

Felids have short carnassials in relation to their body size, and accordingly their equation gives a high predicted body mass when applied to the Sansan *A. major*. On the other hand, canids have long carnassials, and their equation predicts a low body mass for *A. major*. Body size in terms of the lengths of the back and limbs of the studied *A. major* individual are within the range of *P. pardus*. The mustelid equation gives an estimate within this range for the Sansan amphicyonid, but a likely source of error in this equation is the relatively low body mass of recent mustelids. Schmidt-Nielsen (1984, see also Fortelius 1990) has pointed out that allometric equations should not be used for extrapolation in general, although sometimes it is the only alternative (e.g. Fortelius & Kappleman 1993). The size range of the Recent Mustelidae is very narrow, and the fact that the slope of the mustelid regression on the lower carnassial shows positive allometry might suggest that the range is not great enough for a reliable regression analysis.

The total sample regression slope of body mass against carnassial length is isometric, as has been observed in other studies where the regression analyses have been carried out on higher taxonomic units (Fortelius 1984, and references therein). As noted before, some carnivore families display negative (Canidae), and others positive (Mustelidae) allometry. I feel that using the isometric equation, Van Valkenburgh's (1990) regression based on her total sample, is the safest averaging method. The body mass prediction based on this equation for the Sansan skeleton is, however, slightly above the range of *P. pardus*, unlike that based on head-body length, and this suggests that the prediction may be a slight overestimate. Considering the general accuracy of the methods presented by Van Valkenburgh (1990), the prediction based on the total sample equation is still reasonable, however.

Predicted body masses for European Miocene amphicyonids are shown in Table 4. The species for which estimates are calculated separately for each sex are those that show a bimodal distribution in carnassial length, which was interpreted to indicate sexual dimorphism.

4.1.3.6. Increased cursoriality in C. schlosseri

It has been suggested that the genera *Agnotherium* and *Thaumastocyon* had more cursorial and less graviportal limbs than rest of the European Miocene amphicyonids. Cursoriality may have also evolved in other clades. Abundant unpublished material of amphicyonid postcranial bones has been collected from Wintershof West. It was not possible to examine this collection thoroughly for this study, but a set of metapodials attributed to *C. schlosseri* based on their size, was measured and compared to those of *A. major* by means of a ratio diagram (Fig. 15ab). The metapodials of *C. schlosseri* are absolutely longer in comparison to those of *A. major*. This is unexpected, because the estimated body mass of *C. schlosseri* is much smaller than that of *A. major*. They are also longer in relation to their proximal widths than those of *A. major*. The metapodials of *C. schlosseri* shorten from medial to lateral more than those of *A. major*. Both of these features are evident in the Recent wolf, which is a cursorial carnivore.

A trend towards more gracile distal limbs is obvious in *C. schlosseri*. This lends support to the idea that some of the amphicyonids may have evolved towards more cursorial forms and away from the basic robust body plan typical of rest of the Miocene amphicyonids and exemplified by

Table 4. Body mass estimates based on lower carnassial length.

Species/sex	m1 (mm)	Body mass est. (kg)
Pseudarctos bavaricus	12.3	9
Amphicyon major/female	29.8	122
A. major/male	35.3	212
A. giganteus/female	31.9	157
A. giganteus/male	40.4	317
A. pannonicus	34.5	198
A. gutmanni	37.1	246
Pseudocyon sansaniensis	29.6	126
P. steinheimensis	29.4	123
P. caucasicus	30.3	130
P. styriacus	29	118
Cynelos helbingi/female	23	60
C. helbingi/male	26	86
C. schlosseri	16.8	23
Euroamphicyon olisiponensis	31.2	147
"A." castellanus	34.5	198

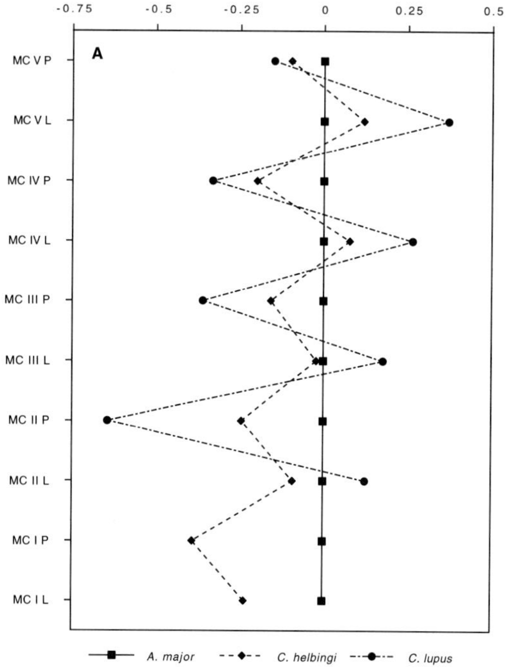

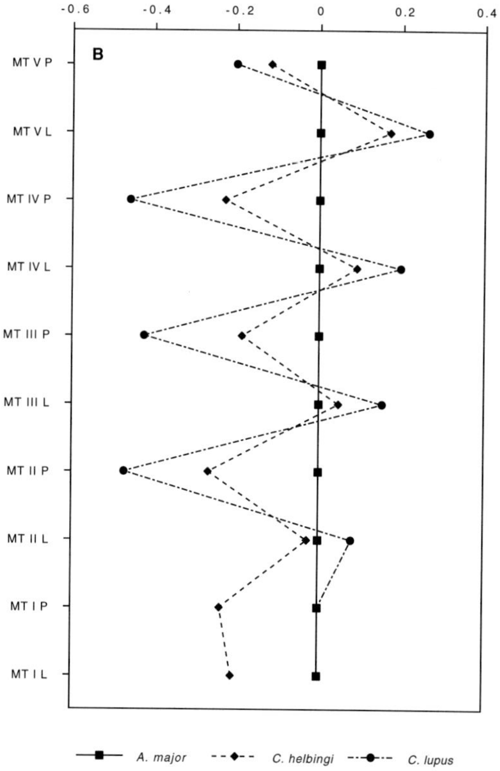

Fig. 15. Ratio diagram comparing metcarpals (A), and metatarsals (B) of *C. helbingi* and *C. lupus* with those of *A. major* (zeroline).

A. major. This may have an effect on the body mass estimate of the species.

4.1.3.7. Body mass estimates for Agnotherium spp. and Thaumastocyon spp.

Observed cursorial traits along with the highly modified dentition of *Agnotherium* spp. and *Thaumastocyon* spp., makes them likely to differ from the regression relations between body parts and body mass of other amphicyonids. For lack of a better method, Van Valkenburgh's (1990) regression on lower carnassial length based on her total sample was also used to estimate body masses for *Agnotherium* spp. and *Thaumastocyon* spp. (Table 5). No test of the reliability of the estimates are available, and they are used here as indications only.

Table 5. Tentative body mass estimates for *Agnotherium* spp. and *Thaumastocyon* spp.

Species	m1 (mm)	Body mass est. (kg)
Agnotherium antiquum	32	158
A. grivense	34	173
Thaumastocyon bourgeoisi	26	86

4.1.4. Discussion

A few remarks must be made regarding the body mass estimates. As emphasized before, postcranial material is known for only a few amphicyonid species, so the possibility remains that some species evolved significantly different morphologies. The trends found in the metapodial morphology in *C. schlosseri* and the cursorial morphologies in *Agnotherium antiquum* and *Thaumastocyon dirus* suggest that this is in fact the case. For such species the proposed estimates would be overestimates. As for the estimate predicted by Legendre and Roth (1988) for this *A. major* female, 630 kg is certainly more than a carnivore with the body dimensions of a leopard could possibly weigh. The result, which was obtained on the basis of crown area of the lower carnassial, is of interest for another reason: it confirms that amphicyonid carnassials were broad in comparison to the carnassials of modern carnivores.

Some amphicyonid species show enormous variation in body mass (or carnassial length), some of which can be explained as sexual dimorphism. A body mass of 355 kg is predicted here for *A. giganteus* males. There is a femur from Lisboa which is larger than that of any known recent grizzly bear femur. (A large Kodiak bear male with a weight of 680 kg had a femur of length of 520 mm, while this amphicyonid femur was approximately 600 mm in length.) This specimen is attributed to *A. giganteus* (Ginsburg & Telles Antunes 1968). Ginsburg and Telles Antunes (1968) estimated the size variation of *A. giganteus* to be 30%. Since they refer to length measurements, this estimate seems reliable also in the light of the present study. Again, among brown bears great variation in size has been observed. While average body masses for the Finnish population are 115 kg and 95 kg (Pulliainen 1973) and for southwestern Yukon for males and females 139 kg and 95 kg respectively, body masses as high as 780 kg have been recorded for the Alaskan grizzlies (Nowak 1991). In the light of this data it seems possible that in favorable conditions giant individuals of *A. giganteus* could have existed.

4.2. Ecology of the European Miocene amphicyonids

4.2.1. Introduction

The mammalian order Carnivora is diagnosed by the presence of carnassial teeth. According to Fox and Youzwyshyn (1994) the common ancestor of the order is Cretaceous in age. They suggested (see also Butler 1946, Van Valen 1969, Springhorn 1980) that this ancestor had an unspecialized, puncture-crushing dentition without shearing blades, unlike the Creodonta or Palaeoryctidae, which have also been implicated in Carnivoran ancestry (see e.g. Lillegraven 1969). In the Carnivora, P4 evolved to serve as a blade, and became the upper carnassial. The lower carnassial, however, is a molar (m1) that maintained a puncture-crushing talonid, while the trigonid formed a blade for slicing. The postcarnassial molars maintained the puncture-crushing function, and the precarnassial premolars were reduced and mainly used for holding the prey.

This occurrence of functionally different postcanine teeth in the dentition has led to adaptive flexibility in Carnivoran evolution (e.g. Werdelin 1987). Recent carnivores occupy a variety of omnivorous and faunivorous niches, as was true of Tertiary carnivores as well. Even closely related species or different populations within a species can be very different in their ecologies. Recent Canidae, Mustelidae and Ursidae are good examples. The polar bear *Ursus maritimus* relies mainly on medium-sized vertebrate prey (seals), while the closely related brown bear (*Ursus arctos*) is omnivorous in most areas in which it occurs. Brown bears may turn to a mainly faunivorous diet in some areas (Nowak 1991).

Due to the fact that many attributes of the ecology of extinct species, such as the preferred prey and its size, remain unknown, their niches cannot be reconstructed precisely. To understand the ecologies of extinct species we must rely on their morphological characters to place them in broad feeding categories. Interpretations based on morphology become more reliable when phylogeny is also taken into account. Without a knowledge of the condition in closely related taxa, it is impossible to judge the importance of the characters to the ecology of a given species. Although all Recent bears have large molars and a lower carnassial with a large talonid basin, the polar bear shows more carnivorous adaptations in its dentition than other bears. Given the short time of separation between the polar and brown bear (less than 0.3 Ma, Kurtén 1964, Cronin *et al.* 1991), the adaptations of the former are remarkable and indicate strong selection pressure for carnivorous traits.

Among Recent carnivores the giant panda (*Ailuropoda melanoleuca*) relies heavily on plant foodstuffs in its diet, and may be classified as herbivorous. The giant panda almost exclusively consumes bamboo. The dentition consists mainly of broad grinding surfaces. Even the posterior premolars are slightly broadened. P3 has two lingual cusps and p4 has one. The molars are large and tuberculate. The m1 paraconid is in a lingual position, the protoconid is short, and thus the trigonid fails to form a blade. In the P4 the protocone is posterior making the tooth triangular in occlusal view. It has no metastyle blade, but the metacone forms a cusp (Davis 1964).

Dominated by crushing features, the dentition of omnivorous carnivores still maintains some of the slicing characters. Recent bears (with the exception of the polar bear) are good examples of omnivorous carnivores. The lower carnassials still maintain a blade-like trigonid, but the talonid is very broad and makes up a large part of the length of the carnassial. Broad, grinding postcarnassial teeth are typical adaptations to a more herbivorous diet. The premolars may be very reduced, the canines are usually robust and the incisors are well-developed.

Faunivorous forms include a large range of ecologies that are separated from the omnivorous forms in that a major part of the diet consists of animal foodstuffs. Faunivores can further be classified on the basis of the quality and quantity of the animal foodstuffs they consume. Initially, faunivores can be separated into two groups: those that primarily eat invertebrates (invertebrativores), and those that primarily eat vertebrates (carnivores). The invertebrativorous forms differ clearly from the carnivorous forms in their dental morphology. Invertebrativory requires the crushing of hard, but fragile items, such as the chitin exoskeletons of insects, and this is realized by individual tapering points rather than long blades (see Strait 1993). In the lower carnassial, pointed talonid cusps and a small paraconid in a lingual position are indications of invertebrativory, as well as pointed cusps on the other posterior molars. I know of no invertebrativorous amphicyonids.

Carnivorous forms include forms that primarily consume meat. Carnivores that despite their preference for meat eating are able to switch to a herbivorous diet have been called "mesocarnivores". Their dentition shows both the shearing and the grinding conditions in roughly equal proportions. They often have well-developed premolars and a lower carnassial with a long blade in the trigonid. The lower carnassial may have a bi- or tricuspid talonid and the molars are often large.

The extreme carnivorous forms have lost the ability to ingest large amounts of plant matter. Their physiology necessitates meat in their diet, and their morphology is best suited to killing vertebrates and consuming their meat. This hypercarnivory is shown in the dentition by lengthening of the cutting blade, at the expense of the grinding area. In the lower carnassial, a long trigonid (in relation to the talonid), reduced metaconid, and reduced talonid width indicate increased slicing function. The postcarnassial molars are often reduced, or even lost (Van Valkenburgh 1989).

An additional feeding behaviour, related to carnivory (either hyper- or mesocarnivory), is that of bone breaking in order to consume the nutritious marrow inside. This can be achieved by at least two different ways, by bone crushing and bone cracking (Werdelin 1989). Bone cracking involves focusing the bite on the bone at a point, often by means of a pair of premolars. In bone crushing the bite force is distributed over a wider area, often the postcarnassial molars. In the Recent faunas, the bone breakers are not only specialized carrion-feeders, but also active hunters. There is some evidence that scavenging specialists may have occurred in Tertiary faunas (Werdelin 1989, Zhou *et al.* 1992).

Many Miocene amphicyonids display hypertrophy of their second molars (m2 and M2) (Kuss 1965). Enlarged molars broaden the grinding and crushing surface of the dentition. This could be interpreted as an adaptation to bone crushing or omnivory. The amphicyonids underwent a major turnover in the Early Miocene and this is when all the later Miocene taxa originated. Their radiation was rapid and their subsequent evolutionary history short. As apparent from the cladogram (Fig. 11) the hypertrophy was inherited from a common ancestor, and thus even slight deviations from this pattern are of interest.

4.2.2. Methods

In this section, general descriptions of the possible ecologies of amphicyonid species are provided. The emphasis is on feeding ecologies and a dietary classification adapted and modified from Ewer (1973) and Van Valkenburgh (1989) is used. Four categories of amphicyonids are recognized: omnivores, mesocarnivores, bone-crushers and hypercarnivores. In principle, two kinds of morphological data are available: morphologies of unworn teeth and bone, and tooth wear. In addition to qualitative methods, a quantitative method is applied. The blade length of the trigonid of the lower carnassial is plotted against the total length of the lower molars. A long blade indicates an emphasis on carnivorous habits (Van Valkenburgh 1989).

The functional morphology of the postcranial skeleton is important to the study of the ecology and evolution of mammals. Among other things, the postcranial skeleton reflects the hunting mode and living substrate of a carnivore. In this study, owing to the scarcity of material of European Miocene Amphicyonidae, only minor emphasis is placed on the postcrania. However, some aspects of modes of locomotion are considered on the basis of data in the literature.

4.2.3. Results

4.2.3.1. Dietary adaptations in the European Miocene amphicyonids

Omnivores

Pseudarctos bavaricus and *Ictiocyon socialis* both show typical omnivore characters in their dentition. They have high-crowned teeth with blunt cusps. The lower carnassial has a short trigonid with a relatively posterior metaconid, and there is practically no cutting blade. It follows that the talonid basin is large. The upper carnassial is short and stout. The large grinding area in the dentition is also retained by means of the large occlusal surfaces of the postcarnassial molars. *P. bavaricus* and *I. socialis* were the smallest amphicyonids (body masses less than 20 kg) in the European Miocene, and probably had very different ecologies from the other species in the family.

As is the case with amphicyonids in general, there are no modern analogues for *P. bavaricus* or *I. socialis*. They may have had diet similar to that of the Recent lesser panda (*Ailurus fulgens*). The lesser panda feeds on bamboo sprouts, grasses, roots, fruits and nuts. It also occasionally takes small vertebrates, eggs and insects (Nowak 1991).

Mesocarnivores

Cynelos spp. have a primitive dentition: they retain the primitive tooth formula, the molar cusps are well developed, the premolar row is quite crowded, and the canines are not especially robust. In fact, functionally, the dentitions of these amphicyonids resemble those found in many mesocarnivorous Recent canids. Based on their moderate sizes and non-specialized dentition, *Cynelos* spp. were most likely mesocarnivorous predators of relatively small prey. High cusps in the two last lower molars may hint at some proportion of invertebrates in diet. On the other hand, similar cusps in the postcarnassial molars are found in the bush dog, *Speothos venaticus,* which is a hypercarnivore (Nowak 1991, Van Valkenburgh 1989). The wear on the teeth of *Cynelos* spp. is both vertical and horizontal, and this is in accordance with the present interpretation.

Hunt (1972) attributed a lower carnassial, first upper molar and proximal end of a radius from Agate Spring Quarries, Nebraska, North America to *Cynelos* sp. He noted that the proximal articular surface of the radius has a spiral configuration, such that the surface rises in height from the lateral to the medial edge. This means that there was a large angle between the long axis of the humerus and that of the radius, so that the elbow stuck out laterally. This allows for greater supination and pronation than in cursorial forms, and rules out the possibility of this species being cursorial. This observation decreases the likelihood that the other representatives of the genus were cursorial. On the other hand, the relationships between North American and Eurasian amphicyonids remain to be resolved. The American *Cynelos* sp. and the European *Cynelos* spp. may or may not represent the same genus.

Springhorn (1977b) studied limb bones of an early *Cynelos* sp. and marked characters that indicate digitigrade mode of locomotion. Also, as noted earlier (Fig. 15) the Wintershof-West collection contains unsorted amphicyonid postcranial material including a complete set of metatarsals and metacarpals. Based on their size these may be attributed to *C. schlosseri*. They are similar in general to those of *A. major*, but appear more gracile. They indicate less graviportal limbs, and probably increased cursoriality. These are also trends suggested to be common in digitigrade carnivores (Ginsburg 1961b). In this light, it seems likely that the *Cynelos* spp. were animals fleeter of foot than *A. major* and other *Amphicyon* spp., but not as swift as the Recent wolf.

Bone-crushing mesocarnivores

Ginsburg (1966) and Ginsburg and Telles Antunes (1968) have, respectively, interpreted *A. major* and

A. giganteus as active and agile predators. I agree with their analysis. There are no modern analogues for the dentitions of these species. They have well-developed molars and a sparsely distributed, complete set of premolars. The heavy carnassials with horizontal abrasion indicate bone-crushing. The posterior molars also have heavy wear, and were probably used in crushing. The high sagittal crest with two layers of bone known for *A. major* and American *Amphicyon* spp. (Olsen 1958, Ginsburg 1961, pers. obs.) also indicates the powerful bite necessary for bone breaking. The modern spotted hyena has a similarly strong sagittal crest. The long snout with a long row of premolars was probably used for catching and holding the prey. The relatively long canines with their double cutting edges could have been used in a neck bite.

The body plan of *Amphicyon* spp. indicates active hunting. Although as far as known the limbs of *Amphicyon* spp. have short distal elements that show no cursorial adaptations like those of Recent pursuit predators, their flexible backs and powerful limbs seem to indicate either group hunting or the solitary stalk-pounce mode of hunting performed by today's large felids and bears. The long tail is useful in balancing the pounce, as seen in the large felids. Group hunting is unlikely in such heavily forested areas as the Early and Middle Miocene of Europe appears to have been.

Of the Recent large-bodied carnivores, bears and felids carry out solitary hunting in closed habitats. Recent brown bears hunt mostly through a short pursuit followed by a powerful attack that either wounds or kills the prey depending on the size-relationship between prey and predator and the body part attacked. Often the prey dies of its wounds in a short time, and is consumed by the bear. This differs from the felid method in that the attack is not necessarily targeted on the neck or snout. The large pantherine felids attack the snout or throat of the prey and suffocate it, while the smaller ones usually attack with a neck bite. An attack by a felid usually leads to more rapid death for the prey. Both of these hunting modes may have been carried out by the bone-crushing amphicyonids. They were probably active predators, that were also able to consume the marrow of larger bones.

A. pannonicus and *A. gutmanni* are both known only from their lower carnassials, which share quite similar morphologies. They are also very similar to other *Amphicyon* spp. Assuming that they were closely related to that group they were probably actively hunting predators with bone-crushing habits.

E. olisiponensis is similar to *A. major* and *A. giganteus* in the general plan of its dentition, and it probably had a similar feeding ecology as well. Postcrania for *E. olisiponensis* are not known, so the mode of locomotion remains unknown.

The dentition of *Pseudocyon sansaniensis* with its truncated m1 talonid and relatively narrow molars, indicates a carnivorous diet and more limited adaptations to bone crushing. The same applies to *P. styriacus* and *P. steinheimensis*. These species probably did some bone-crushing as shown by the horizontal wear on the molars of these species, and were active hunters like *Amphicyon* spp.

P. sansaniensis has long phalanges and a short olecranon process compared with those of *A. major* (Ginsburg 1966). These characters are common in climbing carnivores (Van Valkenburgh 1987), but *P. sansaniensis* was relatively heavy (125 kg) for a climbing animal.

The dentition of *G. macrognathus* differs from that of the other Miocene bone-breaking amphicyonids in Europe. It was a bone-cracker rather than a bone-crusher. As seen in the type specimen (AMNH26597) and in AMNH26601 (Colbert 1939), the premolars have a pyramidal shape and the posterior molars are reduced. The robust symphysis and deep mandibular ramus indicate a powerful bite.

Hypercarnivores

The two known mandibles of *P. caucasicus* show hypercarnivorous trends. They have lost the anterior premolars and have a narrow, slicing carnassial. The two posterior molars are small. The wear on the lower carnassial of the type specimen is predominantly vertical. The Spanish *"A." castellanus* shows a similar reduction of the metaconid and tapering of the talonid in m1. Of the premolars only p4 and P3 are known for *"A." castellanus*. They are both reduced, and especially the P3 is diminutive. In the light of this it seems likely that

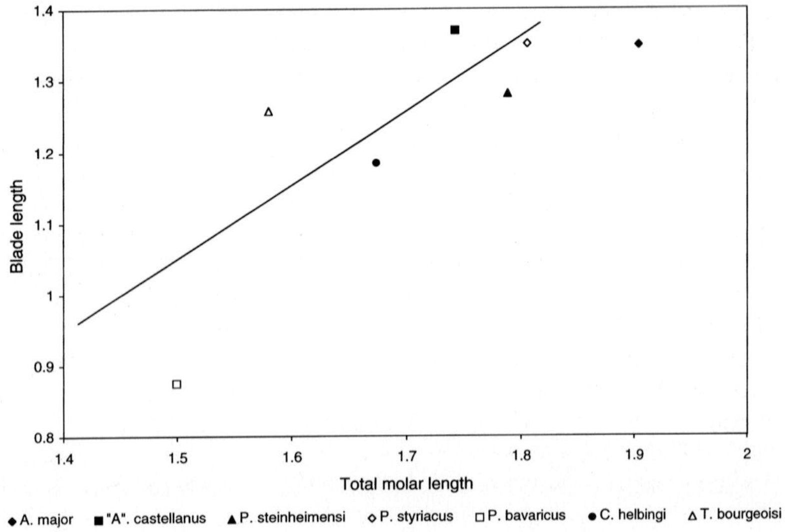

Fig. 16. Blade length of the lower carnassial plotted against total molar length (mm).

the most anterior premolars were absent in this species as well. These two late amphicyonids evidently very much relied on meat in their diet.

Judging by the sparse material known, *Y. valentiana* shows affinities with the hypercarnivorous amphicyonids. With its reduced molars, it probably was hypercarnivorous.

Agnotherium spp. and *Thaumastocyon* spp. have special adaptations to slicing in their dentition. Their vertical tooth wear supports the idea that the primary function of the carnassials in these forms was for slicing. Reduction of the molars has gone so far in this group that any grinding of food might have been difficult. The sparse postcranial material known of *Agnotherium* shows clear cursorial adaptations (see 1.4.). Because of the unclear associations of these specimens it can only be hypothesized that these were cursorial animals, and may have been pursuit hunters.

4.2.3.2. Slicing vs. total molar length in the lower dentition

Complete sets of lower molars are available for seven species. Their blade lengths are here (Fig. 16) plotted against the total lengths of the molars. The distribution of the species in the plot is in accordance with their placement in the feeding categories above. The omnivorous *P. bavaricus* has a short blade characteristic of a less carnivorous mode of life. The bone-crushing mesocarnivores have longer blades than *P. bavaricus*, but shorter than the species that were classified as hypercarnivores. The blade is shorter in *A. major* than in *P. steinheimensis* or *P. styriacus*. *C. helbingi*, which was classified as a mesocarnivore, has a blade length similar to that of *P. steinheimensis*. The two hypercarnivores, "*A*". *castellanus* and *T. bourgeoisi*, have long blade lengths.

4.2.4. Discussion

The Middle and Late Miocene amphicyonids were an ecologically diverse group of carnivores. Because there are no living descendants of amphicyonids, and because of their unique characteristics, reconstructing their exact ecologies is difficult. Most likely many of them lack modern ecological analogues. However, it is fruitful to view these species in the context of the Middle and Late Miocene carnivore fauna as a whole. This is done in the next section.

4.3. Amphicyonid radiation and extinction in the Miocene

4.3.1. Introduction

The fate of the amphicyonids was marked in a turnover event, the mid-Vallesian crisis, when only a few amphicyonid species survived on limited areas in Eastern Europe. The term turnover refers to the rise of new species and the extinction of old (through migration or evolution) in a local fauna (Kurtén 1959). Turnover events are periods marked by enhanced turnover. During mid-Vallesian time extended migrations of new taxa from eastern Eurasia (initially from the New World) took place. Among the immigrants, species of hipparionine horses (Equidae) were the most remarkable. They differed from the earlier Equidae (Anchitheriinae) in having longer limbs and hypsodont teeth. These are both considered to be adaptations to more open environments. At the same time, the climate apparently changed some ways, causing changes in habitats. Both the migrations and the climatic change have been related to the extinctions among the earlier fauna (e.g. Bernor 1983).

In the previous chapters the taxonomy, systematics and ecology of Middle and Late Miocene amphicyonids have been reviewed. In this chapter these are placed in a chronologic and environmental context. This chapter sums up the last radiation of amphicyonids in Europe which was followed by their total extinction, and tries to ascertain the reasons for this extinction using both the ecological and paleontological literature.

4.3.2. Material and methods

The present taxonomic revision of amphicyonids, along with body mass estimates and diet interpretations, were entered into the NOW database (Fortelius *et al.* 1996ab). Temporal and spatial ranges and locality information for the amphicyonid species, as well as their contemporaneous ungulate and carnivore species together with body masses (size classes for pecoran species), diets and tooth heights (for ungulates only) were obtained. Chapters in Bernor *et al.* (1996a) form the main bulk of the data in the NOW, although additional references have been used. Specific references for each locality and species are available in the database and they are not repeated here. Since NOW uses the ETE database structure I refer to Damuth (1993) for further explanations regarding the database itself.

In addition, the existing literature was used for the environmental reconstructions. The three main sources were Werdelin (1996b), and Werdelin and Solounias (1996) for the contemporaneous carnivore fauna and Fortelius *et al.* (1996b) for the potential prey species and general environment.

In order to study the paleoecology of a larger area it is crucial to know whether it forms a uniform faunal area or consists of several provinces. Fortelius *et al.* (1996b) tested faunal provinciality in Western Eurasia by analyzing faunal similarities between different regions in different MN units by means of Dice, Pielou and Simpson indices. The conclusion was that during the Miocene this geographic area sustained two biogeographically distinct regions ("provinces") denoted West and East. West includes what is now Portugal, Spain, France, Italy, Germany, Switzerland, Austria, Poland, Czechia and Slovakia while East includes what is now Hungary, Romania, Moldova, Ukraine, Slovenia, Croatia, Macedonia, Serbia, Bosnia, Albania, Greece, Turkey, Bulgaria, Georgia, Iran, Afghanistan, and Kazachstan.

Because the present study concerns only the amphicyonids of Europe, only a few localities from East are included. These are from Georgia, Hungary and Serbia. It is assumed that rest of the amphicyonid localities are from a general, common region, West as defined by Fortelius *et al.* (1996b).

4.3.3. Results

4.3.3.1. Body size and dietary trends in the European Miocene Amphicyonidae

Three large (body mass more than 50 kg) amphicyonid species, *A. major*, *P. caucasicus* and *G. macrognathus* are known from MN 6 in eastern Europe. *P. caucasicus* is hypercarnivorous species, while *G. macrognathus* is classified as bone-cracking and *A. major* as bone-crushing mesocarnivore. Subsequent to this, only one locality assigned to an MN unit has a record of amphicyonids in eastern Europe. This is Rudabánya with undescribed material of a probable *A. antiquum.* Pécs in Hungary, a locality of uncertain age but probably Upper Miocene, records a possibly bone-crushing species of amphicyonid, *A. pannonicus,* not known from elsewhere.

In West, MN 4 saw a diverse amphicyonid fauna with three smaller species, *C. helbingi*, *C. schlosseri* and *Y. valentiana*, and five larger ones, *A. grivense*, *E. olisiponensis*, *A. major*, *A. giganteus* and *P. sansaniensis*. The small species were carnivorous with one of them being a hypercarnivore, while one of the large ones was a hypercarnivore and rest of them were bone-crushing mesocarnivores. All the smaller species disappeared before MN 5, while the large ones survived until MN 7/8 or MN 9. One small omnivorous species (*P. bavaricus*) appeared in MN 5 and survived until MN 9. It was the last of the smaller amphicyonids in Europe.

The large amphicyonids were diverse in MN 5 with three bone-crushing mesocarnivores and

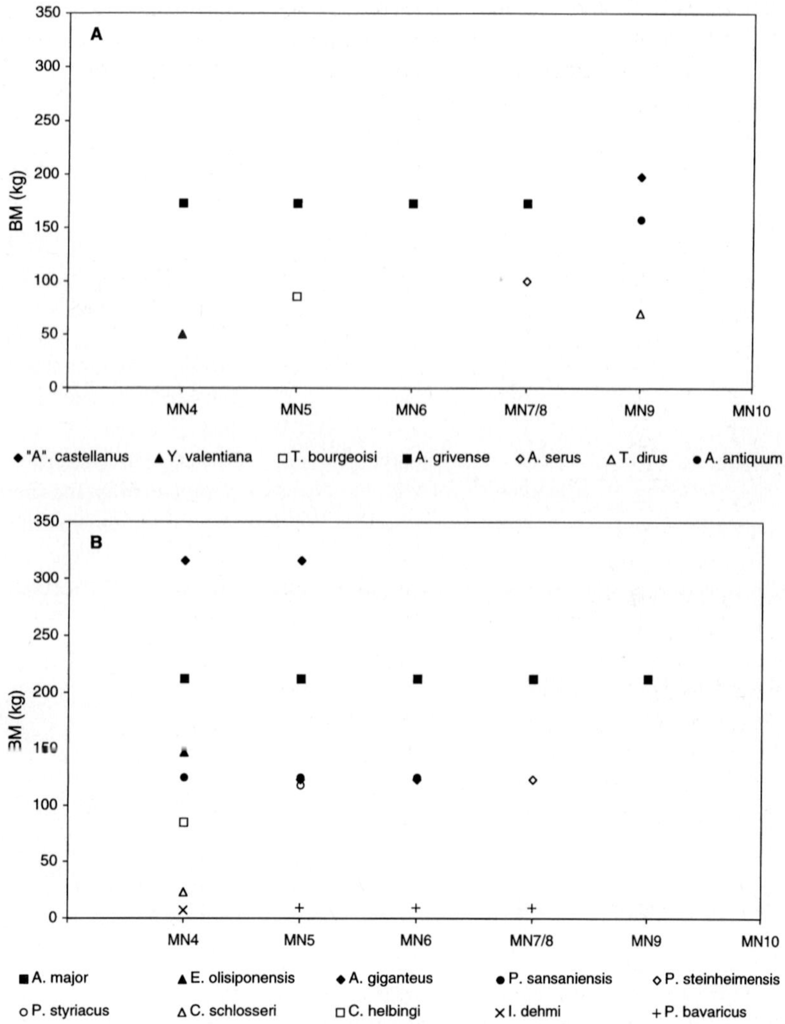

Fig. 17. Amphicyonid diversity by (A) ecomorph hypercarnivore and by (B) ecomorphs bone-crushing mesocarnivore, mesocarnivores and omnivores MN4 to MN10. Each spot refers to a species, and y-axis indicates the estimated body mass. *A. gutmanni*, a species that occured in Germany after MN 9, is not shown in the figure.

two hypercarnivorous species, and in MN 6 with three bone-crushing and one hypercarnivorous species. The decline started in MN 7/8, when there were only three amphicyonid species, of which two were survivors from the previous units and one was a new species. MN 9 marks the probable disappearance of the amphicyonids in West (Fig. 17a and b). There is a possibility that the bone-crushing *A. gutmanni* from Germany survived later than this. There is only one bone-crushing mesocarnivore in MN 9, *A. major,* which is a survivor from MN 5. Interestingly, there are three hypercarnivorous species, *A. antiquum T. dirus* and "*A*". *castellanus*. All three are new species.

4.3.3.2. Climatic change and the extinction of the amphicyonids

Recent work has confirmed the evidence for abrupt changes in climate in the Early Late Miocene and the instability of the climate ever since (Raymo & Ruddiman 1992, and references therein). The drop in paleotemperature may have been as much as 8 degrees on average during MN 9–MN 10 (Burchardt 1978). Direct information on vegetation, such as pollen diagrams or peat formations are rare from the European Miocene. Based on the available evidence Zagwijn and Hager (1987) concluded that data from Germany and the Netherlands show prolonged warm and moist conditions with evergeen forests and many representatives of the oak family (Fagaceae) in the Middle Miocene. The Late Miocene conditions were more temperate with a greater abundance of herbs and deciduous trees.

According to Fortelius *et al.* (1996b), the spatial distribution as well as the dietary guild diversities of mammals in West during MN 4–MN 9 were similar to those found in Recent subtropical, semi-deciduous forests. The situation remained relatively constant in MN 10, although the diversity of brachydont ungulates (typical forest-dwellers) declined dramatically from MN 9 to MN 10. The diversity of some other typical forest inhabitants dropped only in MN 11. These somewhat controversial results suggest that West remained forested with more open areas until MN 11, and that a more serious fragmentation took place in MN 11 (Fortelius *et al.* 1996b). Amphicyonid extinction in West appears to be contemporaneous with the climatic change and precede the fragmentation in MN 11 (although the trend towards landscape opening clearly has started already at the time of amphicyonid extinction, though the process was slow).

4.3.3.3. Extinction of the potential prey

Since most of the amphicyonids seems to have been active predators, the evolution of their ecology might be expected to correlate with that of their prey species. Amphicyonids were clearly not primarily cursorial predators. On the basis of morphology, their probable hunting modes were stalk and pounce as in the large cats of today or laceration as in the Recent brown bear. Stalk and pounce can be used in group hunting in open areas, as in the case of modern lions (*Panthera leo*). However, at least today, it is more commonly carried out by solitary predators of relatively small game in forested areas. Small prey is safer for the predator, which often has to subdue the prey before the kill takes place. As it seems that West mostly consisted of closed habitats in the time of the amphicyonids, group hunting seems unlikely. Amphicyonids were probably solitary hunters and their main prey species can be expected to have had body sizes not greater than the amphicyonids themselves. Fig. 18a and b show pecoran species diversity in different size classes in comparison with the diversity of bone-crushing mesocarnivorous and hypercarnivorous amphicyonids. Fig. 18a shows how both the numbers of pecoran species in body class 36 kg (range 29–43 kg) and bone-crushing amphicyonids decreased MN 6 to MN 9. Diversities in the other pecoran size classes seem to be lower and more stable. A body mass of around 36 kg seems possible for the prey of a solitary stalking amphicyonid weighing less than 100 kg, although it is slightly less than the prey of the similarly sized Recent leopard (*Panthera pardus*) (Nowak 1991).

The suoids represent another possible artiodactylan prey for the amphicyonids. The diversity of suoids was high during MN 4–MN 9 and most of the species had body masses between 10 and 100 kg (Fig. 19a and b). In MN 9 suoids were diverse both in small- and large-bodied species. By MN 10 all the small-bodied forms had become extinct.

If amphicyonids hunted by laceration (as Recent brown bears do), their prey could have included larger ungulates. Laceration is not a very efficient way of hunting, and not very common. The chances that the wounded prey will escape and never be found by the predator are considerable. No large (over 225 kg) pecoran species were available in the environments of West. The pecoran size class 175 kg (115–225 kg) was filled by one or two species during the time period concerned. The family Rhinocerotidae was rich in species during the whole period. Their diversity stayed relatively constant from MN 4 to MN 7/8. Body weights of rhinocerotids varied from 300 kg to 750 kg, except for *Brachypotherium brachypus,* which appeared in MN 6 and weighed some 1 500 kg. In MN 9, the diversity of large-bodied rhinocerotids increased by up to three species, while the diversity of the smaller species remained high. There seems to be no correlation between rhinocerotid diversity and amphicyonid diversity, but the relatively high diversity of the small-sized rhinocerotids allows for the possibility that they were prey of amphicyonids (Fig. 20). No major changes occurred in either equid or chalicotherid diversity. Hipparionine horses appeared in MN 9 with two species, and MN 10 saw four species (Bernor *et al.* 1996b). The diversity of chalico-

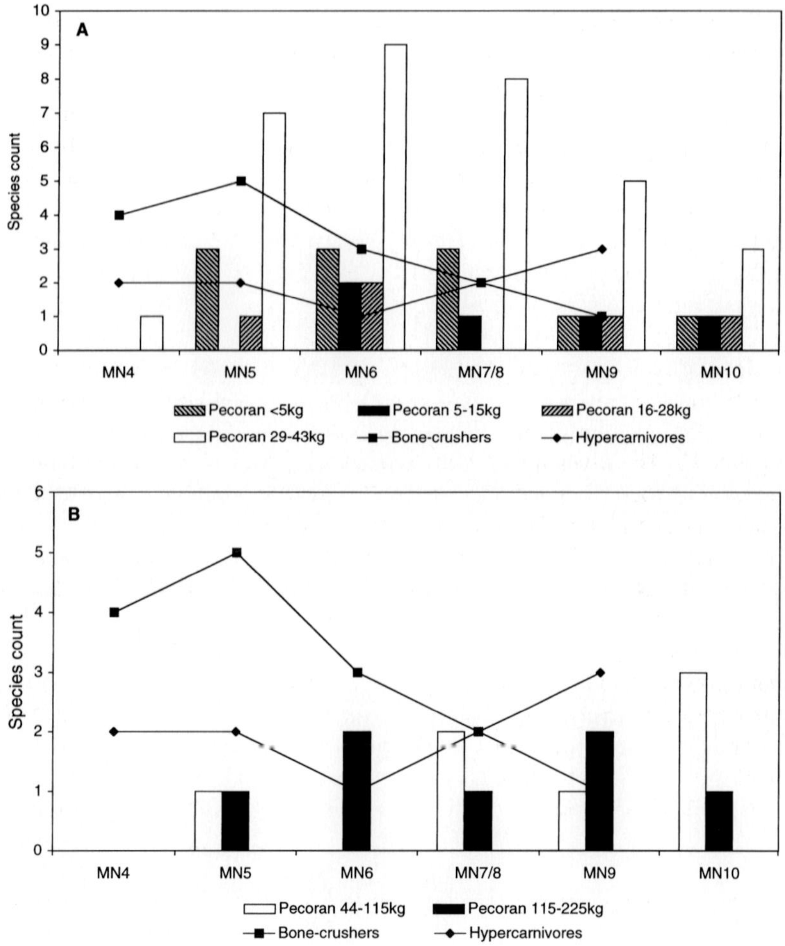

Fig. 18. Amphicyonid species diversity by (A) ecomorphs (only hypercarnivores and bone-crushing mesocarnivores), and (B) pecoran diversity by size classes.

therids remained at two species through MN 9, but dropped to one in MN 10 (Zapfe 1979, 1989). The two anchitheriine horses became extinct in MN 9. The equids and chalicotheres probably played a minor role in the evolutionary history of the Miocene bone-crushing amphicyonids. The hypercarnivorous amphicyonids, on the other hand, increased in diversity in pace with the perissodactylan species until their extinction in MN 9. Perissodactylan, and especially equid, diversity kept increasing to MN 10, and thus the extinction of the hypercarnivorous amphicyonids has no direct connection with perissodactylan evolution.

4.3.3.4. Other carnivores

The possible competitors of actively hunting amphicyonids with bone-crushing abilities in the European Miocene were the hyenids, percrocutids, machairodontine felids, ursids and large mustelids. The hyenids evolved and radiated during the Early and Middle Miocene. The early forms were small insectivorous and omnivorous animals. More carnivorous taxa soon evolved, but not until MN 10 did the first hyenid species with well-developed shearing carnassials and bone-cracking premolars appear. This was *Adcrocuta eximia*. *A. eximia* was geographically widespread, and is very common in fossil faunas from MN 10–MN 13, which indicates that this species was abundant. The resolution of the fossil record is not yet sufficient to determine whether *A. eximia* co-occurred with amphicyonids. However, amphicyonids and *A. eximia* do not co-occur at any locality. It seems that the majority of the amphicyonid species had disappeared by MN 10, when *A. eximia* appeared. However, the latest known amphicyonid species

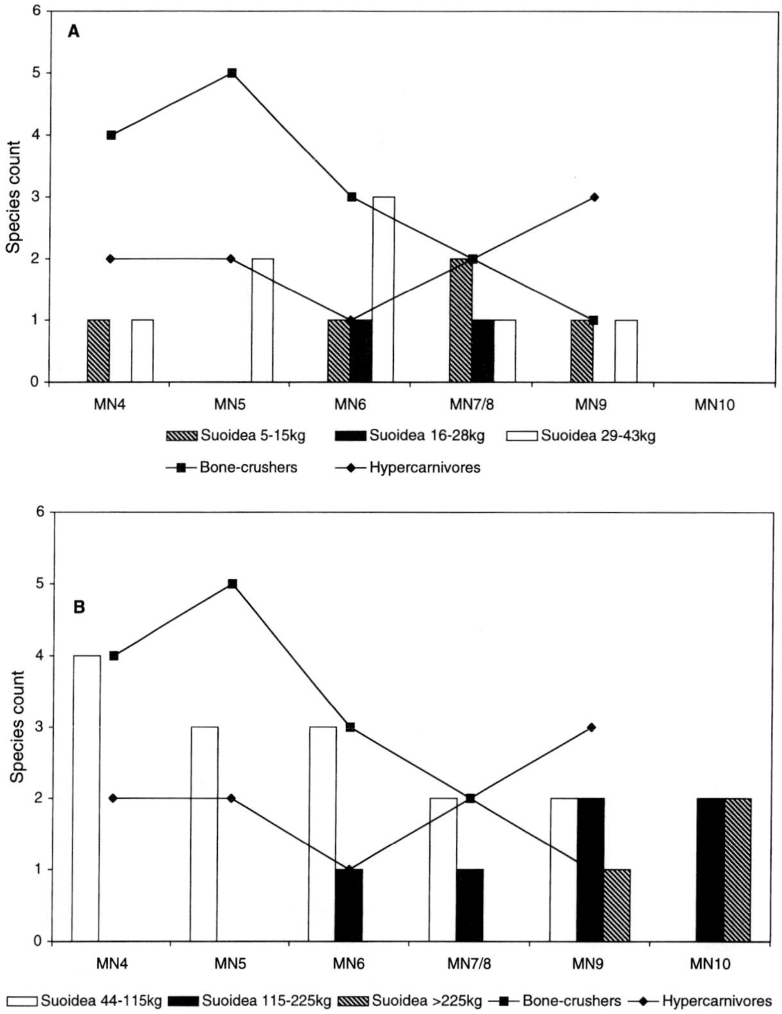

Fig. 19. Amphicyonid species diversity by ecomorphs (only hypercarnivores and bone-crushing mesocarnivores), and suoid diversity by size classes.

are known from rare material from MN 10 or 11 and they may have been outcompeted by *A. eximia*.

A somewhat mysterious family of bone-cracking carnivores, the Percrocutidae, originated in Asia in the Early Miocene, and had spread as far west as Turkey by MN 6 (Howell & Petter 1985). There is a record of a percrocutid in West (from La Grive St. Alban, France) in MN 6, but it is not until MN 12 that a representative of the family is next found in the fossil record of West. Given the good quality of the fossil record of the other carnivore families from West in MN 6–MN 10 it seems unlikely that percrocutids were present there.

The machairodontine felids appear in the fossil record of West in MN 9 with three species. These were all of large body size (100–200 kg), and with their felid shearing dentitions accompanied by long upper canines they were obviously predators on large prey. They were unlikely to be able to consume bones (Marean 1989), although some species have been shown to be effective at carcass disarticulation (Marean & Ehrhardt 1995). Since almost all the last amphicyonids were hypercarnivores, machairodontines were likely competitors with them in MN 9.

Most of the Miocene ursids show more carnivorous adaptations in their dentition than the Recent ones and their body mass range (40 to 300 kg) was similar to that of the amphicyonids. Their diversity, however, shows a pattern very similar to that of the amphicyonids in MN 4-MN 10 (Werdelin 1996b). They differed from amphicyonids in that two genera, *Indarctos* spp. (*I. arc-*

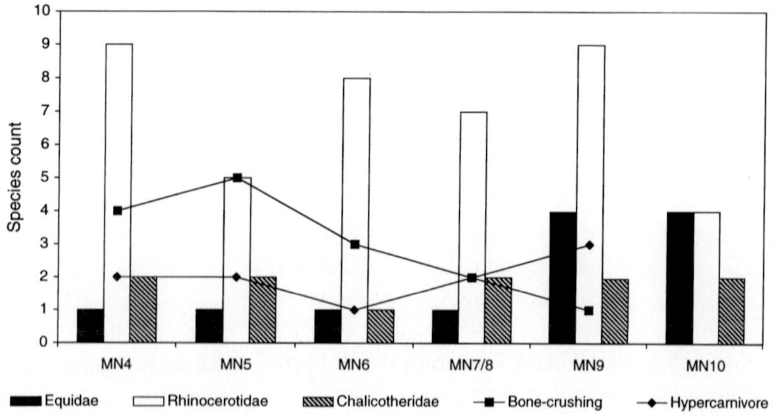

Fig. 20. Amphicyonid species diversity by ecomorphs (only hypercarnivores and bone-crushing mesocarnivores), and perissodactylan diversity.

toides, I. atticus) and *Ursavus* sp. (*U. depereti*), survived into MN 10 in West and at least *Indarctos* spp. flourished in West from MN 11 on. In the Early and Middle Miocene the mesocarnivorous ursids and amphicyonids were apparently able to co-occur.

Two large (over 50 kg) carnivorous mustelid, *Hadrictis fricki* and *Eomellivora hungarica* are known from MN 9. Prior to MN 9 they are unknown in West. *H. fricki* is known from one locality (Gaiselberg), and taking into account the good representation of other carnivores in MN 9 localities, it seems that *H. fricki* was not very common, and probably had no great effect on the evolution of the amphicyonids. The same applies to *E. hungarica* which is only known from Los Valles de Fuentiduena.

4.3.3.5. Opportunities for scavenging

Based on the morphological evidence it was concluded above (4.2) that most of the amphicyonids were active predators with bone-crushing abilities, and none of them can be considered specialist scavengers. However, the possiblity of scavenging is examined here in the light of the Miocene environments.

Most of the carnivores tend to scavenge when an opportunity arises. A primarily scavenging mode of life has two requirements: the presence of efficient hunters (or some other factor inducing high prey mortality), and that the prey large enough is hunted, because small prey is usually consumed in its entirety by the predators themselves. Even if these two requirements are met, in any environment the number of large, primarily scavenging species must be limited. This is because it is a niche with little opportunity for specialization. Since amphicyonids were the only carnivores with bone-crushing adaptations in MN 4–MN 9 of West, they are practically the only possible candidates for the scavenging niche. Hyenids with some adaptations towards bone-cracking appeared in MN 7/8, but these species are still generally characterized by an unspecialized carnivore dentition with only a somewhat greater emphasis on bone-eating (Werdelin & Solounias 1996).

Prior to MN 9, all the hypercarnivores were small (less than 100 kg) except the hypercarnivorous amphicyonids, which were relatively large, but they were not represented by more than two species before MN 9. The other hypercarnivores were primitive felids (*Pseudaelurus* spp.) and nimravids. Ursids were the only large predators (in addition to the amphicyonids, of course). Many Miocene ursids show moderate adaptations to carnivory (*Plithocyon* spp., *Hemicyon* spp.), but their posterior molars still clearly have crushing functions. They were not likely to leave many large, unconsumed carcasses behind.

As already noted, prey of large body size in the form of perissodactyls was abundant. The largest hypercarnivores were nimravids, with bodymasses around 50 kg. Many recent canids of about same size (e.g. *Canis lupus*, *Lycaon pictus*) are able to take down prey as large as 500 kg (e.g. *Alces alces, Equus grevyi*, respectively) by means of group hunting (Nowak 1991). Such a behaviour is not probable for the nimravids. Although mammalian scavengers that exclusively live on carcasses without actively hunting themselves have existed in past ecosystems (Zhou *et al.* 1992), I find it difficult to believe that the amphicyonids occupied such a niche in the European Miocene.

4.3.3.6. Stochasticity in amphicyonid extinction

The MN 9/MN 10 turnover was not random among mammals, but shows certain patterns. First, the Carnivora seems to have a high loss of species. Werdelin (1996b) found relatively low turnover from MN 5 to MN 8 in the European and Western Asian carnivores and an elevated turnover in MN 10. He also found a continuous rise in exit rate (indicates extinctions) in the Carnivora during MN 5–MN 7/8, culminating in MN 9.

The carnivore families most seriously affected were the nimravids which became totally extinct, amphicyonids which apparently declined to one species in West and survived only a little longer in the whole of Eurasia, and the viverrids, which became locally extinct in West. On the other hand, the number of species lost was greatest in the mustelids. While the number of mustelid species was 29 in MN 9, it was 9 in MN 10. All the large mustelids (over 10 kg, 4 species) and also all the piscivorous forms (4 species) disappeared in this transition.

It is also apparent when the whole mammalian fauna is considered that the drop in diversity was greatest in the body size class 10 to 100 kg. There was even a slight increase in diversity of species with a body mass of less than 10 kg (Fortelius *et al.* 1996b). It thus seems that amphicyonids were typical losers in the mid-Vallesian crisis.

4.3.4. Discussion

As shown in paleoecologic works (e.g. Bernor 1983, Fortelius *et al.* 1996b), East was very different from West in the Middle and Late Miocene. According to Fortelius *et al.* (1996b) the environments of East were dry forests with open areas in MN 5 and MN 6 with probably a more forested period from MN 7/8 to MN 9, followed by a new opening up of the tree canopy. It is of interest that the amphicyonids seem to have become extinct in East already by MN 7/8 with the possible exception of *A. pannonicus* in Hungary in the Late Miocene and an undescribed amphicyonid from Rudabánya (MN 9). The taxonomic composition of the carnivore fauna of East differed from that of West at least in the presence of the percrocutids (e.g. Schmidt-Kittler 1976).

It appears that East preceded West in the development of the biotic environment, including the migration of many mammalian groups and species. Although the extinction of the amphicyonids seems to have occurred under these new conditions in both regions, the enigma of their extinction remains unexplained. A comparison with known general patterns in carnivore evolution may assist in approaching this question.

Carnivore evolution has been shown to be iterative, producing similar functional characters in different phylogenetic lineages repetitively under even different conditions (e.g. Martin 1989, Van Valkenburgh 1991, Werdelin & Solounias 1991, Werdelin 1996a). Carnivore communities usually consist of several taxa with similar feeding habits, e.g., hypercarnivory. The coexistence of different species with similar ecologies is possible because of ecological partitioning of niches. Hypercarnivores of the same community differ in body size, in body characters or in hunting mode, and these differences lead to different sets of prey species for each predator. Ecological partitioning has been shown for Recent (e.g. Rosenzweig 1966, Dayan *et al.* 1992, Dayan & Simberloff 1994), and past (Van Valkenburgh 1985, 1989) mammalian carnivore communities, as well as for Mesozoic carnivorous marine reptile communities (Massare 1987).

Carnivore communities appear stable in that they may maintain the same species diversity over many millions of years. During this time there are often turnover events that change the taxonomic composition of the communities, and ecological roles are adopted by different species, often from different families, while the number of species remains the same. This kind of stability is demonstrated in the study by Van Valkenburgh and Janis (1993) of mammalian herbivore and carnivore communities during the Tertiary of North America. Species diversity in carnivore communities remained relatively stable, despite the fact that herbivore diversity fluctuated through time, and even during times of high herbivore diversity there was no marked increase in carnivore diversity. Van Valkenburgh and Janis (1993) suggested that this may reflect the limited number of ways predators can share their prey ("niche") and the more flexible use of habitats by carnivores.

The data on the Vallesian carnivores suggest that increased specialization may occur in a carnivore community as a result of decreased herbivore diversity. These data differ from those of Van Valkenburgh and Janis (1993) in that carnivore diversity dropped dramatically along with herbivore diversity in the Mid-Vallesian crisis. Carnivores went through a dramatic turnover and the post-crisis carnivore community consisted of clearly more specialized forms than the earlier one (Table 6). In many ecological communities trophically similar species have very different overlap in the use of resources during the periods of abundance and scarceness of food (see a review in Schoener 1982). Most commonly the period of the least overlap (i.e. enhanced partitioning) occurs in the lean periods (seasons or years). In other words, members of different populations become specialists when resources are scarce, and return to more generalist's mode of life when resources become more abundant. Such observations have also been made on the birds of prey (Opdam 1975). Accordingly, the numbers of generalist mammalian predators have been shown to decrease from South to North in Fennoscandia (Hanski *et al.* 1991), apparently along with scarcer resources.

Long term environmental changes that cause resource deprivation probably also intensify competition and promote partitioning. In the faunal communities this is likely to result in entry (immigration or speciation) of more specialized species and extinction of generalists. It is possible that this explains the stability of the carnivore diversity observed by Van Valkenburgh and Janis

Table 6. Carnivore species (estimated body mass > 20 kg) of western Eurasia in MN 9 and MN 10.

Family	Species	Feeding ecology	Estimated body mass (kg)
MN 9			
	Simocyon diaphorus	Bone-crusher	55
Amphicyonidae	*Agnotherium antiquus*	Hypercarnivore	158
Amphicyonidae	*"Amphicyon" castellanus*	Hypercarnivore	220
Amphicyonidae	*Amphicyon major*	Bone-crusher	183
Amphicyonidae	*Thaumastocyon dirus*	Hypercarnivore	60 (?)
Ursidae	*Hemicyon goeriachensis*	Mesocarnivore	120
Ursidae	*Indarctos vireti*	Mesocarnivore	175
Ursidae	*Ursavus brevirhinus*	Omnivore	80
Ursidae	*Ursavus primaveus*	Omnivore	90
Mustelidae	*Eomellivora hungarica*	Bone-cracker	50
Mustelidae	*Hadrictis frickii*	Mesocarnivore	55
Nimravidae	*Sansanosmilus jourdani*	Hypercarnivore	80
Hyaenidae	*Hyaenictitherium* sp.	Mesocarnivore	50
Hyaenidae	*Thalassictis montadai*	Mesocarnivore (bone-cracker)	31
Hyaenidae	*Thalassictis robusta*	Mesocarnivore (bone-cracker)	20
Felidae	*Machairodus alberdiae*	Hypercarnivore	100
Felidae	*Machairodus aphanistus*	Hypercarnivore	220
Felidae	*Paramachairodus ogygia*	Hypercarnivore	44
Felidae	*Pseudaelurus quadridentatus*	Hypercarnivore	30
MN 10			
	Simocyon sp.	Bone-crusher	55
Ursidae	*Indarctos arctoides*	Omnivore	160
Ursidae	*Indarctos atticus*	Omnivore	360
Hyaenidae	*Adcrocuta eximia*	Bone-cracker	70
Felidae	*Machairodus aphanistus*	Hypercarnivore	220
Felidae	*Machairodus giganteus*	Hypercarnivore	210
Felidae	*Paramachairodus orientalis*	Hypercarnivore	85
Felidae	*Paramachairodus ogygia*	Hypercarnivore	44

(1993). Maybe carnivores that live with a poorer prey selection are all specialized relying on particular prey or life style. In a community with a higher prey diversity generalist carnivores may even have advantage over the specialists.

The iterative evolution, ecological partitioning, and stability of communities found to be common in carnivore evolution are probably the result of interspecific and intraspecific competition (Van Valkenburgh 1991, 1995). On the other hand, the theory of a strict co-evolution between carnivores and their prey has gained little credence. For example, the evolution of carnivoran hunting methods seems not to have happened as an arms race in pace with ungulate defence methods. Janis and Wilhelm (1994) showed that the evolution of cursorial predators lagged behind that of the ungulates by millions of years. Pennycuick (1979) has suggested that cursorial adaptations in ungulates evolved for economical locomotion rather than as a means to escape predators. Extinctions of the generalist carnivores in the Mid-Vallesian crisis were probably also results of competition within the carnivore community.

In conclusion it seems that the amphicyonids were never able to specialize as far as the new contemporaneous members of the carnivore community, such as sabertooth felids or bone-cracking hyenas. Specialization to large prey was required because of the extinction of small artiodactyl prey. Ursids were probably able to survive by turning into more omnivorous habits, but amphicyonids which were hypercarnivores and bone-crushers in MN 9 were apparently not able to switch to omnivory. Neither were they able to compete with the new kinds of carnivores for the prey that became more homogenous in the Vallesian.

Acknowledgements. Dr. Lars Werdelin and Prof. Mikael Fortelius were advisors for this Ph.D.-thesis; I gratefully acknowledge their help. Prof. Rainer Springhorn and Prof. Ann Forstén provided thorough reviews and useful suggestions on the text. The systematic section was developed in collaboration with Dr. Jyrki Muona. His help is likewise appreciated. I thank Dr. Jorge Agustí, Dr. Luis Alcalá, Dr. Burhart Engesser, Prof. Volker Fahlbusch, Dr. Léonard Ginsburg, Prof. Kurt Heissig, Dr. Elmar P. J. Heizmann, Dr. Gudrun Höck, Prof. Laszlo Kordos, Dr. Jorge Morales, Mag. Doris Nagel, Prof. Rabeder and Dr. G. Plodowski for the access to specimens in their care and useful discussions. I thank Dr. Brigitte Lange-Badre, Dr. F. O. Neuffer, Prof. Karlheinz Rothausen, Prof. M. Telles Antunes, Prof. Norbert Schmidt-Kittler and Prof. Rainer Springhorn for informative and helpful correspondence. This work has also benefited from discussions with Dr. Jukka Jernvall, Dr. Tracy Popowics, M.Sc. Lluis Rius, and M.Sc. Pirkko Ukkonen. Ms. Marjaana Hakala kindly helped with Russian texts and Dr. Lars Werdelin checked the English. Xavier Herbozo drew the life reconstruction on the front cover. I received financial support from the Finnish Cultural Foundation, the Academy of Finland, and Naisten Tiedesäätiö.

References

Alcalá, L., Morales, J. & Soria, D. 1989–1990: El registro fósil neógeno de los carnívoros (Creodonta y Carnivora, Mammalia) de España. — Paleontologia i Evolució 23: 55–66.

Antunes, M. T. & Ginsburg, L. 1977: Notes sur la géologie et la paléontologie du miocène de Lisbonne XIX – Sur un Amphicyon (Mammalia, Ursidae) du Burdigalien. — Comunicacoes dos Servicos Geologicos de Portugal LXI: 335–342.

Anyonge, W. 1993: Body mass in large extant and extinct carnivores. — J. Zool. Lond. 231: 339–350.

Bakker, R. T. 1983: The deer flees, the wolf pursues. — In: Futuyma, D. J. & Slatkin, M. (eds.), Coevolution: 350–382. Sinauer associates Inc., Sunderland, Massachusetts.

Barry, J. & Flynn, L. 1989: Key stratigraphical events in the Siwaliks sequence. — In: Lindsay, E. H., Fahlbusch, V. & Mein, P. (eds.), European Neogene mammal chronology: 557–573. Plenum Press, New York.

Belinchón, M. & Morales, J. 1989: Los carnívoros del mioceno inferior de Buñol (Valencia, España). — Revista Española de Paleontologia 4: 3–8.

Bergounioux, F. & Crouzel, F. 1973: Amphicyon major Blainville du Miocène moyen de Sansan (Gers). — Annales Paléontology (Vertébrés) 59: 27–52.

Bernor, R. L. 1983: Geochronology and zoogeographic relationships of Miocene Hominoidea. — In: Ciochon, R. L. & Corrucini, R. S. (eds.), New Interpretations of ape and human ancestry: 21–64. Plenum Press, New York & London.

— 1984: A zoogeographic theatre and biochronologic play: the time/biofacies phenomena of Eurasian and African Miocene mammal provinces. — Paléobiologie continentale, Montpellier, XIV: 121–142.

Bernor, R., Fahlbusch, V. & Mittmann, W. (eds.) 1996a: The evolution of western Eurasian Miocene mammal faunas. — Columbia University Press. (In press.)

Bernor, R. L., Koufos, G. D., Woodburne, M. O. & Fortelius, M. 1996b: The evolutionary history of European and Southwestern Asian late Miocene amd Pliocene hipparionine horses. — In: Bernor, R., Fahlbusch, V. & Mittmann, W. (eds.), The evolution of western Eurasian Miocene mammal faunas. — Columbia University Press. (In press.)

Blainville, H. de 1841: Ostéographie et déscription iconographique des Mammifères récents et fossiles (Carni-

vores), Vols. 1 & 2. Paris.
Burchardt, B. 1978: Oxygen isotope paleotemperatures from the Tertiary period in the North Sea area. — Nature 275: 121–123.
Butler, P. M. 1946: The evolution of carnassial dentitions in the Mammalia. — Proc. zool. Soc. Lond. 116: 198–220.
Carpenter, J. M. 1996: Phylogeny and biogeography of Polistes. — In: Turillazzi, S. & West-Eberhard, M. J. (eds.), Natural history and evolution of paper-wasps: 18–27. Oxford University Press, Oxford.
Coddington, J. & Scharff, N. 1994 Problems with zero-length branches. — Cladistics 10: 415–423.
Colbert, E. D. 1939: Carnivora of the Tung Gur formation of Mongolia. — Bulletin of the American Museum of Natural History LXXVI: 47–81.
Creighton, G. K. 1980: Static allometry of mammalian teeth and the correlation of tooth size and body size in contemporary mammals. — J. Zool., Lond. 191: 435–443.
Cronin, M. A., Amstrup, S. C. & Garner, G. W. 1991: Interspecific and intraspecific mitochondrial DNA variation in North American bears (Ursus). — Can. J. Zool. 69: 2985–2992.
Crusafont, M. & Kurtén, B. 1976: Bears and bear-dogs from the Vallesian of the Vallés-Penedés basin, Spain. — Acta Zool. Fennici 144: 1–29.
Crusafont, M., de Villalta, J. F. & Truyols, Y. J. 1955: El Burdigaliense continental de la cuena del Vallés-Penedés. — Memorias y comunicaciones del Instituto Geológico XII: 7–273.
Damuth, J. 1990: Problems in estimating body masses of archaic ungulates using dental measurements. — In: Damuth, J. & MacFadden, B. J. (eds.), Body size in mammalian paleobiology: estimation and paleobiological implication: 229–255. Cambridge University Press, New York.
— 1993: ETE database manual. Evolution of Terrestrial Ecosystems Consortium: Washington.
Damuth, J. & Macfadden, B. J. 1990: Introduction: body size and its estimation. — In: Damuth, J. & MacFadden, B. J. (eds.), Body size in mammalian paleobiology: estimation and paleobiological implication: 1–11. Cambridge University Press, New York.
Davis, D. D. 1964: The Giant Panda, a morphological study over of evolutionary mechanisms. — Fieldiana Zoology Memoirs 3: 1–339.
Dayan, T., Simberloff, D., Tchernov, E. & Yom-Tov, Y. 1992: Canine carnassials-character displacement in the wolves, jackals and foxes of Israel. — Biological Journal of Linnean society 45: 315–331.
Dayan, T. & Simberloff, D. 1994: Character displacement, sexual dimorphism, and morphological variation among British and Irish mustelids. — Ecology 75: 1063–1073.
De Beaumont, G. 1964a: Un crane d'Amphicyon ambiguus (Filhol) (Carnivora) des Phosphorites du Quercy. — Archives des Sciences, Geneve 17: 332–342.
— 1964b: Essai sur la position taxonomique des genres Alopecocyon Viret et Simocyon Wagner (Carnivora). — Eclogae geologicae Helveticae 57: 829–836.
— 1984: Des dents d'Amphicyon (Mammifère, Carnivore, Ursidé) du Turolien basal de Kohfidisch, Burgenland, Autriche. — Archives des Sciences, Geneve 37: 77–83.
De Bonis, L. 1966: Arrière-Crâne et Moulages endocrâniens de Carnivores fossiles. — Annalés Paléontology (Vertébrés) 53: 143–162.
— 1969: Remarques sur la position systématique des Amphicyon. — C. R. Acad. Sci. (Paris) Ser. D. 269: 1748–1750.
De Bruijn, H., Daams, R., Daxner-Höck, G., Fahlbusch, V., Ginsburg, L., Mein, P. & Morales, J. with the contribution of Heizmann, E., Mayhew, D. F., van der Meulen, A. J., Schmidt-Kittler, N. & Telles-Antunes, M. 1992: Report of the RCMNS working group on fossil mammals, Reisensburg 1990. — Newletters on Stratigraphy 26: 65–118.
Dehm, R. 1950: Die Raubtiere aus dem Mittel-Miocän (Burdigalium) von Wintershof-West bei Eichstätt in Bayern. — Abhandlungen der Bayerischen Akademie der Wissenschaften 58: 4–141.
Ewer, R. F. 1973: The carnivores. — Cornell University Press, Ithaca, New York.
Farris, J. S. 1972: Estimating phylogenetic trees from distance matrices. — American Naturalist 106: 645–668.
— 1983: The logical basis of phylogenetic analysis. — In: Platnick, N. I. & Funk, V. A. (eds.), Advances in cladistics, 2. Proceedings of the second Meeting of the Willi Hennig Society: 7–36. Columbia University Press, New York.
— 1988: Hennig86 reference, version 1.5. Computer program and documentation. — Port Jefferson Station, New York.
Filhol, H. 1874: Sur les Vertébrés fossiles des dépôts de phosphate de chaux du Quercy. — Bull. Soc. philom. Paris, Ser. 6, 11: 16–20.
Filhol, H. 1891: Etudes sur les Mammifères fossiles de Sansan. — Masson & Cie, Paris. 314 pp.
Flynn, J. J., Neff, N. N. & Tedforf, R. H. 1988: Phylogeny of the Carnivora. — In: Benton, M. J. (ed.), The phylogeny and classification of tetrapods, Vol 2: 73–116. Clarendon Press, Oxford.
Fortelius, M. 1984: Ungulate cheek teeth: developmental, functional and evolutionary interrelations. — Acta Zool. Fennica 180: 1–76.
— 1990: Problems with using fossil teeth to estimate body sizes of extinct mammals. — In: Damuth J. & MacFadden, B. J. (eds.), Body size in mammalian paleobiology: estimation and paleobiological implication: 207–229. Cambridge University Press, New York.
Fortelius, M. & Kappelman, J. 1993: The largest land mammal ever imagined. — Zool. J. Linn. Soc., London 108: 85–101.
Fortelius, M., Andrews, P., Bernor, R. L., Viranta, S. & Werdelin, L. 1996a: Preliminary analysis of taxonomic diversity, turnover and provinciality in a subsample of large land mammals from the later Miocene of Western Eurasia. — Acta Zoologica Cracoviensia. (In press.)

Fortelius, M., Werdelin, L., Andrews, P., Bernor, R. L., Gentry, A., Humphreys, L., Mittmann, W. & Viranta, S. 1996b: Provinciality, diversity, turnover and paleoecology in land mammal faunas of the later miocene of western Eurasia. — In: Bernor, R., Fahlbusch, V. & Mittmann, W. (eds.), The evolution of western Eurasian Miocene mammal faunas. Columbia University Press. (In press.)

Fox, R. C. & Youzwyshyn, A. W. 1994: New primitive carnivorans (Mammalia) from the Paleocene of western Canada and their bearing on relationships of the order. — Journal of Paleontology 14: 382–404.

Fraas, C. 1885: Beiträge zur Fauna von Steinheim. — Jh. Ver. vaterl. Naturk. Württemberg, Stuttgart 41: 313–326.

Gabunia, L. 1973: (Fossil vertebrates of the fauna of Bjelometschkaya). — Akademia Nauk, Georgian SSR, Tbilisi. (In Russian.)

Gabunia, L. K. 1981: (New data on the genus Gobicyon (Canidae) in Asia). — Paleontologkheskij zhurnal 2: 131–134. (In Russian.)

Ginsburg, L. 1961a: La faune des Carnivores Miocènes de Sansan. — Memoires du Muséum National D'Histoire Naturelle Sér. C Tome IX.

— 1961b: Plantigradie et digitigradie chez les carnivores fissipèdes. — Mammalia 25: 1–21.

— 1965: L'"Amphicyon" ambiguus des phosphorites du Quercy. — Bulletin du Muséum National d'Histoire Naturelle 37: 724–730.

— 1966: Les Amphicyons des Phosphorites du Quercy. — Annales Paléontology (Vert.), Paris 52: 23–64.

— 1977: Cynelos lemanensis (Pomel), carnivore ursidé de l'Aquitanien d'Europe. — Annales de Paléontologie, Vertébrés 42: 97–117.

— 1992: Les genres Pseudarctos et Ictiocyon, Amphicyonidae (Carnivora, Mammalia) du Miocène européen. — Bull. Mus. natl. Hist. nat. Paris 4: 301–317.

Ginsburg, L. & Antunes, M. T. 1968: Amphicyon giganteus carnassier Géant du Miocène. — Annales de Paléontologie (Vertébrés) LIV: 1–32.

— 1995: Les carnivores du Miocène de Lisbonne (Portugal). — Annales de Paléontologie 81: 125–165.

Ginsburg, L., Morales, J. & Soria, D. 1981: Nuevos datos sobre los carnivoros de los Valles de Fuentiduena. — Estudios geol. 37: 382–415.

Goloboff, P. 1993a: NONA. Noname ver. 1.0 (a bastard son of Pee-Wee 32-bit version). Program and documentation. — Computer program distributed by J. M. Carpenter, Dept. Entomology, American Museum Natural History, New York.

— 1993b: Pee-Wee (P)arsimony and (I)mplied (W)eights ver. 2.0 (32 bit version). Program and documentation. — Computer program distributed by J. M. Carpenter, Dept. Entomology, American Museum Natural History, New York.

Gürbüz, M. 1974: Amphicyon major Blainville discovered in the Middle Miocene beds of Candir. — Bulletin of the Mineral Research and Exploration Institute of Turkey 83: 109–111.

Hanski, I., Hansson, L. & Henttonen, H. 1991: Specialist predators, generalist predators, and the microtine rodent cycle. — Journal of Animal Ecology 60: 353–367.

Hedrick, A. V. & Temeles, E. J. 1989: The evolution of sexual dimorphism in animals: hypotheses and tests. — Trends in Evolution and Ecology 4: 136–138.

Heizmann, E. P. J. 1973: Die tertiären Wirbeltiere des Steinheimer Beckens. — Palaeontographica VIII(V): 1–95.

Helbing, H. 1928: Carnivoren des oberen Stampien. — Abhandlungen der Schweizerischen Palaeontologischen Gesellschaft XLVII: 1–83.

Hough, J. R. 1948a: The auditory region in some members of the Procyonidae, Canidae, and Ursidae: its significance in the phylogeny of the Carnivora. — American Museum of Natural History Bulletin 92: 67–118.

— 1948b: A systematic revision of Daphoenus and some allied genera. — Journal of Paleontology 22: 573–600.

Howell, F. C. & Petter, G. 1985: Comparative observations on some Middle and Upper Miocene hyaenids. Genera Percrocuta Kretzoi, Allohyaena Kretzoi, Adcrocuta Kretzoi (Mammalia, Carnivora, Hyaenidae). — Géobios 18: 419–476.

Hunt, R. M. Jr. 1972: Miocene amphicyonids (Mammalia, Carnivora) from the Agate Spring Querries, Sioux County, Nebraska. — American Museum Novitates 2506: 1–39.

— 1974a: The auditory bulla in Carnivora: an anatomical basis for reappraisal of carnivore evolution. — Journal of Morphology 143: 21–76.

— 1974b: Daphoenictis, a cat-like carnivore (Mammalia, Amphicyonidae) from the Oligocene of North America. — Journal of Paleontology 48: 1031–1048.

— 1977: Basicranial anatomy of Cynelos Jourdan (Mammalia: Carnivora) an Aquitanian carnivore from the Allier Basin, France. — Journal of Paleontology 51: 826–843.

— 1996: Biogeography of the order Carnivora. — In: Gittleman, J. (ed.), Carnivore behaviour, ecology and evolution, Vol. II: 485–542. Cornell University Press, Ithaca, New York.

Hunt, R. M. Jr., Xiang-Xu, X. & Kaufman, J. 1983: Miocene Burrows of Extinct Bear Dogs: Indication of Early Denning Behavior of Large Mammalian Carnivores. — Science 221: 364–366.

Hürzeler, J. 1940: Über felinoide Caniden des europäischen Miocaens. — Verh. Schweizer. Naturforsch. Ges.: 150–151.

Janis, C. M. 1990: Correlation of cranial and dental variables with body size in ungulates and macropodoids. — In: Damuth, J. & MacFadden, B. J. (eds.), Body size in mammalian paleobiology: estimation and paleobiological implication: 255–301. Cambridge University Press, New York.

Janis, C. M. & Wilhelm, P. B. 1993: Were there mammalian pursuit predators in the Tertiary? Dances with wolf avatars. — Journal of Mammalian Evolution 1: 103–126.

Jourdan, M. 1862: La déscription de restes fossiles de grands Mammifères 2. Les terrains sidérolithiques. — Rev. Soc. sav., publ. Min. instr. publ. et desc. 1: 126–130.

Kaup, J. J. 1832: Description d'Ossements fossiles de deux grands Mammifères inconnus jusquà présent, qui se trouvent au Muséum grand-ducal de Darmstadt. — H. 2, Darmstadt.
Kittl, E. 1891: Die jungtertiären Säugethierfunde in der Mannersdorfer Zieglei bei Angern. — Annalen des K. K. Naturhistorischen Hofmuseums VI: 92–97.
Kluge, A. G. & Farris, J. S. 1969: Quantitative phyletics and the evolution of anurans. — Systematic Zoology 18: 1–32.
Kretzoi, M. 1937: Die Raubtiere von Gombaszög nebst einer Übersicht der Gesamtfauna. — Annales Musei Nationalis Hungarici XXXI: 88–157.
Kretzoi, M. 1985: New Amphicyonid from the Early Pannonian of Pécs, South Hungary. — Annales Historico-Naturales Musei Nationales Hungarici. 77: 65–68.
Kurtén, B. 1959: On the longevity of mammalian species in the Tertiary. — Commentationes Biologicae 21: 1–14.
— 1964: The evolution of the polar bear Ursus maritimus Phipps. — Acta Zool. Fennica 108: 1–30.
— 1976: Fossil carnivora from the late Tertiary of Bled Douarah and Cherichira, Tunisia. — Notes du Service de Tunisie 42: 177–214.
Kuss, S. E. 1962: Problematische Caniden des europäischen Tertiärs. — Ber. Naturf. Ges. Freiburg i Br. 52: 123–172.
— 1965: Revision der europäischen Amphicyoninae (Canidae, Carnivora, Mamm.) — Sitzungsberichte der Heidelberger Akademie der Wissenschaften 1: 1–169.
Lartet, E. 1836: Nomenclature des mammifères et des coquilles qu'il a trouvés dans un terrain d'eau douce près de Simorre et des Sansan (Gers). — Bull. Soc. Géol. Fr. 7: 217–220.
— 1851: Notice sur la colline de Sansan. — Auch.
Legendre, S. & Roth, C. 1988: Correlation of carnassial tooth size and body weight in recent carnivores (Mammalia). — Historical Biology 1: 85–98.
Lillegraven, J. A. 1969: Latest Createaceous mammals of upper part of the Edmonton Formation of Alberta, Canada, and review of marsupial-placental dichotomy in mammalian evolution. — The University of Kansas Paleontological Contributions 50 (Vertebrata 12): 1–122.
Marean, C. W. 1989: Sabertooth cats and their relevance for early hominid diet and evolution. — J. hum. Evol. 18: 559–582.
Marean, C. W. & Ehrhardt, C. L. 1995: Paleoanthropological and paleoecological implications of the taphonomy of a sabertooth's den. — J. hum. Evol. 29: 515–547.
Martin, L. D. 1989: Fossil history of the Terrestrial Carnivora. — In: Gittleman, J. L. (ed.), Carnivore behavior, ecology and evolution: 536–569. Chapman & Hall, London.
Massare, J. A. 1987: Tooth morphology and prey preference of Mesozoic marine reptiles. — J. Vert. Paleont. 7: 121–137.
Matthew, W. D. 1924: Third contribution to the Snake Creek fauna. — Bulletin of the American Museum of Natural History L: 61–210.
Mayet, L. 1908: Etude des mammiferes miocenes des sables de l'Orleanais et des faluns de la Touraine. — Ann. Univ. Lyon 24: 1–336.
Mein, P. 1975: Résultats du groupé de travail des vertébrés. — In: Report on activity of the RCMNS Working Groups (1971–1975), Bratislava: 78–81.
Morales, J., Alcalá, L., Hoyos, P., Montoya, P., Nieto, M., Pérez, B. & Soria, D. 1993: El yacimiento del Aragoniense medio de La Retama (Depresión Intermedia, Provincia de Cuenca, España): significado de las faunas con Hispotherium. — Scripta Geologia 103: 23–39.
Muona, J. 1996: The phylogeny of Elateroidea, or which tree is best today? — Cladistics. (In press.)
Nixon, K. C. 1995: Clados, ver. 1.49. Program and documentation. — Trumansburg, New York.
Nixon, K. C. & Carpenter, J. M. 1993: On outgroups. — Cladistics 9: 413–426.
— 1996: On consensus, collapsibility, and contention. — Cladistics. (In press.)
Nixon, K. C. & Davis, J. 1991: Polymorphic taxa, missing values and cladistic analysis. — Cladistics 7: 233–241.
Noulet, J. B. 1876: Notes sur des gisements nouveaux de Canis palaeolycos et du Cadurcotherium cayluxi. — Mém. Acad. Sci. Insc, et B.-L. Toulouse, Sér. VII 8: 3–6.
Nowak, R. M. 1991: Walker's Mammals of the World, Vol. II. — The John Hopkins University Press, Baltimore.
Olsen, S. J. 1958: The fossil carnivore Amphicyon intermedius from the Thomas Farm Miocene Part 1. Skull and dentition. — Bulletin of the Museum of Comparative Zoology 118: 157–172.
— 1960: The fossil carnivore Amphicyon longiramus from the Thomas Farm Miocene. — Bulletin of the Museum of Comparative Zoology 123: 1–44.
Opdam, P. 1975: Inter- and intraspecific differentation with respect to feeding ecology in two sympatric species of the genus Accipterus — Ardea 63: 30–54.
Pavlović, M. & Thenius, E. 1959: Gobicyon macrognathus (Canidae, Mammalia) aus dem Miozän Jugoslawiens. — Sitzung der mathematisch-naturwissenschaftlichen Klasse.
Pennycuick, C. J. 1979: Energy costs of locomotion and the concept of "foraging radius". — In: Sinclair, A. R. E. & Northorn-Griffths, M. (eds.), Serengeti: Dynamics of an ecosystem: 64–184. University of Chicago Press.
Pilgrim, G. E. 1931: Catalogue of the pontian carnivora of Europe. — BM(NH).
Platnick, N. I., Griswold, C. E. & Coddington, J. A. 1991: On missing entries in cladistic analysis. — Cladistics 7: 337–343.
Pomel, A. 1846: Note sur des animaux fossiles découverts dans le Départment de l'Allier (Addition au Mémoire sur la géologie paléontologie etc.). — Bull. Soc. géol. France 4: 2.
Pulliainen, E. 1973: Suomen suurpedot. — Otava, Porvoo (Finland).
Radinsky, L. B. 1980: Endocasts of amphicyonid Carnivorans. — American Museum Novitates 2694: 1–11.
Raymo, M. E. & Ruddiman, W. F. 1992: Tectonic forcing of late Cenozoic climate. — Nature 359: 117–1122.
Rosenzweig, M. L. 1966: Community structure in sympatric

carnivores. — J. Mamm. 47: 602–620.

Roth, C. H. 1989: Die Raubtierfauna (Carnivora, Mamm.) der untermiozänen Spaltenfüllung von Erketshofen 2 bei Eichstätt/Bayern. — Mitt. Bayer. Staatsslg. Paläont. hist. Geol. 29: 163–205.

Rothausen, K. 1988: Carnivoren im Kalktertiär (Oberoligozän-Untermiozän) des Mainzer Beckens (1. Amphicyonidae). — Geol. Jb. A110: 241–260.

Schintz, R. H. 1925: Cuviers Thierreich. IV.

Schlosser, M. 1888: Die Affen, Lemuren, Carnivoren des europäischen Tertiärs. — Beiträge zur Palaeontologie von Oesterreich-Ungarn. Neumayr und Mojsisovics.

— 1899: Über die Bären und bärenähnlichen Formen des europäischen Tertiärs. — Palaeontographica 46: 95–146.

— 1904: Notizen über einige Säugetiere aus dem Miocän von Württemberg und Bayern. — Neues. Jb. Miner. B19: 485–502.

Schmidt-Kittler, N. 1981: Zur Stammesgeschichte der marderverwandten Raubtiergruppen (Musteloidea, Carnivora). — Eclogae Geologicae Helvetiae 74: 753–801.

Schmidt-Nielsen, K. 1984: Scaling – why is animal size so important. — Cambridge University Press, Cambridge.

Schoener, T. W. 1982: The controversy over interspecific competition. — American Scientist 70: 586–595.

Simpson, G. G. 1941: Large Pleistocene felines of North America. — American Museum Novitates 1136: 1–27.

— 1945: The principles of classification and a classification of mammals. — Bulletin of the American Museum of Natural History 85.

Springhorn, R. 1976: Zur Craniologie südfranzösischer oligozäner Amphicyonidae. — Palaeontographica 152: 1–13.

— 1977a: Revision of Paleogene European Amphicyonidae (Carnivora, Mammalia). — Palaeontographica 158: 26–113.

— 1977b: Zum Extremitätensskelett von Cynelos crassidens vireti (Kuss, 1965) (Amphicyonidae, Carnivora, Mammalia). — Ber. Naturforsch. Ges. Freiburg i. Br. (Pfannenstiel-Gedenkband) 67: 325–333.

— 1980: Radiation der Hyaenodonta (Mammalia: Deltatheridia) unter gebisstrukturellen Gesichtspunken und ein Vergleich mit fissipeden Carnivora. — Ber. Naturforsch. Ges. Freiburg i. Br. (Pfannenstiel-Gedenkband) 70: 97–109.

Stehlin, H. G. & Helbing, H. 1925: Catalogue des Ossesments de Mammifères Tertiaires de la Collection Bouregeois. — Extrait du Bulletin no. 18 de la Société d'Histoire naturelle d'Anthropologie de Loir-et-Cher: 77–277.

Steininger, F. F., Bernor, R. L. & Fahlbusch, V. 1989 European Neogene marine/continental chronologic correlations. — In: Lindsay, E. H., Fahlbusch, V. & Mein, P. (eds.), European Neogene mammal chronology: 15–46. Plenum Press, New York.

Strait, S. 1993: Molar morphology and food texture among small-bodied insectivorous mammals. — J. Mamm. 74: 391–402.

Thenius, E. 1949: Die Carnivoren von Göriach (Steiermark). — Sitzungsberichte der Österr. Akademie der Wissenschaften, Mathem.-naturw. Kl., Abt. I, 158: 696–762.

Trouessart, E. L. 1885: Conspectus systematicus et geographicus mammalium tam viventium quam fossil. — Catalogue des Mammifères vivants et fossilies. Bull. Soc. Edts. sci. Angers, Paris 15: 4.

Van Valen, L. 1969: Evolution of dental growth and adaptation in placental carnivores. — Evolution 23: 96–117.

Van Valkenburgh, B. 1985: Locomotor diversity within past and present guilds of large predatory mammals. — Paleobiology 11: 406–428.

— 1987: Skeletal indicators of locomotor behavior in living and extinct carnivores. — J. Vert. Paleont. 7: 162–182.

— 1989: Carnivore dental adaptations and diet: a study of trophic diversity within guilds. — In: Gittleman, J. (ed.), Carnivore behavior, ecology, and evolution: 410–436. Cornell University Press, Ithaca, New York.

— 1990: Skeletal and dental predictors of body mass in carnivores. — In: Damuth, J. & MacFadden, B. J. (eds.), Body size in mammalian paleobiology: estimation and paleobiological implication: 1–11. Cambridge University Press, New York.

— 1991: Iterative evolution of hypercarnivory in canids (Mammalia: Carnivora): evolutionary interactions among sympatric predators. — Paleobiology 17: 340–362.

— 1995: Tracking ecology over geological time: evolution within guilds of vertebrates. — Trends in Ecology and Evolution 10: 71–76.

Van Valkenburgh, B. & Janis, C. A. 1993: Historical diversity patterns in North American large herbivores and carnivores. — In: Ricklefs, R. E. & Schluter, D. (eds.), Species diversity in ecological communities: 331–341. The University of Chicago Press, Chicago & London.

Viranta, S. 1994: Limb bone proportion and body mass of the Cave bear (Ursus spelaeus). — Historical Biology 7: 239–250.

Viret, J. 1929a: Tomocyon grivensis nov. gen. n. sp. et les Canidés de La Grive Saint-Alban (Isère). — B. S. G. F. 4 serie t XXIX: 217.

— 1929b: Les faunes de Mammifères de l'oligocène supérieur de la Limagne Bourbonnaise. — Ann. Univ. Lyon Sci. Méd. 47: 1–328.

— 1951: Catalogue critique de la faune des mammifères miocènes de La Grive-St. Alban (Isère) I. Chiroptères, carnivores, édentés pholidontes. — Nouv. Arch. Mus. Hist. Nat. Lyon, Fasc 3: 1–102.

Vrana, P. B., Milinkovitch, M. C., Powell, J. R. & Wheeler, W. C. 1994: Higher level relationships of the arctoid Carnivora based on sequence data and "total evidence". — Molecular phylogenetics and evolution 3: 47–58.

Vrba, E. S. 1996: On the connections between paleoclimate and evolution. — In: Vrba, E. S., Denton, G. H., Partridge, T. C. & Burckle, L. H. (eds.), Paleoclimate and evolution, with emphasis on human origins: 24–45. Yale University press, New Haven.

Wang, X. 1996: New cranial material of Simocyon from China, and its phylogenetic relationship to the red panda (Ailurus). — J. Vert. Paleont. (In press.)

Wang, X. & Tedford, R. H. 1994: Basicranial anatomy and

phylogeny of primitive canids and closely related miacids (Carnivora: Mammalia). — American Museum Novitates 3094: 1–34.

Webb, S. D. 1977: A history of savanna vertebrates in the New World. — Ann. Rev. Ecol. Syst. 8: 355–380.

Weitzel, K. 1930: Neue Amphicyoniden aus dem Mainzer Becken. — Notizbl. Ver. u. d. hess. geol. L.-A. Darmstadt 5: 11–158.

Werdelin, L. 1987: Jaw geometry and molar morphology in marsupial and placental carnivores: analysis of constraint and its macroevolutionary consequences. — Paleobiology 13: 342–350.

— 1989: Constraint and adaptation in the bone-cracking canid Osteoborus (Mammalia: Canidae). — Paleobiology 15: 387–401.

— 1996a: Carnivoran ecomorphology: a phylogenetic perspective. — In: Gittleman, J. L. (ed.), Carnivores, behavior, ecology and evolution, Vol II: 582–625. Cornell University Press, Ithaca, New York.

— 1996b: Carnivores, exclusive of Hyaenidae, from the later Miocene of Europe and western Asia. — In: Bernor, R., Fahlbusch, V. & Mittmann, W. (eds.), The evolution of western Eurasian Miocene mammal faunas. Columbia University Press. (In press.)

Werdelin, L. & Solounias, N. 1991: The Hyaenidae: taxonomy, systematics and evolution. — Fossils and Strata 30: 1–104.

— 1996: The evolutionary history of hyaenas in Europe and western Asia during the Miocene. — In: Bernor, R., Fahlbusch, V. & Mittmann, W. (eds.), The evolution of western Eurasian Miocene mammal faunas. Columbia University Press. (In press.)

Wilkinson, M. 1995: Arbitrary resolutions, missing entries and the problem of zero-length branches in parsimony analysis. — Syst. Biol. 44: 108–111.

Wolsan, M. 1993: Phylogeny and classification of early European Mustelida (Mammalia: Carnivora). — Acta Theriol. 38: 345–384.

Wozengraft, W. C. 1989: The phylogeny of the recent carnivora. — In: Gittleman, J. L. (ed.), Carnivore behavior, ecology and evolution: 495–535. Cornell University Press, Ithaca, New York.

Wyss, A. R. & Flynn, J. J. 1993: A phylogenetic analysis and definition of the Carnivora. — In: Szalay, F. S., Novacek, M. J. & McKenna, M. C. (eds.), Mammal phylogeny: 32–52. Springer-Verlag, New York.

Zagwijn, W. H. & Hager, H. 1987: Correlations of continental and marine Neogene deposits in the south-eastern Netherlands and the Lower Rhine district. — Meded. Werkgr. Tert. Kwart. Geol. 24: 59–78.

Zapfe, H. 1979: Chalicotherium grande (BLAINV.) aus der miozänen Spaltenfüllung von Neudorf and der miozänen Spaltenfüllung von Neudorf and der March (Devinska Nova Ves), Tscheckoslowakwi. — Ferdinand Berger and Söhne, Wien.

— 1989: Chalicotherium goldfussi KAUP aus dem Vallesien vom Höwenegg im Hegau (Südwestdeutschland). — Andrias 6: 117–126.

Zdansky, O. 1924: Jungtertiäre Carnivoren Chinas. — Palaeont. Sinica C, 2: 1–146.

Zhanxiang, Q., Defa, Y., Hang, J. & Bo, S. 1986: The large-sized ursid fossils from Shanwang, Shandong. — Vertebrata PalAsiatica 24: 182–192.

Zhou, X., Sanders, W. J. & Gingerich, P. D. 1992: Functional and behavioral implications of vertebral structure in Pachyaena ossifraga (Mammalia, Mesonychia). — Contributions from the Museum of Paleontology, the University of Michigan 28: 289–319.

OXFORD MONOGRAPHS IN INTERNATIONAL LAW

Extraterritorial Application of Human Rights Treaties

Law, Principles, and Policy

OXFORD MONOGRAPHS IN INTERNATIONAL LAW

The aim of this series is to publish important and original pieces of research on all aspects of international law. Topics that are given particular prominence are those which, while of interest to the academic lawyer, also have important bearing on issues which touch the actual conduct of international relations. Nonetheless, the series is wide in scope and includes monographs on the history and philosophical foundations of international law.

RECENT TITLES IN THE SERIES

The International Minimum Standard and Fair and Equitable Treatment
Martins Paparinskis

The Margin of Appreciation in International Human Rights Law Deference and Proportionality
Andrew Legg

Individual Criminal Responsibility in International Law
Elies van Sliedregt

Formalism and the Sources of International Law A Theory of the Ascertainment of Legal Rules
Jean d'Aspremont

Extraterritorial Application of Human Rights Treaties Law, Principles, and Policy
Marko Milanovic

State Responsibility for International Terrorism
Kimberley N. Trapp

Disobeying the Security Council Countermeasures against Wrongful Sanctions
Antonios Tzanakopoulos

Maritime Security and the Law of the Sea
Natalie Klein

Extraterritorial Use of Force against Non-State Actors
Noam Lubell

Extraterritorial Application of Human Rights Treaties

Law, Principles, and Policy

MARKO MILANOVIC

OXFORD
UNIVERSITY PRESS

OXFORD
UNIVERSITY PRESS

Great Clarendon Street, Oxford, OX2 6DP,
United Kingdom

Oxford University Press is a department of the University of Oxford.
It furthers the University's objective of excellence in research, scholarship,
and education by publishing worldwide. Oxford is a registered trade mark of
Oxford University Press in the UK and in certain other countries

First Edition published 2011
First published in paperback 2013

Impression: 1

British Library Cataloguing in Publication Data
Data available

Library of Congress Cataloguing in Publication Data
Data available

ISBN 978–0–19–969620–8
ISBN 978–0–19–968225–6 (pbk.)

Printed in Great Britain
on acid-free paper by
CPI Group (UK) Ltd. Croydon CR0 4YY

General Editors' Preface

The actions of States are not confined to their territory, but frequently affect the lives of individuals beyond their borders. In some cases these actions involve killings, torture or indefinite detention, and this raises the question of the extraterritorial application of a States's human rights obligations. Many human rights treaties, notably the European Convention of Human Rights, limit the scope of their application to everyone within the 'jurisdiction' of the Contracting Party concerned. There is considerable uncertainty as to what the term 'jurisdiction' means in this context. Depending on the standard applied, the ensuing human rights protection might be over-inclusive, and thus unrealistically burdensome for States to comply with, or too limited to secure adequate international human rights protection.

Dr Milanovic's important and original study explores the issue of 'jurisdiction' with clarity and vigour, both within the field of human rights treaties and within the wider framework of general international law. It is the quality of this analysis which merits careful consideration of Dr Milanovic's proposed way forward in this sensitive and difficult area of the law.

AVL, DS, ST
Oxford, April 2011

Preface

This book is an expanded version of the doctoral thesis that I defended at the University of Cambridge in 2010. I am most grateful to David Feldman for his guidance and input in the course of the preparation of the thesis; I could not have wished for a better supervisor. I say this not (just) with the utter relief of someone who is finally over and done with his PhD, or as an obligatory platitude in a post-PhD monograph. He was truly the Goldilocks supervisor, neither forcing me heavy-handedly into whatever direction he thought best, nor letting me drift through the PhD all on my own; he was *just right*, and his future students are lucky to have him.

I also wish to express my thanks to my two examiners, James Crawford and Colin Warbrick. Their great knowledge and good humour made my viva not only challenging and intellectually stimulating, but genuinely fun. Few are so lucky, and for that, again, I am thankful, as I am thankful for their comments. I am likewise indebted to Dapo Akande, Ken Anderson, Charles Garraway, Vidan Hadzi-Vidanovic, Rebecca Jenkin, Francesco Messineo, Tatjana Papic, Jelena Pejic, Bruno Simma, Sandesh Sivakumaran, and Tobias Thienel who were gracious enough with their time to read all or parts of this study and provide me with most helpful suggestions and comments. My gratitude also extends to John Louth, Merel Alstein, Bethan Cousins, and the whole OUP production team for their work on this book, as well as to Zeljko Djuric, Dusan Kanazir, Milica Kostic, Ines Lasic, Zarko Markovic, and Petar Zmak for preparing the tables of authorities.

Finally, I am of course grateful to my family and friends for their continuing support—particularly so to a certain circle of friends who made my three years in Cambridge not only bearable, but happy. The Tapp Fund of Gonville & Caius College provided me with a more material kind of support, for which I am most thankful.

Chapter II of the study is a revised version of an article I published as 'From Compromise to Principle: Clarifying the Concept of State Jurisdiction in Human Rights Treaties', (2008) 8 *HRLR* 411, while Chapter V is similarly a revision of a piece I originally published as 'A Norm Conflict Perspective on the Relationship between International Humanitarian Law and Human Rights Law', (2009) 14 *JCSL* 459, and then in a more expanded form as 'Norm Conflicts, International Humanitarian Law and Human Rights Law', in O. Ben-Naftali (ed.), *Human Rights and International Humanitarian Law* (Oxford University Press, 2011). The remainder of this study has not been published before, although some sections were reworked as posts at *EJIL: Talk!*, the blog of the European Journal of International Law.

It is always a danger to write a book in what is a fast-developing field, as the risk of obsolescence exponentially increases. There are always new cases and developments on the horizon when it comes to extraterritoriality and human rights—as I write, *Al-Skeini* is pending before the Grand Chamber of the European Court of Human Rights. One must, however, stop somewhere; the book takes into account all cases (to the best of my knowledge) decided until December 2010.

Author's Preface to the Paperback Edition

I am very grateful to OUP for producing this book in paperback; the book itself is unchanged when compared to the hard cover edition which came out in 2011. As I had mentioned in the preface to the hard cover, the book's coverage of relevant cases and materials ends with December 2010, and at the time it was due to come out the European Court of Human Rights' *Al-Skeini* judgment was on the horizon.

For good luck or bad, *Al-Skeini* duly came out as the book itself was hitting the shelves. The Grand Chamber used the case as a vehicle for fixing some of the major faults in its prior case law and did so with some success, even going so far as to partially (if half-heartedly) overturn *Bankovic*, its previous attempt at drawing a grand design of Article 1 of the European Convention.

Thankfully (for me, at least), while more human rights-friendly than *Bankovic*, *Al-Skeini* was still the product of a judicial balancing act between universality and effectiveness, a major theme of this book, and is still far from where the case law on the extraterritorial application of human rights treaties needs to be. While closing the doors to some uncertainties it opened others—consider, for instance, the question whether and how the Convention applies to the actions of European troops in Afghanistan, and note just how unhelpful *Al-Skeini* is in providing an answer.

In short, *Al-Skeini* notwithstanding, please continue reading this book! For my detailed views on that judgment please see '*Al-Skeini* and *Al-Jedda* in Strasbourg,' (2012) 23 *European Journal of International Law* 121. And *Al-Skeini* was of course not the only important—or instructive—case decided the past two years; for a running commentary of sorts I would ask the reader to take a look at my posts on *EJIL: Talk!* at http://www.ejiltalk.org/category/human-rights/extraterritorial-application/.

Nottingham, January 2013

Contents

List of Abbreviations

ACHR	American Convention on Human Rights
ADHR	American Declaration on Human Rights
AJIL	American Journal of International Law
ASR	Articles on State Responsibility
BYIL	British Yearbook of International Law
CAT	Convention Against Torture
CEDAW	Convention on the Elimination of All Forms of Discrimination Against Women
CERD	Convention on the Elimination of All Forms of Racial Discrimination
CRC	Convention on the Rights of the Child
ECHR	European Convention on Human Rights
EJIL	European Journal of International Law
EU	European Union
ICCPR	International Covenant on Civil and Political Rights
ICESCR	International Covenant on Economic, Social, and Cultural Rights
ICJ	International Court of Justice
ICLQ	International and Comparative Law Quarterly
ICRC	International Committee of the Red Cross
ICTY	International Criminal Tribunal for the former Yugoslavia
IRRC	International Review of the Red Cross
IHL	International Humanitarian Law
ILC	International Law Commission
JCSL	Journal of Conflict & Security Law
JICJ	Journal of International Criminal Justice
NATO	North Atlantic Treaty Organization
OJLS	Oxford Journal of Legal Studies
RdC	Recueil des Cours
UDHR	Universal Declaration on Human Rights
UK	United Kingdom
UN	United Nations
US	United States

NOTE ON METHOD OF CITATION

Citations to authorities will generally conform to the OSCOLA style. However, citations to cases will conform to the style of their own jurisdiction, as will standard abbreviations for journal titles. This study is divided into five chapters, which form separate units with regard to citation and footnote numbering. Cross-references within each chapter will cite the section number only, while cross-references to other chapters will indicate both the chapter number and the section number.

Because of frequency of use, the following works will be cited in abbreviated form throughout this study, except for their first appearance:

F. Coomans and M. Kamminga (eds), *Extraterritorial Application of Human Rights Treaties* (Intersentia, 2004), hereinafter Coomans and Kamminga;

M. Gondek, *The Reach of Human Rights in a Globalising World: Extraterritorial Application of Human Rights Treaties* (Intersentia, 2009), hereinafter Gondek.

Table of Cases

INTERNATIONAL

European Commission of Human Rights

European Court of Human Rights

Inter-American Commission of Human Rights

Inter-American Court of Human Rights

International Court of Justice

International Criminal Tribunal for the Former Yugoslavia

UN Treaty Bodies: Human Rights Committee

Consideration of state party reports and general comments

Individual cases

UN Treaty Bodies: Committee Against Torture

Consideration of state party reports and general comments

Individual cases

UN Treaty Bodies: Committee on the Elimination of Racial Discrimination

UN Treaty Bodies: Committee on the Rights of the Child

National case law

Canada

Israel

United Kingdom

United States

Table of Treaties

I

Introduction

1. Outline and Scope of the Study

If a state affects the lives of individuals outside its sovereign borders, when does it owe them obligations pursuant to the human rights treaties to which it is a party? Answering this question is the object of this study. In recent years, the issue of the extraterritorial application of human rights treaties has truly come to the fore. Not only is it now the subject of a growing literature, but more and more actual cases are being litigated. Courts have involved themselves on human rights grounds in such controversies as Turkey's invasion of northern Cyprus, NATO's use of force against Serbia, Russia's involvement in Georgia, or the US and UK invasion of Iraq. No longer are such cases examined solely from the more orthodox standpoints of the *jus ad bellum* and the *jus in bello*, or state sovereignty more generally. Now it is increasingly the individuals directly affected by extraterritorial state action who are pursuing the avenues open to them under international human rights law. And because the impact of human rights treaties in an extraterritorial context is growing, states need to take it into account in their policy-making.

Many of the controversies surrounding the extraterritorial application of human rights treaties which will be explored in this study have been pushed to their limit by the actions of certain states engaged in the ongoing, if rebranded, 'Global War on Terror', particularly by the United States under the administration of George W. Bush, which was adept at creating and exploiting purported gaps in international law to further its purposes.[1] Thus the United States has argued in international fora that human rights treaties to which it is a party do not apply to persons it has detained outside its sovereign territory as 'unlawful enemy combatants' in Guantanamo Bay, Cuba, or in its bases in Iraq and Afghanistan, and that even if they did, they do not apply in times of war.[2] Both of these claims have been strongly rejected by UN treaty bodies, though without any immediate effects on

[1] For a general discussion and more background, see R. Wilde, 'Legal "Black Hole"? Extraterritorial State Action and International Treaty Law on Civil and Political Rights', (2005) 26 *Mich. J. Int'l L.* 739.

[2] See, e.g., the Opening Remarks by John Bellinger, Legal Adviser, US Department of State, before the UN Committee against Torture, 5 May 2006, available at <http://www.state.gov/g/drl/rls/68557.htm> and the Opening Statement of Matthew Waxman, Head of US Delegation before the UN Human Rights Committee, 17 July 2006, available at <http://2001-2009.state.gov/g/drl/rls/70392.htm>.

US compliance.[3] Though the Obama administration has made some changes to US anti-terrorism strategy and has dropped the 'global war' moniker, its overall approach to international terrorism—and, consequently, to the relevance of human rights—has remained broadly the same as that of the Bush administration, at least in its somewhat less unilateralist second term. In fact, the Obama administration has actually escalated some of the policies that directly implicate the extraterritorial application of human rights treaties, most notably the targeted killing of suspected terrorists through the use of unmanned aerial vehicles, or drones, in Pakistan, Afghanistan, and Yemen.

Most of these controversies are to an extent counterintuitive. Human rights are, after all, supposed to be universal—why should it matter whether a state violates a person's rights through killing, torture, indefinite detention, or unfair trials by acting within its territory or outside it? Indeed, when a state acts against an individual outside its territory, there is almost a human rights reflex to immediately venture into the substantive issue of whether the person's rights were violated. For example, in November 2002, on Yemeni soil and with the apparent consent of the Yemeni government, a US Predator drone fired a missile killing the suspected mastermind of the USS Cole terrorist bombing—an action condemned, among others, by the late Anna Lindh, then the foreign minister of Sweden, as a 'summary execution that violates human rights'.[4] Yet, as a matter of law, this statement simply begs the question of whether the target of this attack even had rights vis-à-vis the United States, or whether he had such rights only in relation to Yemen, which permitted the United States to assassinate him.[5]

At the legal level, the question whether a Yemeni national, who is living in Yemen when he is killed by the United States with the consent of the Yemeni government, has rights vis-à-vis the United States is a matter of treaty interpretation. As we shall see, the scope of application of many major human rights treaties is defined by a very similar clause: the persons concerned must fall within the state's *jurisdiction* for that person to be able to raise his or her rights against the state. This introductory chapter will define more precisely what is meant by the extraterritorial application of human rights treaties, explain that the law of treaties sets no general rules on extraterritorial application, and outline the basic normative framework of the human rights treaties which are the object of this study.

Having said that, of course one might well ask what a human rights treaty is to begin with. I will not attempt to provide an abstract definition of this notion, nor try to divine some supposedly unique characteristics that these treaties might have

[3] See the Conclusions and Recommendations of the Committee Against Torture: United States of America, UN Doc. CAT/C/USA/CO/2, 25 July 2006, paras 14 and 15 and the Concluding Observations of the Human Rights Committee: United States of America, UN Doc. CCPR/C/USA/CO/3, 15 September 2006, para. 10.

[4] 'Killing probes the frontiers of robotics and legality', *The Guardian*, 6 November 2002, available at <http://www.guardian.co.uk/usa/story/0,12271,834311,00.html>.

[5] The Yemeni government still frequently consents to the US use of force on its territory, albeit furtively, so as to avoid a domestic political backlash—see 'WikiLeaks cables: Yemen offered US "open door" to attack al-Qaida on its soil', *The Guardian*, 3 December 2010, available at <http://www.guardian.co.uk/world/2010/dec/03/wikileaks-yemen-us-attack-al-qaida>.

in relation to other international agreements. Much has already been written on this topic,[6] to which there is little useful to add. It is, of course, possible to broadly identify human rights treaties by their subject-matter as those treaties which have the object of safeguarding those rights of individuals which are somehow perceived as being inherent in their human dignity. However, the specific treaties which will be analysed in this thesis will be set out in the following section.

I should also note at this point that this study will *not* be dealing with the extraterritorial application of *customary* human rights law. Many scholars have argued that the rules announced in the Universal Declaration of Human Rights (UDHR) and later implemented in legally binding treaties have now attained customary status.[7] As an exhortative instrument, the UDHR, much like the American Declaration on Human Rights, has no clause defining its territorial scope.[8] Some authors have thus suggested that although the territorial scope of human rights treaties may be limited, no such limitation attaches to human rights arising under custom.[9] However, bearing in mind the many difficulties which as a general matter surround the inference of customary rules from the practice of states acting pursuant to a treaty obligation,[10] it does not seem at all possible to disentangle the territorial limitations on human rights as prescribed in treaties from any customary substantive rules of human rights law. In other words, it is quite unlikely that states have assumed more extensive obligations under customary human rights law than they have done under treaty law. Even if they did, there is rarely any forum for enforcing such obligations directly. Customary human rights law will therefore generally be outside the purview of this study.

Among the relevant human rights treaties, a greater emphasis will be put on treaties protecting civil and political rights, for the sole reason that there is much more case law and other material to work with in respect of these treaties than those

[6] See, e.g., M. Craven, 'Legal Differentiation and the Concept of the Human Rights Treaty in International Law', (2000) 11 *EJIL* 489; B. Simma, 'International Human Rights and General International Law: A Comparative Analysis', (1995) 4 *Collected Courses of the Academy of European Law* 153. See also J. Crawford, 'Multilateral Rights and Obligations in International Law', (2006) 319 *RdC* 325.

[7] See, e.g., H. Hannum, 'The Status of the Universal Declaration of Human Rights in National and International Law'(1995–1996) 25 *Ga. J. Int'l.& Comp. L.* 28, at 317–52; V. Dimitrijevic, 'Customary Law as an Instrument for the Protection of Human Rights', (2006) *ISPI* 7 *Working Papers*, available at <http://www.ispionline.it/it/documents/wp_7_2006.pdf>.

[8] However, the last paragraph of the Preamble of the UDHR does make a reference to territorial application, as it speaks of the UDHR

> as a common standard of achievement for all peoples and all nations, to the end that every individual and every organ of society, keeping this Declaration constantly in mind, shall strive by teaching and education to promote respect for these rights and freedoms and by progressive measures, national and international, to secure their universal and effective recognition and observance, both among the peoples of Member States themselves and among the peoples of *territories under their jurisdiction*. (emphasis added)

[9] See, e.g., N. Lubell, *Extraterritorial Use of Force Against Non-State Actors* (Oxford University Press, 2010), at 232–5.

[10] See, e.g., the Final Report of the International Law Association's Committee on the Formation of Customary (General) International Law (London, 2000), at 42 *et seq.*

protecting socio-economic rights.[11] That said, this study will proceed from the assumption that there is something to be gained from focusing on problems common to all or most human rights treaties, and will view these treaties as a whole. Hence, it is generally structured thematically, issue by issue, not chronologically or treaty by treaty.

Of all the treaties, I will focus the greatest attention on the European Convention on Human Rights, for two reasons. First, the ECHR system is by far the strongest of all human rights regimes (if far from perfect) in its ability to effectively secure compliance and have a direct impact on state policy. The stakes are highest in Strasbourg, because the Court will be listened to. Secondly, it is precisely because the stakes are highest in Strasbourg that the jurisprudence of the European Court of Human Rights on extraterritorial application is the richest and the most developed. At the same time, it is the most problematic, suffering from rampant casuistry and conceptual chaos. It is a jurisprudence of (at times quite unprincipled) compromise, caused mostly by the Court's understandable desire to avoid the merits of legally and politically extremely difficult cases by relying on the preliminary issue of extraterritorial application.

Chapter II of this study will try to clear up some of this conceptual confusion. It will examine the notion of state jurisdiction in human rights treaties, and will attempt to place it within the framework of international law. Is this notion the general concept of jurisdiction one finds in public international law, as contemplated by the European Court in *Bankovic*,[12] or is it a distinct, autonomous concept, which is a part of a self-contained human rights regime?[13] Is it a simple admissibility requirement for an application,[14] or a test of attribution in the framework of state responsibility, as assumed by the International Criminal Tribunal for the Former Yugoslavia in the *Tadic* case,[15] or is it in fact a threshold criterion

[11] For a detailed examination of the extraterritorial reach of treaties on socio-economic rights, see M. Gondek, *The Reach of Human Rights in a Globalising World: Extraterritorial Application of Human Rights Treaties* (Intersentia, 2009), at 291 *et seq*. See also F. Coomans, 'Some Remarks on the Extraterritorial Application of the International Covenant on Economic, Social and Cultural Rights,' in F. Coomans and M. Kamminga (eds), *Extraterritorial Application of Human Rights Treaties* (Intersentia, 2004), 183; R. Künnemann, 'Extraterritorial Application of the International Covenant on Economic, Social and Cultural Rights', in Coomans and Kamminga 201; S. Skogly, *Beyond National Borders: States' Human Rights Obligations in International Cooperation* (Intersentia, 2006), as well as the contributions in M. Gibney and S. Skogly, *Universal Human Rights and Extraterritorial Obligations* (University of Pennsylvania Press, 2010).

[12] *Bankovic and Others v. Belgium and Others* [GC] (dec.), App. No. 52207/99, 12 December 2001. The applicants in *Bankovic* were the victims of the bombing of the RTS television station in Belgrade by NATO airplanes during the 1999 NATO campaign against Serbia. They filed the application against all NATO member states that were also parties to the ECHR, claiming that these states were jointly responsible for the violation of their human rights, above all the right to life. The Grand Chamber of the European Court declared the application inadmissible, holding that the applicants were outside the jurisdiction of the respondent states in the sense of Article 1 ECHR, and accordingly had no rights under the ECHR that they could enforce. *Bankovic* will be discussed in detail throughout this study, but in particular in Chapters II and IV.

[13] See M. Scheinin, 'Extraterritorial Effect of the International Covenant on Civil and Political Rights', in Coomans and Kamminga 73.

[14] See M. Nowak, *CCPR Commentary* (Engel, 2nd revised edn, 2005), at 858–62.

[15] *Prosecutor v. Tadić*, IT-94-1, Judgment, Appeals Chamber, 15 July 1999.

determining whether a human rights obligation exists in the first place?[16] These are the questions I hope to answer.

Chapter II begins the inquiry into the semantic, ordinary meaning of the jurisdiction clauses in human rights treaties, and of their construction into workable legal concepts and rules. Their interpretation cannot be complete, however, without examining the object and purpose of these treaties, and the various policy considerations which influence courts in their decision-making. This will be the object of Chapter III, which will focus on the tension between universality and effectiveness; this is, I submit, the prime cause of methodological and conceptual inconsistencies in the case law. Chapter IV will conclude the interpretative inquiry into the meaning of the jurisdiction clauses, and will elaborate on the several possible models of extraterritorial application of human rights treaties. Finally, Chapter V will explore the relationship between international humanitarian law and international human rights law from a norm conflict perspective, as the interaction between these two bodies of law is frequently a concurrent issue with that of extraterritorial application, implicating the same policy considerations.

Most of the case law that we will be examining is of fairly recent extraction. It appears that the problem of the extraterritorial application of human rights treaties has been growing progressively more acute in the past decade or so. It is indeed rather startling that such a fundamental issue regarding the scope of application of these treaties has not been definitively resolved much earlier during their life-span. One, almost trite, response to this observation would be that in the age of globalization states are increasingly affecting the human rights of individuals outside their borders, and that this explains both the increase in litigated cases on extraterritorial application and the growing importance of the issue generally.

There is some truth in this remark, particularly with regard to socio-economic rights and transnational criminal law enforcement. There is also, however, something profoundly mistaken in suggesting that most of the situations which today involve the extraterritorial application of human rights treaties are truly novel. States, especially powerful states, have *always* acted outside their borders and have *always* affected the lives of foreigners. They have moreover continued to do so even in the period after the Second World War, in which the modern human rights instruments were created. It seems that the better explanation for the increasing urgency of this topic is that society at large has changed and is changing still. Our culture has been permeated by law generally and human rights specifically to such a level that even those state acts that have hitherto been considered as the ultimate expressions of sovereign prerogative have become exposed to human rights scrutiny, in public discourse as well as in courts. We live in an age of rights,[17] and the rhetoric of rights is no longer solely the province of increasingly aggressive lawyers and human rights activists, but is employed by policy-makers and actors of all stripes.

[16] See R. Lawson, 'Life after Bankovic: On the Extraterritorial Application of the European Convention on Human Rights', in Coomans and Kamminga 83.

[17] L. Henkin, *Age of Rights* (Columbia University Press, 1996).

Hence, people complain to human rights bodies more frequently and they do so in situations undreamed of even fairly recently.[18] If, for example, an international lawyer had predicted even just a decade or two ago that Saddam Hussein, of all people, would soon be lodging an application with the European Court of Human Rights, he would have been thought eccentric at best. Yet this is exactly what happened after the 2003 invasion of Iraq, and that particular case turned on the territorial scope of application of the European Convention.[19]

In another example straight out of the pages of a spy novel, the family of Alexander Litvinenko, a former high-ranking officer of the Russian security services who was fatally poisoned in London in November 2006 with polonium, a highly radioactive substance, is reported to have lodged an application against Russia with the European Court.[20] On his deathbed, Mr Litvinenko accused the Kremlin of involvement in his death and his family now argues that Russia violated several articles of the European Convention. The case is still pending, and clearly the applicants' prospects depend on whether Russia's obligations under the European Convention can be interpreted to extend to a person killed in London.

There is no little irony in the fact that today even deposed dictators,[21] former KGB officers, or Marxist-Leninist revolutionaries *cum* notorious terrorists[22] try to avail themselves of the protections granted by international human rights law. Indeed, one could also view this phenomenon as a corollary of the widespread 'humanization' that international law has been subjected to under the influence of human rights.[23] That humanizing effect is furthermore not confined to international law and international courts, as similar issues have arisen before domestic courts as well. For example, US courts have grappled with the question of the extraterritorial application of the US Constitution to detainees in Guantanamo or in US bases in Afghanistan, while UK courts have dealt with the acts of UK armed forces in Iraq under the ECHR and the Human Rights Act 1998. In sum, human rights and their universalist premise have become *internalized* to such an extent that their extraterritorial application is no longer merely a theoretical issue.

At this point I must say that this study is a product of the same process. It is not just an attempt to solve a doctrinal puzzle, or to bring together the disparate strands of the case law. It necessarily has an ideological bent. Human rights treaties are themselves not value-neutral instruments, and the process of their interpretation cannot be completely value-neutral either. I start from the assumption that human rights grounded in universal human dignity are a *good thing*, but this is not an assumption that I am able, or wish, to defend here. What matters is that this

[18] See in that regard G. Simpson, 'The Death of Baha Mousa', (2007) 8 *Melbourne JIL* 340.

[19] *Saddam Hussein v. 21 Countries* (dec.), App. No. 23276/04, 14 March 2006.

[20] See 'Litvinenko widow takes case to European Court', *Reuters*, 22 May 2007, available at <http://www.reuters.com/article/idUSL2264003120070522>.

[21] See also *Milošević v. The Netherlands* (dec.), App. No. 77631/01, 19 March 2002.

[22] *Ramirez Sanchez v. France* [GC], App. No. 59450/00, Judgment, 4 July 2006.

[23] See generally T. Meron, *The Humanization of International Law* (Nijhoff, 2006); M. Kamminga and M. Scheinin (eds), *The Impact of Human Rights Law on General International Law* (Oxford University Press, 2009).

assumption is not just my own personal view, but the normative premise of the entire body of law that I will be analysing. Thus, to the extent that value judgments prove to be necessary, this study is part of a project—an academic and not an *activist* project, but a project nonetheless—with the general aim of furthering the humanization of both international law and the reality of international relations.

To conclude this introduction, this is a study on human rights which is unconcerned with the actual substantive content of human rights treaties, but is concerned instead with the preliminary conditions for their application. However, although the interpretation of the jurisdiction clauses is conceptually distinct from the substantive application of a treaty to a specific issue, we will see that this is simply not the case in practice. Rather, the preliminary question of application is frequently used as a proxy for dealing with the merits, as nothing more than a judicial avoidance technique. My central argument is that the only way that the case law on the threshold issue of extraterritorial application can be sensible and coherent is if it is divorced from such an unstated assessment of the merits. This, however, will only be possible if due regard is paid to considerations of effectiveness, so that the actual substantive application of a human rights treaty in an extraterritorial context does not appear to be hopelessly unrealistic or utopian.

2. Defining Extraterritorial Application

Before venturing any further in explaining the normative framework that regulates the territorial scope of application of human rights treaties, I must first briefly define what, in fact, *is* the extraterritorial application of a human rights treaty. In its commentaries to the Draft Articles on the Law of Treaties (later to become the Vienna Convention on the Law of Treaties (VCLT), the International Law Commission (ILC) stated that

> [c]ertain types of treaty, by reason of their subject matter, are hardly susceptible of territorial application in the ordinary sense. Most treaties, however, have application to territory and a question may arise as to what is their precise scope territorially.[24]

Human rights treaties fall within the latter category. As the ILC explains, it is by looking at the subject-matter of a treaty—the content of the rights and obligations that it creates—that we can tell whether and how these rights and obligations apply territorially. With respect to human rights treaties specifically, we must note that they only impose obligations on their states parties, and do not do so for third states or private individuals. Moreover, they create obligations not only between the states parties themselves, but also between states and individuals—indeed, that is their whole purpose. The *application* of a human rights treaty to a particular individual thus requires that a state owes that individual some legal obligation under the treaty.

[24] ILC, 'Draft Articles on the Law of Treaties with Commentaries', (1996) 2 *Yearbook of the International Law Commission* 187, at 213.

Extraterritorial application simply means that at the moment of the alleged violation of his or her human rights the individual concerned is not physically located in the territory of the state party in question, a geographical area over which the state has sovereignty or title. Extraterritorial application of a human rights treaty is an issue that will most frequently arise from an *extraterritorial state act*, i.e. conduct attributable to the state, either of commission or of omission, performed outside its sovereign borders—for example, the killing of a suspected terrorist in Pakistan by a US drone. However—and this is a crucial point—extraterritorial application does not *require* an extraterritorial state act, but solely that the individual concerned is located outside the state's territory, while the injury to his rights may as well take place inside it. For instance, if we accept that the ECHR applies to the taking within the UK of the property of a UK national living in Monaco in tax haven bliss, this would also be an instance of extraterritorial application, since the individual concerned is not himself within the UK's territory even if his property is.

It can be somewhat misleading to talk about the extraterritorial application of human rights treaties as a useful category, which is why I sought to define it. First, with the possible exception of the International Covenant on Civil and Political Rights (ICCPR),[25] the actual text of the relevant treaties makes no mention of *title* over territory as the threshold for their applicability. Rather, the treaties require that the individual be within or subject to the state's *jurisdiction*.[26] As I will shortly explain, this is a concept revolving around *de facto* control over territory, and perhaps also individuals. It is not about title or sovereignty over territory, which is, or should be, perfectly irrelevant for the treaties' territorial scope. Jurisdiction is the *actual* exercise of control and authority by a state, while title or sovereignty establishes the state's *right* in international law to exercise such authority within a specific territory. Secondly, an emphasis on the extraterritorial application of human rights treaties may give the mistaken impression that no issue of applicability can arise *intra*-territorially.[27] A state may have title over territory, but not have jurisdiction, i.e. *de facto* control, over it. Thus, it is Cuba, not the United States, that has title over Guantanamo Bay, yet it would to my mind be absurd to say that it is Cuba, rather than the United States, which has the obligation to ensure the human rights of persons detained there.[28]

Extraterritorial application should not be confused with other phenomena. Most notably, in *Soering*[29] the European Court incorporated into Article 3 ECHR the principle of *non-refoulement* first developed in refugee law, ruling that the UK could not extradite a murder suspect to the US since there was a serious risk that if convicted he would be sentenced to death and subjected to years of waiting on death row for the sentence to be carried out, which was in the Court's view

[25] On which see further below, Chapter IV, Section 5.

[26] Note also that the text of the relevant treaties is quite clear that it is the *individual*, rather than the state *agent* affecting his rights, who has to be within, under, or subject to the state's jurisdiction.

[27] On this issue, see particularly K. Mujezinovic Larsen, '"Territorial Non-Application" of the European Convention on Human Rights', (2009) 78 *Nord. J. Int'l L.* 73.

[28] See further below, Chapter III, Section 3.

[29] *Soering v. United Kingdom*, App. No. 14038/88, Judgment, 7 July 1989.

tantamount to inhuman treatment.[30] Such cases do not involve extraterritorial application, since the individual concerned is located within the territory of the extraditing state. As the European Court itself said in *Bankovic*:

However, the Court notes that liability is incurred in such cases by an action of the respondent State concerning a person while he or she is on its territory, clearly within its jurisdiction, and that such cases do not concern the actual exercise of a State's competence or jurisdiction abroad.[31]

The violation in such cases is *not* the inhuman treatment that the individual would suffer in the state seeking extradition—after all, the US is not even a party to the ECHR, and is incapable of violating it—but the decision of the extraditing state to actually proceed with the extradition, while being aware of the risk that the individual will be subjected to inhuman treatment.[32] The violation, in other words, consists of the state knowingly exposing an individual to harm at the hands of third parties, intra- or extraterritorially, even though the harm itself may or may not materialize.[33] Note that a *Soering*-type scenario *can* arise in an extraterritorial context—for example, with the transfer of detainees held by the US or UK in Iraq to Iraqi authorities—but only then would such a scenario be one of extraterritorial application.[34]

3. The Law of Treaties Provides No General Answer

Having explored what is meant by the extraterritorial application of human rights treaties, it is necessary to establish whether public international law has something

[30] Similarly, see *Judge v. Canada*, UN Doc. CCPR/C/78/D/829/1998, 5 August 2002 and *Kindler v. Canada*, UN Doc. CCPR/C/48/D/470/1991, 11 November 1993.

[31] *Bankovic*, para. 68.

[32] See also Gondek, at 6; M. Nowak, 'Obligation of States to Prevent and Prohibit Torture in an Extraterritorial Perspective,' in Gibney and Skogley, above note 11, 11, at 19; Nowak, above note 14, at 861–2; D. McGoldrick, 'Extraterritorial Application of the International Covenant on Civil and Political Rights', in Coomans and Kamminga 41, at 52–3; M. O'Boyle, 'The European Convention on Human Rights and Extraterritorial Jurisdiction: A Comment on "Life After Bankovic"', in Coomans and Kamminga 125, at 126–7.

[33] A commentator suggests that it is the extraterritorial harm that would itself be a violation of the ECHR, because of the prior territorial connection to the extraditing state: 'if a signatory state managed to expel an individual without establishing a sufficient certainty that he would not be subject to torture or cruel, inhuman, or degrading treatment and he proceeded to suffer such harms at the hand of the receiving country's authorities, the individual would still be "within the jurisdiction" of the Convention for the purposes of Article 1'—see S. Miller, 'Revisiting Extraterritorial Jurisdiction: A Territorial Justification for Extraterritorial Jurisdiction under the European Convention', (2010) 20 *EJIL* 1223, at 1243. But this is simply not what the Court said in *Soering*, where it quite clearly considered the extraterritorial harm subsequent to extradition to be 'outside the jurisdiction' of the extraditing state—see *Soering*, paras 85 and 86. See also *Kindler*, para. 13.1: 'If a State party extradites a person within its jurisdiction in circumstances such that as a result there is a real risk that his or her rights under the Covenant will be violated *in another jurisdiction*, the State party itself may be in violation of the Covenant' (emphasis added).

[34] Extraterritorial *non-refoulement* is also implicated in the 'war on terror' practice of so-called extraordinary renditions—see generally M. Satterthwaite, 'Rendered Meaningless: Extraordinary Rendition and the Rule of Law', (2007) 75 *Geo. Wash. L. Rev.* 1333.

to say about the extraterritorial application of treaties generally. It has been suggested in that regard that Article 29 VCLT, which reads 'unless a different intention appears from the treaty or is otherwise established, a treaty is binding upon each party in respect of its entire territory', creates a presumption against extraterritoriality.[35]

This argument is unfounded. Article 29 VCLT deals with a specific problem—treaty-making by federal states or states with overseas dependencies—and creates a presumption in favour of the applicability of the treaty to the whole territory of the state, i.e. to all geographical areas over which it has title, unless a contrary intention is established.[36] In order to rebut this presumption states have historically employed a variety of colonial or federal clauses in treaties that were meant to limit the territorial scope of the treaties that they concluded—indeed, there is one such colonial clause in Article 56 ECHR, which I will address in more detail below. Not only is there a considerable logical and linguistic leap in interpreting Article 29 VCLT as saying that a treaty binds a state party in respect of *nothing but* its entire territory, but the ILC commentaries to Article 25 of the Draft Articles on the Law of Treaties, which with a slightly different formulation became Article 29 VCLT, make it clear that this provision has nothing to say on the extraterritorial application of treaties:

In [the ILC's] view, the law regarding the extra-territorial application of treaties could not be stated simply in terms of the intention of the parties or of a presumption as to their intention; and it considered that to attempt to deal with all the delicate problems of extra-territorial competence in the present article would be inappropriate and inadvisable.[37]

There is no default rule of international law, no presumption against extraterritoriality which we can turn to in the absence of a clear norm in the treaty itself regulating its extraterritorial applicability. Conversely, there is also no presumption in *favour* of extraterritoriality.[38] The only guidance can be found in the text, object,

[35] This was, for instance, the position of the Russian Federation before the ICJ with regard to the extraterritorial application of the CERD—see *Case concerning Application of the International Convention on the Elimination of All Forms of Racial Discrimination* (*Georgia v. Russia*), CR 2008/23, available at <http://www.icj-cij.org/docket/files/140/14713.pdf>, at 39 *et seq.* Israel espoused the same interpretation with regard to the occupied Palestinian territories—see O. Ben-Naftali and Y. Shany, 'Living in Denial: The Application of Human Rights Treaties in the Occupied Territories', (2003) 37 *Israel L Rev* 17, at 66–8.

[36] See generally A. Aust, 'Treaties, Territorial Application', in *Max Planck Encyclopedia of Public International Law*, available at <http://www.mpepil.com>, and the sources cited therein.

[37] ILC, above note 24, at 214.

[38] The ICJ's somewhat expansive dictum in its provisional measures order in *Georgia v. Russia* that, absent a textual limitation, the 'provisions of CERD generally appear to apply, like other provisions of instruments of that nature, to the actions of a State party when it acts beyond its territory' should not be interpreted to say otherwise—see Order on the Indication of Provisional Measures, 12 October 2008, para. 109. As the Court rightly notes, it is the *nature* of human rights treaties, their foundation in universality, that requires a justification for a territorial limit on their scope, but as a consequence of their object and purpose, not of some kind of formal presumption. See Chapter III below; see also *Legal Consequences of the Construction of a Wall in the Occupied Palestinian Territory,* Advisory Opinion, 9 July 2004, ICJ Reports 2004, at 136, para. 109, and see further below, Chapter III, Section 2. For further comment on the Court's order, see T. Thienel, 'Provisional Measures in the "*Case concerning Application of the International Convention on the Elimination of All Forms of Racial Discrimination*"

and purpose of each particular treaty.[39] In that regard, provisions governing the territorial scope of human treaties can be classified in several broad, yet flexible and overlapping categories.

4. Basic Normative Framework

A. Jurisdiction clauses

The first, and most interesting, of these are treaties containing a jurisdiction clause. The first human rights treaties proper—though actually not the first treaties generally[40]—to have such a clause are the ECHR and the ICCPR. Though these two treaties were for the most part drafted over the same period, the ECHR was adopted more than fifteen years before the ICCPR, since, as is well known, due to ideological differences states were unable to agree on a comprehensive UN human rights convention protecting both civil and political and socio-economic rights. Article 1 ECHR contains the prototype jurisdiction clause: 'The High Contracting Parties shall secure to everyone *within their jurisdiction* the rights and freedoms defined in Section I of this Convention' (emphasis added).

Though this phrase is repeated in many treaties, no two jurisdiction clauses are exactly alike. The wording of Article 2(1) ICCPR differs significantly from that of the ECHR: 'Each State Party to the present Covenant undertakes to *respect and to ensure* to all individuals *within its territory and subject to its jurisdiction* the rights recognized in the present Covenant, without distinction of any kind, such as race, colour, sex, language, religion, political or other opinion, national or social origin, property, birth or other status' (emphasis added)—the key difference here being the seemingly conjunctive territorial requirement ('within its territory') which is missing from Article 1 ECHR and from other human rights treaties, including the Second Optional Protocol to the ICCPR abolishing the death penalty,[41] and which the Human Rights Committee has interpreted disjunctively.[42] Article 7 of the Migrants Workers Convention is the only treaty provision which uses the ICCPR formula, but with the 'and' from Article 2(1) ICCPR explicitly turned into an 'or'.[43]

(Georgia v. Russian Federation)', (2009) 1 *Goettingen JIL* 143, esp. at 148–52; T. Thienel, 'The Georgian Conflict, Racial Discrimination and the ICJ: The Order on Provisional Measures of 15 October 2008', (2009) 9 *HRLR* 465, esp. at 469–72.

[39] See also Gondek, at 11.

[40] See below, Chapter II, Section 2.E.

[41] Article 1 of which reads: '1. No one within the jurisdiction of a State Party to the present Protocol shall be executed. 2. Each State Party shall take all necessary measures to abolish the death penalty within its jurisdiction.'

[42] See Human Rights Committee, General Comment 31, Nature of the General Legal Obligation on States Parties to the Covenant, U.N. Doc. CCPR/C/21/Rev.1/Add.13 (2004), para. 11. For a more extensive discussion of Art. 2(1), see Chapter IV, Section 5 below.

[43] 'States Parties undertake, in accordance with the international instruments concerning human rights, to respect and to ensure to all migrant workers and members of their families within their territory or subject to their jurisdiction the rights provided for in the present Convention.'

Unlike the ICCPR, the American Convention on Human Rights (ACHR)[44] and the UN Convention on the Rights of the Child (CRC)[45] alter the ECHR jurisdiction formula only slightly, but like the ICCPR, they do so with the addition of a non-discrimination clause.

Other treaties have jurisdiction clauses only in respect of specific obligations arising under the treaty, instead of a single applicability clause for the treaty as a whole. The first of these treaties was the 1969 Convention on the Elimination of All Forms of Racial Discrimination (CERD), which provides in its Article 3 that 'States Parties particularly condemn racial segregation and apartheid and undertake to prevent, prohibit and eradicate all practices of this nature in territories under their jurisdiction'. The only other provisions of the CERD with a similar jurisdiction clause are Article 6, which guarantees the right to an effective remedy against racial discrimination, and Article 14, which regulates the submission of individual petitions to the CERD Committee. No other provision of the CERD, particularly Articles 2 and 5 which protect a wide range of substantive rights, has any sort of territorial limitation.

This particular approach is found in several other treaties.[46] The most developed example of this type is the Convention against Torture (CAT), which contains no less than nine jurisdiction clauses (in Articles 2(1), 5(1)(a), 5(2),[47] 7(1), 11, 12, 13, 16, and 22(1)), with the obligations of a contracting state under these articles generally being confined to 'any territory under its jurisdiction'.

The final type of jurisdiction clause is the one which deals exclusively with the scope of application of the supervisory mechanism under a particular treaty, most notably with the competence of a treaty body to examine individual petitions. Such clauses can either be found in the treaty itself, as in Article 14 CERD and Article 22(1) CAT, or in a supplementary protocol to the treaty, as with Article 1 of the (First) Optional Protocol to the ICCPR. A variation on this type of jurisdiction clause can be found in treaties for the prevention of torture which create special inspection mechanisms. For instance, Article 4(1) of the Optional Protocol to the CAT stipulates that '[e]ach State Party shall allow visits, in accordance with the present Protocol, by the mechanisms referred to in articles 2 and 3 *to any place*

[44] Art. 1(1) of which reads: 'The States Parties to this Convention undertake to respect the rights and freedoms recognized herein and to ensure to all persons subject to their jurisdiction the free and full exercise of those rights and freedoms, without any discrimination for reasons of race, color, sex, language, religion, political or other opinion, national or social origin, economic status, birth, or any other social condition.'

[45] Art. 2(1) of which reads: 'States Parties shall respect and ensure the rights set forth in the present Convention to each child within their jurisdiction without discrimination of any kind, irrespective of the child's or his or her parent's or legal guardian's race, colour, sex, language, religion, political or other opinion, national, ethnic or social origin, property, disability, birth or other status.'

[46] See Arts I, IV, VI, Inter-American Convention on Forced Disappearance of Persons; Arts 9, 11, 31, 34, International Convention for the Protection of All Persons from Enforced Disappearance; Arts 6(1), 6(3), 8(1), 8(2), 12, 14, Inter-American Convention to Prevent and Punish Torture.

[47] Art. 5 of the CAT uses several different notions of 'jurisdiction'. The difference between these will be analysed in Chapter II, Section 2 below.

under its jurisdiction and control where persons are or may be deprived of their liberty' (emphasis added).[48]

It is important to note that this type of jurisdiction clause is distinct in that it does not affect the scope of state obligations under a treaty, but creates a standalone condition for admissibility. The territorial scope of the principal treaty in question is independent from such provisions. For instance, as we have seen, state obligations under Article 2(1) of the ICCPR extend to all individuals 'within its territory and subject to its jurisdiction', while Article 1 of the Optional Protocol to the ICCPR provides that '[a] State Party to the Covenant that becomes a Party to the present Protocol recognizes the competence of the Committee to receive and consider communications from individuals subject to its jurisdiction who claim to be victims of a violation by that State Party of any of the rights set forth in the Covenant'. The latter provision omits the arguably more restrictive territorial requirement from Article 2(1) ICCPR, but that does not mean that it extends the applicability of the ICCPR itself.[49] On the other hand, the Convention for the Elimination of Discrimination against Women (CEDAW) has no jurisdiction clause whatsoever limiting its scope of application, yet Article 2 of the Optional Protocol to the CEDAW provides that communications to the CERD Committee 'may be submitted by or on behalf of individuals or groups of individuals, under the jurisdiction of a State Party'. Accordingly, with regard to this particular treaty, its scope of application is not limited by a jurisdiction clause, yet the right to individual petition under the treaty is.

B. Treaties with dedicated provisions on territorial application

Though human rights treaties with one sort of jurisdiction clause or another are by far the most common, there are also a few with dedicated provisions governing their territorial application. As a general matter, these provisions have little or nothing to say about the *extra*-territorial application of a treaty, but are meant to address the two specific problems that we have already seen with respect to Article 29 VCLT: the application of the treaty to a state's colonies and dependencies, or, euphemistically, 'territories for whose international relations it is responsible', on the one hand, and the application of the treaty to a state with several autonomous or federal territorial units, usually with the purpose of limiting the federal state's liability, on the other. Both of these two types of provisions—colonial and federal clauses—are common in treaty practice.

Since colonies are historically primarily a European problem, it makes sense that only European human rights treaties have colonial clauses. Certain European states wished to retain the option of not applying the specific human rights treaty to their dependent territories, and they had the negotiating power to push for the inclusion

[48] See also Arts 2 and 14, European Convention for the Prevention of Torture and Inhuman or Degrading Treatment or Punishment.

[49] See also McGoldrick, above note 32, at 43–4, 49.

of a colonial clause. The prototype colonial clause can be found in Article 56 ECHR:[50]

1. Any State may at the time of its ratification or at any time thereafter declare by notification addressed to the Secretary General of the Council of Europe that the present Convention shall, subject to paragraph 4 of this Article, extend to all or any of the territories for whose international relations it is responsible.
2. The Convention shall extend to the territory or territories named in the notification as from the thirtieth day after the receipt of this notification by the Secretary General of the Council of Europe.
3. The provisions of this Convention shall be applied in such territories with due regard, however, to local requirements.
4. Any State which has made a declaration in accordance with paragraph 1 of this article may at any time thereafter declare on behalf of one or more of the territories to which the declaration relates that it accepts the competence of the Court to receive applications from individuals, non-governmental organisations or groups of individuals as provided by Article 34 of the Convention.

As explained by Moor and Simpson in a masterful article, the colonial clause was inserted into the ECHR due to the insistence of the British government, which professed that it could not extend the application of the Convention to its overseas territories without consulting the governments of these respective territories and obtaining their consent. Thus, the principal justification for the colonial clause is the colonial power's respect for local self-governance in its colonies. The subtext of this explanation, however, was that some circles within the British government, especially the Colonial Office, were reluctant to extend the ECHR's many human rights protections to large swathes of its yet to be dismantled colonial empire, particularly when it came to the right of individual petition to the European Court.[51]

A colonial clause can be found in every European human rights treaty, including the protocols to the major treaties which add further substantive rights, as well as in a large number of other treaties concluded under the auspices of the Council of Europe. The clause in Article 56 ECHR has been reproduced in the territorial applicability clauses of Protocols No. 1 and No. 4 to the Convention, in Articles 4 and 5 respectively, which, however, add another twist. They allow the states parties not only to specify to which territories, for whose international relations the states are responsible, the protocols will apply, but also to what *extent* they will apply. In other words, the states parties were given the freedom to pick and choose the provisions of the two additional protocols that they will apply in their colonies—which makes perfect sense when one considers the provisions of Protocol

[50] Art. 63 before the amendments introduced by Protocol No. 11.

[51] See L. Moor and A.W.B. Simpson, 'Ghosts of Colonialism in the European Convention on Human Rights', (2006) 76 *BYIL* 121, at 136–58.

No. 1 which guarantee basic democratic rights, and the reluctance of colonial states to extend such rights to their overseas dominions.

The colonial clauses in other Council of Europe treaties differ from that of the ECHR and Protocols No. 1 and No. 4 thereto and indeed vary greatly among themselves. For example, Article 2 of the 1953 European Convention on Social and Medical Assistance provides that it will apply to the territory of the contracting parties, and that they may define this territory in a subsequent declaration.[52] On the other hand, the 1955 European Convention on Establishment was originally meant to apply to the metropolitan territories of states parties alone, but in the final stages of drafting it was decided to give the states parties the option of extending the applicability of the treaty to their colonies.[53] This same solution was adopted in Article L (former Article 34) of the European Social Charter, and several other treaties.[54]

In 1980 the Council of Ministers of the Council of Europe adopted the Model Final Clauses for conventions and agreements concluded within the Council.[55] When it comes to territorial applicability, model clause (d) deleted any reference to embarrassing anachronisms such as metropolitan territories, colonies, and territories for whose international relations a state is responsible, by providing that '[a]ny State may at the time of signature or when depositing its instrument of ratification, acceptance, approval or accession, specify the territory or territories to which [the treaty] shall apply'. This clause has been employed in subsequent treaties, including Protocols Nos 6, 7, 12, and 13 to the ECHR and the European Convention for the Prevention of Torture.

What then is the relationship between, for example, the colonial clause in Article 56 ECHR and the ECHR's jurisdiction clause in its Article 1? The first possible reading would be that Article 56 allows for the *extension* of the ECHR to territories which the states parties did *not* consider to be 'within their jurisdiction' pursuant to Article 1—most of all those territories belonging to the category of protected states. The second possible reading is that, on the contrary, Article 56 allows states to *exclude* the applicability of the ECHR to territories which are most certainly within their jurisdiction for the purposes of Article 1. Finally, the third possible reading is that Article 56 allows for *both* of the previous possibilities, i.e. that it can be used by states both to extend the applicability of the ECHR to a territory to which it would not normally apply, and to exclude the ECHR's applicability to territories to which it would otherwise apply.[56]

There is little in the *travaux préparatoires* to the ECHR that indicates which of these three readings the states parties favoured, or whether they even considered the

[52] See at <http://conventions.coe.int/Treaty/en/Treaties/Html/014-IV.htm>.

[53] See the Explanatory Report to this treaty, available at <http://conventions.coe.int/Treaty/en/Reports/Html/019.htm>, para. 86.

[54] See, e.g., Art. 27 of the European Convention on Extradition; Art. 80 of the European Code of Social Security.

[55] Model Final Clauses for Conventions and Agreements concluded within the Council of Europe, as adopted by the Committee of Ministers of the Council of Europe at its 384th meeting, in February 1980, available at <http://conventions.coe.int/Treaty/EN/Treaties/Html/ClausesFinales.htm>.

[56] Moor and Simpson, above note 51, at 126–32.

relationship between the jurisdiction clause and the colonial clause. This relationship has furthermore never been definitively clarified by the European Court. However, the second reading does seem the most plausible, as evidenced by the subsequent practice of the colonial powers, as well as by the general treaty practice within the Council of Europe and most of the case law, though the third reading is not to be excluded. The colonial clauses in European treaties appear to be designed to give the contracting states the freedom to designate those parts of *their own territories* to which the treaties will apply, thereby avoiding the presumption in favour of territorial application to all of the territories of a state party stipulated in Article 29 of the VCLT.[57] Indeed, internal memoranda of the British government show that this presumption, which was at the time yet to be codified in the VCLT but was considered by the British to be a customary rule, was the primary reason for the British insistence on inserting a colonial clause into the ECHR and all subsequent treaties.[58]

There is some contradiction between the inclusion of colonial clauses into some human rights treaties and the case law on their extraterritorial application developed by the relevant supervisory bodies. For instance, per a ruling of the European Court, Turkey must apply the ECHR to the area of northern Cyprus which it occupied, as it falls 'within its jurisdiction' under Article 1 ECHR. However, the United Kingdom can choose, pursuant to Article 56 ECHR, whether or not to apply the ECHR to the Isle of Man, the Falklands, or some of its other dependencies which it in fact controls entirely as a sovereign and over which its jurisdiction (whatever 'jurisdiction' might exactly mean) is exclusive and has lasted for centuries.[59]

Non-European human rights treaties do not contain any colonial clauses due to the unease that other states, among them mainly former colonies, have had with such relics of European imperialism. Nonetheless, colonial states, particularly the United Kingdom, have used reservations or interpretative declarations, for instance with the ICCPR, in order to limit their obligations in relation to some of their dependencies.[60]

[57] See also P. van Dijk et al., *Theory and Practice of the European Convention on Human Rights* (Intersentia, 4th edn, 2006), at 17–18.

[58] Moor and Simpson, above note 51, at 151–2.

[59] See *Thanh v. United Kingdom* (dec.), App. No. 16137/90, 12 March 1990, in which the Commission declared inadmissible a *Soering*-type case filed by a detainee who was to be returned to Vietnam from Hong Kong, on the grounds that the UK has not extended the Convention to Hong Kong under then Art. 63 ECHR. See also *Yonghong v. Portugal* (dec.), App. No. 50887/99, 25 November 1999, in which the Court declared inadmissible a similar case dealing with Portugal and Macao; *R. (Quark Fishing Ltd) v. Secretary of State for Foreign and Commonwealth Affairs* [2005] UKHL 57, [2006] 1 AC 529, where the House of Lords held that no claim could arise under Art. 1 of Protocol No. 1 for a violation in a British dependency when that Protocol was not extended to it by a declaration, a ruling affirmed by the European Court in *Quark Fishing Ltd v. United Kingdom* (dec.), App. No. 15305/06, 19 September 2006; as well as *R. (Bancoult) v. Secretary of State for Foreign and Commonwealth Affairs* [2008] UKHL 61, [2009] 1 AC 453, where the House of Lords held that the Chagos Islanders expelled from their islands by the UK in order to make room for a US military base had no ECHR rights since the UK declaration extending the ECHR to Mauritius, of which the islands once formed part, lapsed with the independence of Mauritius.

[60] See at <http://treaties.un.org/pages/ViewDetails.aspx?src=TREATY&mtdsg_no=IV-4&chapter=4&lang=en>.

Human rights treaties also generally disfavour federal clauses, which would have the purpose of limiting the responsibility of a federal state in matters which belong to the jurisdiction of its constituent units. Indeed, the ICCPR in its Article 50 explicitly states that its provisions 'shall extend to all parts of federal States without any limitations or exceptions'. The only human rights treaty proper to have such a clause is the ACHR, in its Article 28, though states can still employ reservations to limit their responsibility.[61]

C. Treaties with no provisions on territorial application

Finally, there are those treaties which have no jurisdiction clause, nor any other clause defining their territorial scope of application. The first such universal human rights treaty was the Convention against Discrimination in Education, while the most notable are the International Covenant on Economic, Social and Cultural Rights (ICESCR) and the Convention on the Elimination of All Forms of Discrimination Against Women (CEDAW), as well as the recent Convention on the Rights of Persons with Disabilities. The application of these treaties, at least at first glance, is consequently neither confined to the territory of a state party nor to the territories or persons over which the state party has jurisdiction. It should be noted, however, that the optional protocols to the last two treaties, as well as the Optional Protocol to the ICESCR, do indeed contain a jurisdiction clause limiting the right to individual petition. Moreover, the ICESCR does indeed contain a sort of a jurisdiction clause for one specific right in its Article 14, which stipulates that

> [e]ach State Party to the present Covenant which, at the time of becoming a Party, has not been able to secure in its metropolitan territory or other territories under its jurisdiction compulsory primary education, free of charge, undertakes, within two years, to work out and adopt a detailed plan of action for the progressive implementation, within a reasonable number of years, to be fixed in the plan, of the principle of compulsory education free of charge for all.

Of the regional treaties, there is no jurisdiction clause to be found in the African Charter on Human and Peoples' Rights, or in the Inter-American Convention on the Prevention, Punishment and Eradication of Violence Against Women.

D. Concluding remarks

As we have seen, in some of the most important human rights treaties, especially those protecting civil and political rights, it is the jurisdiction clauses which determine their scope of application. There are no such clauses in humanitarian law treaties, for example. Common Article 1 of the 1949 Geneva Conventions provides that 'the High Contracting Parties undertake to respect and to ensure

[61] The permissibility of such reservations is beyond the scope of this study. For an account of the intense debates regarding the inclusion of a federal clause, pushed for by the United States, and of a colonial clause, pushed for by the United Kingdom, into the then single International Covenant, see B. Simpson, *Human Rights and the End of Empire* (Oxford University Press, 2004), at 470 *et seq.*

respect for the present Convention *in all circumstances* (emphasis added)'. It is after all only natural that treaties which protect certain categories of persons in times of armed conflict are not territorially confined. Indeed, some parts of the law of armed conflict, namely the law of belligerent occupation, apply *only* extraterritorially, as a state by definition cannot occupy its own territory.[62]

Finally, it should also be noted that the provisions defining the scope of applicability of human rights treaties frequently differentiate between two kinds of state obligations. On the one hand, there is the negative obligation of contracting states to *respect* the human rights of persons within their jurisdiction, which commands states to *refrain* from acts capable of violating the rights of individuals. On the other, there is the positive obligation to *secure* or *ensure* the respect of their rights, which requires states to take various steps to fulfil and protect the rights of individuals, even from third parties.[63] These two sets of obligations were first distinguished in a French proposal during the drafting of the then single International Covenant, by which the words 'respect and' were inserted between the words 'undertake to' and 'ensure' in what was to become Article 2(1) of the ICCPR and the ICESCR.[64] This approach was then followed in numerous other human rights treaties, including the ACHR and the CRC, while a similar phrase can be found in Common Article 1 of the Geneva Conventions, cited above.

What is the relevance of the distinction between negative and positive state obligations to the question of extraterritorial application of human rights treaties? Simply put, the ability of a state to comply with—or violate—these obligations is different, since a state needs little by way of means in order to violate a negative obligation, while the state's agents are by definition under its control. On the other hand, a state needs actual or effective control over a territory or a population in order to be able to fulfil its positive obligations. For example, on the facts of the *Bankovic* case before the European Court, it could certainly be said that the NATO countries bombing Serbia had the power to (dis)respect the right to life of the inhabitants of Serbia, but could hardly be said to have had the power to secure that right to life from violation by private parties, as they were not in control of Serbia as such. It is surprising that this distinction has had such a remarkably small role to play in the jurisprudence on the extraterritorial application of human rights treaties, as we shall see, even though it could have major implications.[65] Moreover, the distinction between positive and negative obligations under human rights treaties can also be of help in distinguishing between the concept of state jurisdiction and that of state responsibility, as will be shown in the following chapter of this study.

[62] One the other hand, Common Article 3 might be seen as imposing a territorial limitation when it speaks of 'armed conflict not of an international character *occurring in the territory of one of the High Contracting Parties*', an issue of great practical relevance to the qualification of 'spill-over' internal armed conflicts.

[63] See generally Nowak, above note 14, at 37–41.

[64] See UN Doc. E/CN.4/365, at 16.

[65] See Chapter IV, Section 4 below.

II

From Compromise to Principle

1. A Threshold Criterion: Jurisdiction of a State, Not the Jurisdiction of a Court

The purpose of this chapter will be to unravel and clarify the notion of state jurisdiction found in the various applicability clauses of human rights treaties. As will be seen, the chief problem arising from the use of the word 'jurisdiction' is that it has several meanings. A number of concepts hide themselves behind this single word, and its different meanings contribute to the confusion found both in the jurisprudence and in academic commentary.

The first point of misunderstanding that needs to be dealt with is almost trivial, but it is quite bewildering to see how often it surfaces.[1] The notion of 'jurisdiction' in human rights treaties refers to the jurisdiction of a *state*, not to the jurisdiction of a *court*, even though this latter use of the word is otherwise the most frequent. This is readily apparent from the text of the relevant treaties. For instance, according to Article 1 ECHR, '[t]he High Contracting Parties shall secure to everyone within their jurisdiction the rights and freedoms defined in Section I of this Convention'; under Article 2(1) of the ICCPR '[e]ach State Party to the present Covenant undertakes to respect and to ensure to all individuals within its territory and subject to its jurisdiction the rights recognized in the present Covenant', while Article 2(1) UNCAT provides that '[e]ach State Party shall take effective legislative, administrative, judicial or other measures to prevent acts of torture in any territory under its jurisdiction'.

There is no doubt that the word 'jurisdiction' in all of these clauses refers to the states parties to the treaty, not to the court or other supervisory body established by the treaty. It is a threshold criterion, which must be satisfied in order for the treaty obligations (at least some of them) to arise in the first place.[2] In this way human rights treaties resemble humanitarian law treaties, which also have a trigger for their

[1] See, e.g., G. Ress, 'Problems of Extraterritorial Human Rights Violations: The Jurisdiction of the European Court of Human Rights: the Bankovic case', (2002) 12 *Italian YB Int'l L.* 51; J. Ross, 'Jurisdictional Aspects of International Human Rights and Humanitarian Law in the War on Terror', in Coomans and Kamminga 9, at 23–4; E.C. Gillard, 'International Humanitarian Law and Extraterritorial State Conduct, in Coomans and Kamminga 25, at 37; M. Happold, 'Bankovic v. Belgium and the Territorial Scope of the European Convention on Human Rights', (2003) 3 *HRLR* 77, at 82–4.

[2] See M. O'Boyle, 'The European Convention on Human Rights and Extraterritorial Jurisdiction: A Comment on "Life After Bankovic"', in Coomans and Kamminga 125.

application—the existence of either an international or a non-international armed conflict.[3] In the same sense in which states have no IHL obligations if they are not engaged in armed conflict,[4] they arguably have no treaty obligations to secure or ensure human rights if they do not have jurisdiction over a person, if the treaty in question has a jurisdiction clause.

Of course, if a particular human rights treaty does not apply in the absence of a state's jurisdiction, this would automatically lead to the treaty body in question lacking jurisdiction *ratione materiae*, in the same way that the treaty body would lose jurisdiction *ratione personae* if it found that the wrongful act complained of was not attributable to the defendant state.[5] In other words, since the subject-matter jurisdiction of the European Court or of the Human Rights Committee is limited to interpreting and applying the ECHR and the ICCPR, respectively, they will invariably lack jurisdiction if the treaty itself does not apply. Nonetheless, it is not *their* jurisdiction which is the object of inquiry for the purposes of Article 1 ECHR or Article 2(1) ICCPR.

Moreover, even though the issue of state jurisdiction is by its nature preliminary and may (but need not) be resolved at the admissibility stage of the proceedings, it is still not a simple, technical admissibility requirement,[6] like that of the exhaustion of effective domestic remedies, or the six-month rule. These admissibility requirements do not touch upon the substantive rights enshrined in a human rights treaty, but are a matter of the court's or treaty body's own rules of procedure. For instance, failing to exhaust domestic remedies, or filing an application after the expiration of an admissibility deadline does not mean that the applicant does not have substantive human rights under the treaty—it simply means that his or her rights are unenforceable before the international body, either temporarily or permanently.

The only type of jurisdiction clauses that can conceivably be said to set out a procedural admissibility requirement are those which deal only with the scope of the competence of a treaty body to examine individual petitions, as in Article 14 CERD and Article 22(1) UNCAT, or the (First) Optional Protocol to the ICCPR.[7] As explained above, these jurisdiction clauses are independent of the scope of application of the human rights treaty itself and the substantive rights that it enshrines. Only in that very limited procedural sense can one speak of jurisdiction clauses as directly regulating the compatibility *ratione personae* or *ratione loci* of an individual petition.[8]

[3] See Common Articles 2 and 3 of the four Geneva Conventions of 1949; Article 1 of Additional Protocol I of 1977; Article 1 of Additional Protocol II of 1977.

[4] There are, of course, some IHL obligations which apply outside of armed conflict, most notably during belligerent occupation, but also in peace time—see, e.g., Art. 127 of the Third Geneva Convention.

[5] See, e.g., *Saddam Hussein v. Albania and others* (dec.), App. No. 23276/04, 14 March 2006.

[6] See M. Nowak, *CCPR Commentary* (Engel, 2nd revised edn, 2005), at 858–62.

[7] See above Chapter I, Section 4.A.

[8] See also D. McGoldrick, 'Extraterritorial Application of the International Covenant on Civil and Political Rights', in Coomans and Kamminga 41, at 43–4.

2. Jurisdiction's Many Meanings

A. A spurious assumption

In their case law on the extraterritorial application of human rights treaties, the European Court of Human Rights and to a lesser extent the International Court of Justice proceeded from the assumption that the concept of 'jurisdiction' in human rights treaties is the same concept of jurisdiction which exists in general international law. The European Court, moreover, drew major inferences from this approach. For example, the Court started off its discussion in *Bankovic* with the quite correct, and for a human rights body admirable, sentiment that the ECHR is to be interpreted in accordance with the general rules of treaty interpretation laid down in the VCLT, Article 31(3)(c) of which requires the Court to take into account 'any relevant rules of international law applicable in the relations between the parties'. According to the Court, '[t]he Convention should be interpreted as far as possible in harmony with other principles of international law of which it forms part'.[9] Nothing to quibble with there—but then the Court proceeds to say the following:

As to the 'ordinary meaning' of the relevant term in Article 1 of the Convention, the Court is satisfied that, from the standpoint of public international law, the jurisdictional competence of a State is primarily territorial. While international law does not exclude a State's exercise of jurisdiction extra-territorially, the suggested bases of such jurisdiction (including nationality, flag, diplomatic and consular relations, effect, protection, passive personality and universality) are, as a general rule, defined and limited by the sovereign territorial rights of the other relevant States (Mann, '*The Doctrine of Jurisdiction in International Law*', RdC, 1964, Vol. 1; Mann, '*The Doctrine of Jurisdiction in International Law, Twenty Years Later*', RdC, 1984, Vol. 1; Bernhardt, *Encyclopaedia of Public International Law*, Edition 1997, Vol. 3, pp. 55–59 '*Jurisdiction of States*' and Edition 1995, Vol. 2, pp. 337–343 '*Extra-territorial Effects of Administrative, Judicial and Legislative Acts*'; Oppenheim's *International Law*, 9th Edition 1992 (Jennings and Watts), Vol. 1, § 137; P.M. Dupuy, *Droit International Public*, 4th Edition 1998, p. 61; and Brownlie, *Principles of International Law*, 5th Edition 1998, pp. 287, 301 and 312–314).

Accordingly, for example, a State's competence to exercise jurisdiction over its own nationals abroad is subordinate to that State's and other States' territorial competence (Higgins, *Problems and Process* (1994), at p. 73; and Nguyen Quoc Dinh, *Droit International Public*, 6th Edition 1999 (Daillier and Pellet), p. 500). In addition, a State may not actually exercise jurisdiction on the territory of another without the latter's consent, invitation or acquiescence, unless the former is an occupying State in which case it can be found to exercise jurisdiction in that territory, at least in certain respects (Bernhardt, cited above, Vol. 3 at p. 59 and Vol. 2 at pp. 338–340; Oppenheim, cited above, at § 137; P.M. Dupuy, cited above, at pp. 64–65; Brownlie, cited above, at p. 313; Cassese, *International Law*, 2001, p. 89; and, most recently, the '*Report on the Preferential Treatment of National*

[9] *Bankovic and Others v. Belgium and Others* [GC] (dec.), App. No. 52207/99, 12 December 2001, para. 57.

Minorities by their Kin-States' adopted by the Venice Commission at its 48th Plenary Meeting, Venice, 19–20 October 2001).

The Court is of the view, therefore, that Article 1 of the Convention must be considered to reflect this ordinary and essentially territorial notion of jurisdiction, other bases of jurisdiction being exceptional and requiring special justification in the particular circumstances of each case (see, *mutatis mutandis* and in general, Select Committee of Experts on Extraterritorial Criminal Jurisdiction, European Committee on Crime Problems, Council of Europe, '*Extraterritorial Criminal Jurisdiction*', Report published in 1990, at pp. 8–30).[10]

There is little to be disputed with the first two paragraphs that I have quoted. It is indeed true that, 'from the standpoint of public international law, the jurisdictional competence of a State is primarily territorial'—true, though a major oversimplification, as the entire point of having a law of jurisdiction is precisely to regulate the *exceptions* from territoriality, as we shall soon see. But, no matter—the Court's assessment is still essentially correct.

What is most certainly *not* correct is that what the Court said in the third quoted paragraph flows from the previous two. The word 'therefore' in the first sentence of this paragraph is nothing but a lead-up to a *non sequitur*, since it is based on the *assumption* that the concept of 'jurisdiction' in Article 1 ECHR is the *same concept* of jurisdiction as the one in general international law to which the Court refers. It is from this assumption and from a furtive pseudo-recapitulation of its earlier case law which it used to restate things that it had actually never said before,[11] that the Court drew its ultimate conclusion that

[i]n sum, the case-law of the Court demonstrates that its recognition of the exercise of extraterritorial jurisdiction by a Contracting State is exceptional: it has done so when the respondent State, through the effective control of the relevant territory and its inhabitants abroad as a consequence of military occupation or through the consent, invitation or acquiescence of the Government of that territory, exercises all or some of the public powers normally to be exercised by that Government.[12]

Not only is the assumption that the 'jurisdiction' from Article 1 ECHR is the same jurisdiction as the one in general international law entirely unsupported by anything produced by the Court, but, as will be shown, following that assumption to its logical conclusion would mean accepting completely absurd results, indeed results which would be in total contradiction with the Court's own established jurisprudence.

It seems that the ICJ also proceeded from this assumption in its *Wall* Advisory Opinion, where it first laconically observed, without any analysis, that the 'jurisdiction of states is primarily territorial'.[13] However, unlike the European Court in

[10] *Bankovic*, paras 59–61.

[11] See R. Lawson, 'Life after *Bankovic*: On the Extraterritorial Application of the European Convention on Human Rights', in Coomans and Kamminga 83, at 111, who well notes that the Court silently added the reference to 'public powers' in the quotation that follows, which was never mentioned in its previous case law.

[12] *Bankovic*, para. 71.

[13] *Legal Consequences of the Construction of a Wall in the Occupied Palestinian Territory*, Advisory Opinion, 9 July 2004, ICJ Reports 2004, para. 109.

Bankovic, the ICJ gave no special significance to this supposedly primarily territorial notion of jurisdiction as warranting a restrictive approach to Article 2(1) ICCPR. On the contrary, relying on the practice of the Human Rights Committee, the ICJ found both the ICCPR and the ICESCR, as well as the CRC, applicable to the occupied Palestinian territories.[14]

B. Jurisdiction in general international law

Before showing exactly why this assumption is mistaken, it is first necessary to clearly establish what the concept of jurisdiction in general international law actually is and what it does. For that one may as well rely, *inter alia*, on the various sources cited by the European Court in *Bankovic*.

The definition of that notion of 'jurisdiction' to which the Court refers in *Bankovic* is uncontroversial: that 'jurisdiction' is the authority of the state, based in and limited by international law, to regulate the conduct of persons, both natural and legal, by means of its own domestic law.[15] In essence, a state's jurisdiction is an emanation or an aspect of its sovereignty, its right to regulate its own public order, and limitations on it flow from the equal sovereignty of other states.[16] That 'jurisdiction' in general international law is not a unitary concept, as it encompasses at least two, and possibly three, distinct sets of powers.[17] The first of these is the *jurisdiction to prescribe*—also termed legislative jurisdiction or *compétence normative*—the state's authority to make or prescribe legal rules. On the other hand, the *jurisdiction to enforce*—executive jurisdiction or *compétence d'exécution*—is the state's authority to apply or enforce the rules that it has previously prescribed. Finally, there is the state's *adjudicatory, curial, or judicial jurisdiction*, which refers to the power of its courts to settle legal disputes, though this type of jurisdiction may safely be subsumed under the state's prescriptive and enforcement jurisdiction.[18]

When a state promulgates a criminal statute making murder a crime, it exercises its prescriptive jurisdiction. When the state authorities apprehend a murder suspect, the state exercises its enforcement jurisdiction. When that suspect is brought before a criminal court for trial, the state exercises its adjudicative jurisdiction. Similarly, a state exercises its jurisdiction to prescribe when it passes an antitrust law prohibiting excessive market concentration, while it exercises its enforcement

[14] Ibid., paras 109–11.

[15] See generally V. Lowe, 'Jurisdiction', in M. Evans (ed.), *International Law* (Oxford University Press, 2nd edn, 2006), at 335; M. Shaw, *International Law* (Cambridge University Press, 6th edn, 2008), at 645; M. Akehurst, 'Jurisdiction in International Law', (1972–1973) 46 *BYIL* 145; C. Ryngaert, *Jurisdiction in International Law* (Oxford University Press, 2008), at 5 *et seq.*

[16] See, e.g., I. Brownlie, *Principles of Public International Law* (Oxford University Press, 6th edn, 2003), at 297; A. Cassese, *International Law* (Oxford University Press, 2nd edn, 2005), at 49.

[17] See R. O'Keefe, 'Universal Jurisdiction: Clarifying the Basic Concept', (2004) 2 *JICJ* 735, at 736.

[18] See Lowe, above note 15, at 338–9; Cassese, above note 16, at 49–51; Shaw, above note 15, at 649 *et seq.*; O'Keefe, above note 17, at 736–7; *Restatement (Third) of the Foreign Relations Law of the United States*, s. 401.

jurisdiction when it actually dismembers or otherwise penalizes a company which fails to abide by its antitrust regulations.

The distinction between the jurisdiction to prescribe and the jurisdiction to enforce is not merely academic. The latter is regulated by a very simple rule: a state may not exercise its enforcement jurisdiction on the territory of another state, absent that state's consent.[19] An example of an unlawful exercise of enforcement jurisdiction by a state would be the abduction of Adolf Eichmann, one of the chief organizers of the Holocaust, by Israeli security agents on Argentine soil and without the consent of the Argentine government.[20] Lawful examples would be the consular jurisdiction of states over their nationals abroad, or jurisdiction over military personnel in overseas bases, both of which are grounded in the consent of the territorial state.

Prescriptive jurisdiction, on the other hand, is also territorial, in the sense that a state by definition has the prerogative to legislate for persons present in its own territory.[21] A national of the United States who visits the United Kingdom is subject to the laws of the United Kingdom, and vice versa. However, as stated above, it is in some ways deeply misleading to think of prescriptive jurisdiction as being primarily territorial. Even though the vast majority of situations and relationships that a state will seek to regulate will take place in its own territory, it still may exercise its prescriptive jurisdiction extraterritorially and may do so without the consent of other states.[22]

Indeed, practically the entirety of the *law* of (prescriptive) jurisdiction is about the *exceptions* to territoriality. If particular conduct that the state seeks to regulate is taking place outside its territory (including therein its territorial sea, as well as ships and airplanes flying its flag),[23] it may nonetheless regulate it if there is an additional basis for its jurisdiction, a connecting factor.[24] There are several bases of prescriptive jurisdiction that are generally recognized in international law and these were enumerated by the European Court in *Bankovic*:[25] the nationality (or active personality) principle, according to which a state may regulate the conduct of its nationals, even when they are abroad; the passive personality principle, according to which a state may, within certain limits, prohibit conduct which directly harms its nationals, even if the perpetrator of the harm is not its national and the conduct takes place outside its territory; the protective principle, according to which a state

[19] See Cassese, above note 16, at 50; Lowe, above note 15, at 338; O'Keefe, above note 17, at 740; Ryngaert, above note 15, at 9–10.

[20] This action was unanimously condemned by the UN Security Council, in Resolution 138 (1960). See also Shaw, above note 15, at 651; Lowe, above note 15, at 357; J.E.S. Fawcett, 'The *Eichmann* Case', (1962) 38 *BYIL* 181.

[21] Brownlie, above note 16, at 297.

[22] For examples of extraterritorial prescriptive jurisdiction in fields as diverse as antitrust, taxation, and environmental law, see generally the contributions in K. Meessen (ed.), *Extraterritorial Jurisdiction in Theory and Practice* (Kluwer, 1996).

[23] Though this is not to say, of course, that ships and airplanes are floating and flying pieces, respectively, of the state's own territory.

[24] Lowe, above note 15, at 342.

[25] *Bankovic*, para. 59.

may punish persons who seek to harm its most vital interests—for example, persons plotting to forge its currency; the universality principle, according to which the state may criminalize conduct without any direct connection to it if that conduct harms the international community as a whole, such as piracy and crimes against international law.[26] Certain states have also asserted more controversial bases for jurisdiction—though passive personality and universal jurisdiction have also had their share of controversies. The most notable case is that of the United States, which extended the application of its antitrust laws to intentional conduct which produces economic effects within the territory of the United States.[27]

All of these principles concern the *extraterritorial* exercise of prescriptive jurisdiction, which these days can hardly be said to be truly exceptional. All states do it all the time. Moreover, none of these principles is necessarily subordinate to another, nor is the territorial principle at the apex of some sort of jurisdictional hierarchy. State jurisdictions can and do overlap, and more than one system of municipal law can apply to the same conduct or situation. It is therefore somewhat strange that the European Court in *Bankovic* does not differentiate at all between prescriptive and enforcement jurisdiction when it claims that Article 1 ECHR embodies the notion of jurisdiction from general international law, seemingly implying that the former is subject to equally stringent conditions as the latter, when it most certainly is not.[28]

It is also important to note that the jurisdiction to enforce is necessarily the actualization of a previously made prescription, an assertion that a particular rule of municipal law applies to a concrete situation at hand and that this law is to be enforced.[29] Moreover, a state may lawfully extend its prescriptive jurisdiction, only to unlawfully attempt to enforce it, and vice versa. For example, a state can tax its nationals even when they are abroad, but it must not attempt to seize any of their assets located outside its territory without the consent of the territorial state. If it did so, it would have engaged in an exorbitant exercise of its jurisdiction to enforce, which would have violated the territorial state's sovereignty. Or, a state might arrest a foreign national inside its own territory and demand of him that he pay tax on income earned abroad which has no nexus whatsoever with the state itself. In this case there would be an abuse of the jurisdiction to prescribe, even though the state has not exceeded its enforcement jurisdiction.[30]

To sum up, just as international law delimits the territories of states, so it delimits the spheres of their municipal law through the doctrine of jurisdiction.

[26] See O'Keefe, above note 17, at 738–9; Lowe, above note 15, at 340–53.

[27] See Lowe, above note 15, at 344–5.

[28] *Bankovic*, paras 59 and 60. Unfortunately, as well shown by O'Keefe, above note 17, the European Court is not the only international court to fail to appreciate the fundamental significance of the distinction between the jurisdiction to prescribe and the jurisdiction to enforce. See *Arrest Warrant of 11 April 2000 (Congo v, Belgium)*, Judgment, ICJ Reports 2000, at 3.

[29] See *Restatement (Third) of the Foreign Relations Law of the United States*, s. 401(c) (referring to enforcement jurisdiction as state authority 'to enforce or compel compliance or to punish noncompliance with its laws or regulations, whether through the courts or by use of executive, administrative, police, or other nonjudicial action').

[30] O'Keefe, above note 17, at 741.

Jurisdiction 'concerns essentially the *extent* of each state's *right* to regulate conduct or the consequences of events',[31] this right being limited by the equal rights and sovereignty of other states.[32]

C. Jurisdiction: an absurdity

But what does this concept of jurisdiction from general international law have to do with the one found in human rights treaties? In short, precisely *nothing*. To prove this we must go back to the first truly landmark case on the extraterritorial application of the ECHR, *Loizidou*, in which the Court held that

> ... although Article 1 sets limits on the reach of the Convention, the concept of "jurisdiction" under this provision is not restricted to the national territory of the High Contracting Parties. According to its established case-law, for example, the Court has held that the extradition or expulsion of a person by a Contracting State may give rise to an issue under Article 3, and hence engage the responsibility of that State under the Convention (see the *Soering v. the United Kingdom* judgment of 7 July 1989, Series A no. 161, pp. 35–36, para. 91; the *Cruz Varas and Others v. Sweden* judgment of 20 March 1991, Series A no. 201, p. 28, paras. 69 and 70, and the *Vilvarajah and Others v. the United Kingdom* judgment of 30 October 1991, Series A no. 215, p. 34, para. 103). In addition, the responsibility of Contracting Parties can be involved because of acts of their authorities, whether performed within or outside national boundaries, which produce effects outside their own territory (see the *Drozd and Janousek v. France and Spain* judgment of 26 June 1992, Series A no. 240, p. 29, para. 91).
>
> Bearing in mind the object and purpose of the Convention, the responsibility of a Contracting Party may also arise when as a consequence of military action—*whether lawful or unlawful*—it exercises *effective control of an area* outside its national territory. The obligation to secure, in such an area, the rights and freedoms set out in the Convention, derives from the fact of such control whether it be exercised directly, through its armed forces, or through a subordinate local administration.[33]

This ruling was reaffirmed at the merits stage of *Loizidou*, as well as in the subsequent case of *Cyprus v. Turkey*, in which Turkey was again held liable for human rights violations committed in the territory over which it had 'effective overall control'.[34] Indeed, even *Bankovic* did not purport to overrule *Loizidou*, but merely to clarify it (or not, as it turned out). Note, however, that the authorities

[31] R. Jennings and A. Watts (eds), *Oppenheim's International Law* (9th edn, 1992), at 456 (emphasis added).

[32] F. Mann, 'The Doctrine of International Jurisdiction Revisited After Twenty Years', (1984-III) 186 *RdC* 9, at 20.

[33] *Loizidou v. Turkey*, App. No. 15318/89, Judgment (preliminary objections), 23 February 1995, para. 62 (citations omitted, emphasis added). The applicant was a Greek Cypriot who originally lived in northern Cyprus. After the Turkish invasion and the establishment of the so-called Turkish Republic of Northern Cyprus (TRNC), the applicant was unable to access her property in northern Cyprus. In its preliminary objections judgment the European Court held that Turkey had obligations under the ECHR towards the applicant. In its merits judgment the Court found that the applicant's right to property had been violated by Turkey.

[34] *Loizidou v. Turkey*, App. No. 15318/89, Judgment (merits), 28 November 1996, paras 52–7; *Cyprus v. Turkey*, App. No. 25781/94, Judgment, 10 May 2001, paras 77 and 78. The northern

that the Court cites to give it effective control of an area pronouncement a patina of age in fact hardly support it.[35] As we have seen above, *Soering* and its progeny are not really about extraterritorial application, since the victim of the human rights violation is clearly located within the state's jurisdiction.[36] The *Drozd* case is likewise inapposite. It concerned the acts of French and Spanish judges in Andorra, at whose disposal they were put, and accordingly their acts were not attributable to France and Spain.[37] The *Drozd* quotation on acts of state authorities which 'produce effects outside their own territory', to which the Court in *Loizidou* somewhat surreptitiously added the words 'whether performed within or outside national boundaries', was in fact based solely on the Commission's case law, which regarded Article 1 jurisdiction not as control over *territory*, but as control over *individuals*. It was probably for that reason that this case law was not even cited by the Court in *Loizidou*, which for the first time applied a *spatial*, rather than *personal*, model of jurisdiction.[38]

Loizidou's flimsy foundations in prior case law aside, neither the Commission nor the Court in its pre-*Bankovic* case law based their interpretation of Article 1 ECHR on the general international law doctrine of jurisdiction. No Oppenheims, Brownlies, Casseses, or Pellets were ever cited by the Court in *Loizidou*, and for good reason—exercising 'effective overall control' over a territory does not mean that the state is necessarily exercising its 'jurisdiction'—as general international law speaks of the term—over the inhabitants of that territory. As explained above, this type of jurisdiction requires the state to extend the application of its *domestic law* so that it purports to regulate the conduct of a person. The purpose of the doctrine of jurisdiction in international law is precisely to establish whether a claim by a state to regulate some conduct is *lawful or unlawful*.[39] Conversely, 'effective overall control of an area' is a question of *fact*, of actual physical power that a state has over a territory and its people. Indeed, none of the numerous authors cited above makes any mention of 'effective overall control of an area' as a legal basis of either the jurisdiction to prescribe or the jurisdiction to enforce the rules thus prescribed.

In other words, when Turkey invaded Cyprus, it did not exercise its jurisdiction within the meaning of general international law over the parts of Cyprus that it had conquered. It did not, for example, say that the Turkish Criminal Code applied to Cyprus, and that the inhabitants of Cyprus were bound, as a matter of law, to obey these rules of conduct. On the contrary, Turkey created a puppet regime, the Turkish Republic of Northern Cyprus, which it (and it alone) recognized as an independent state. Turkey did not even *claim* to have jurisdiction in the

Cyprus line of cases employ a spatial conception of state jurisdiction, which will be examined in more detail in Chapter IV, Section 2 below.

35 See Lawson, above note 11, at 97.

36 See Chapter I, Section 2 above.

37 See also Section 3.D below.

38 On the Commission's case law, see further below, especially Chapter IV, Section 3.

39 See B. Oxman, 'Jurisdiction of States' in *Max Planck Encyclopedia of Public International Law*, available at <http://www.mpepil.com>, paras 1 and 9.

classical sense.[40] What it most certainly did have was actual power to affect the lives of the inhabitants of the territory that it occupied.

A defender of *Bankovic* might respond to this argument in the following way. First, as the Court itself said in *Bankovic*, international law does recognize the jurisdiction, if in a limited sense, of a state which engages in a belligerent occupation of another state. Secondly, even if Turkey did not exercise its prescriptive jurisdiction, as it did not set out any rules of its domestic law that the inhabitants of northern Cyprus were mandated to follow, it did exercise its jurisdiction to enforce.

The first of these objections is unfounded. Belligerent occupation is a legal regime that arises only when a certain factual condition is met—that a territory of one state is 'actually placed under the authority' of the army of the hostile state.[41] When this state of affairs occurs, international law tries to limit any potential abuses by setting out a certain number of rights, and a greater amount of obligations, which are incumbent upon the occupying state. As prescribed by Article 43 of the Hague Regulations, '[t]he authority of the legitimate power having in fact passed into the hands of the occupant, the latter shall take all the measures in his power to restore, and ensure, as far as possible, public order and safety, while respecting, unless absolutely prevented, the laws in force in the country'. In theory at least, belligerent occupation is a transitory regime, in which the occupying state must still respect to an extent the authority of the sovereign that it has displaced. Among the rights granted by international law to an occupying power is certainly not the right to extend its own laws to the occupied territory as it pleases, as Article 43 of the Hague Regulations amply demonstrates. Any laws that it passes for the occupied territory—if it does so at all—end with the occupation itself.[42] In other words, a state which occupies the territory of another state has not *ipso facto* exercised its jurisdiction, as that term of general international law is referred to in *Bankovic*, over the persons living in that territory.

The objection that Turkey might not have exercised its prescriptive jurisdiction, but that it did exercise its enforcement jurisdiction over the people of northern Cyprus would also be misplaced. Enforcement jurisdiction presupposes a previous exercise of prescriptive jurisdiction,[43] and not every coercive action that a state might take is an exercise of its jurisdiction to enforce. It is a *legal* rule that is being

[40] For example, in the first *Cyprus v. Turkey* case to be heard by the former European Commission on Human Rights, Turkey argued that

> ... the Commission had no jurisdiction ratione loci to examine the application as Cyprus did not fall under Turkish jurisdiction. Turkey had not extended her jurisdiction to the island of Cyprus since she had not annexed a part of the island nor established a military or civil government there. The administration of the Turkish Cypriot community had absolute jurisdiction over part of the island. Moreover, Turkey could not be held liable under Art. 63 of the Convention since she was not responsible for the international relations of either the whole or a part of Cyprus.

Cyprus v. Turkey (dec.), App. Nos 6780/74 and 6950/74, 26 May 1975, 2 D.R. 125, at 130.

[41] Art. 42 of the Hague Regulations.

[42] See generally M. Sassoli, 'Legislation and Maintenance of Public Order and Civil Life by Occupying Powers', (2005) 16 *EJIL* 661.

[43] See, e.g., Mann, above note 32, at 34, who states that '[n]o State may exercise enforcement jurisdiction outside its own territory in the absence of its own legislature authorizing it to do so, that is to say, in the absence of legislative jurisdiction'. *Restatement (Third) Foreign Relations Law of the United*

enforced, or a *legal* authority that is being exercised, under the guise of domestic law and in at least an asserted conformity with international law—it is not merely a display of naked power, even if it is a 'public' power, whatever that particular concept might mean.[44] If, on the other hand, enforcement jurisdiction were to be *decoupled* from domestic law, it would in essence be reduced to the concept of jurisdiction as factual power as set out in this study, thereby rendering this objection meaningless.[45]

Hence, the concept of jurisdiction in general international law serves a completely different purpose from that developed by the European Court in *Loizidou*—it sets out limits on the domestic legal orders of states, so that they do not infringe upon the sovereignty of others. One possible answer to the inconsistency between the general international law notion of jurisdiction and the one developed by the European Court in *Loizidou* and reaffirmed in later cases, *Bankovic* included, is that *Loizidou* and its progeny are completely wrong and should be overruled. If the 'effective overall control of an area' test to establish jurisdiction for the purposes of Article 1 ECHR is incompatible with the classical doctrine of jurisdiction, then it is this test, not the classical doctrine, that should be discarded. In other words, the answer might be that *Bankovic* did not go *far enough* in bringing the Court's case law back into conformity with general international law.

This argument presumes, however, that problems with interpreting Article 1 ECHR arise only when it is being applied extraterritorially, as everyone who is within a state's territory is *ipso facto* within its jurisdiction, as international law speaks of the term. That is simply not the case. While the domestic law of a state generally applies to all persons within its territory, who are all in that sense within the state's (prescriptive) jurisdiction, not every act done by a state is committed in furtherance of a rule of its domestic law. The state may kill, maim, or persecute people without any guise of legal authority. If, for example, the use of torture by a state agent was authorized by the state's law, and torture was conducted in an institutional setting with torture warrants *à la* Court of Star Chamber, one could well speak of the state exercising its jurisdiction to prescribe and to enforce against the unfortunate victim of such treatment. But that is, of course, not how torture is

States, s. 401: 'jurisdiction to enforce [is the power of the state] to induce or compel compliance or to punish noncompliance with its laws or regulations'. See also Oxman, above note 39, para. 5.

44 *Bankovic*, para. 71.

45 See, in that regard, G. Goodwin-Gill, 'The Extra-Territorial Reach of Human Rights Obligations: A Brief Perspective on the Link to Jurisdiction', in L. Boisson de Chazournes and M. Kohen (eds), *International Law and the Quest for its Implementation/Le droit international et la quête de sa mise en œuvre, Liber Amicorum Vera Gowlland-Debbas* (Brill/Martinus Nijhoff, 2010), 293, esp. at 306 and 307:

> as Brownlie notes, jurisdiction is also understood to include enforcement or prerogative jurisdiction, namely, '... the power to take executive action in pursuance of or consequent on the making of decisions or rules'. It is this sense of 'jurisdiction' which stands out in human rights claims arising out of extra-territorial actions.... 'Jurisdiction', if it is to have a coherent meaning in the context of human rights protection, should therefore be seen to reflect that aspect of the term in general international law which recognizes that every State has the sovereign capacity to act outside its territory, while leaving open the question whether such exercise of jurisdiction is lawful vis-a-vis other States, or consistent with the human rights obligations of the acting State.

done in the modern world, since even the worst dictatorships at least pretend not to torture, and when they in fact do so they do not purport to be enforcing domestic law. Indeed, some human rights violations are *designed* precisely so as to avoid any semblance of legal process. It would be simply bizarre to suggest that, for example, enforced disappearances are an instance of a state exercising its prescriptive or enforcement jurisdiction. They are an exercise of *power*, pure and simple, not of any sort of legal competence—unless, again, the concept of jurisdiction is reduced to one of purely factual power.

Bankovic has been criticized because it created a perverse incentive for states acting outside their boundaries. If state agents detain someone pursuant to a warrant, and then kill him, he would be within the state's jurisdiction. If, however, the state's agents shoot first and ask questions later, the state would presumably not be deemed to have had jurisdiction.[46] However, this same perverse incentive applies both within and outside a state's territory, if one consistently applies the classical international law notion of jurisdiction. If the Royal Air Force were to drop a bomb or two on Bristol or Belfast, it would not be an exercise of the United Kingdom's jurisdiction any more than it was when it bombed Belgrade, since the act itself would not be meant to enforce a previously prescribed rule of domestic law.[47] The extraterritorial application of human rights treaties notwithstanding, interpreting the notion of jurisdiction in these treaties as being identical to the one in general international law would lead to manifestly absurd results even in the domestic sphere, which the states parties to these treaties could not possibly have intended.

D. Jurisdiction: a homonym

What, then, is this notion of jurisdiction in human rights treaties? Is it an autonomous concept specific to these treaties, as some judges of the European Court have suggested in the *Ilascu* case,[48] or, to disguise the uncertainty with a Latin phrase, a concept *sui generis*? We need not go that far. State practice, especially state treaty-making practice, shows that more than one *ordinary* meaning of the word 'jurisdiction' exists in international law. To illustrate this, let us take a look at Article 9(1) of the International Convention for the Protection of All Persons from Enforced Disappearance:

> Each State Party shall take the necessary measures to establish its jurisdiction over the offence of enforced disappearance:

[46] See, e.g., McGoldrick, above note 8, at 72, n. 132.

[47] See also Lawson, above note 11, at 114–15.

[48] *Ilascu and others v. Moldova and Russia* [GC], App. No. 48787/99, Judgment, 8 July 2004; Partially Dissenting Opinion of Judge Bratza, joined by Judges Rozakis, Hedigan, Thomassen, and Pantiru, para. 8.

(a) When the offence is committed in any territory under its jurisdiction or on board a ship or aircraft registered in that State;
(b) When the alleged offender is one of its nationals;
(c) When the disappeared person is one of its nationals and the State Party considers it appropriate.

An article such as this one can be found in many treaties which create specific international offences that states are obliged to criminalize and prosecute.[49] Article 9(1) of this Convention requires states to do so for the crime of enforced disappearance on the basis of the territorial principle, the nationality principle, and the passive personality principle—a classical example of a treaty obligation mandating the extension of the prescriptive jurisdiction of states parties for some specific conduct. But take a look at how the territorial principle is formulated in Article 9(1)(a): 'Each State Party shall take the necessary measures to establish its jurisdiction . . . [w]hen the offence is committed in any territory under its jurisdiction.' Here we have two uses of the word 'jurisdiction', but their meaning is not the same. If it were, a state would be required to establish its jurisdiction when it *already* had jurisdiction.

Quite similar are Article 5(1)(a) CAT, which provides that '[e]ach State Party shall take such measures as may be necessary to establish its jurisdiction over the offences referred to in article 4 . . . [w]hen the offences are committed in any territory under its jurisdiction', and Article 12(1)(a) of the Inter-American Convention to Prevent and Punish Torture, which stipulates that '[e]very State Party shall take the necessary measures to establish its jurisdiction over the crime described in this Convention . . . [w]hen torture has been committed within its jurisdiction'. What possible purpose could these provisions have if the first and the second uses of the word 'jurisdiction' were the same? States generally do not go through the effort of negotiating and adopting legally binding treaties so they can fill them with tautologies.

Article 9(2) of the Disappearances Convention is an even better example of the myriad of meanings of the word 'jurisdiction':

Each State Party shall likewise take such measures as may be necessary to establish its jurisdiction over the offence of enforced disappearance when the alleged offender is present in any territory under its jurisdiction, unless it extradites or surrenders him or her to another State in accordance with its international obligations or surrenders him or her to an international criminal tribunal whose jurisdiction it has recognized.

Three times does this article use the word 'jurisdiction' and every time it does so with a different meaning. The first 'jurisdiction' is the jurisdiction of the state to prescribe an offence in its criminal law, in this particular case a treaty-based universal jurisdiction for the crime of enforced disappearance. The third use of the word 'jurisdiction' refers to the jurisdiction of an international criminal court, which is based on the state's consent. But it is the *second* use of the word

[49] See, e.g., Arts 6 and 7, International Convention for the Suppression of Terrorist Bombings; Art. 7, International Convention for the Suppression of the Financing of Terrorism; Art. 9, International Convention for the Suppression of Acts of Nuclear Terrorism; Art. 4, United Nations Convention against Illicit Traffic in Narcotic Drugs and Psychotropic Substances.

'jurisdiction' which interests us here, as that is the one which can also be found in human rights treaties—it refers to a particular kind of factual power, authority, or control that a state has over a *territory*, and consequently over persons in that territory.[50] So does, for instance, Article 12(2) of the Inter-American Torture Convention: 'Every State Party shall also take the necessary measures to establish its jurisdiction over the crime described in this Convention when the alleged criminal is *within the area under its jurisdiction*' (emphasis added).

This is precisely the kind of concept that the European Court developed in *Loizidou* with its 'effective overall control of an area' test, as sensible a definition of 'jurisdiction' as any. It is *this* notion of jurisdiction—*not* the jurisdiction to prescribe rules of domestic law and to enforce them, but control over a territory and persons within it—that pervades international human rights treaties. Indeed, one can even glimpse it in the Universal Declaration of Human Rights, which in the final clause of its preamble says that

> . . . as a common standard of achievement for all peoples and all nations, to the end that every individual and every organ of society, keeping this Declaration constantly in mind, shall strive by teaching and education to promote respect for these rights and freedoms and by progressive measures, national and international, to secure their universal and effective recognition and observance, both among the peoples of Member States themselves and among the peoples of *territories under their jurisdiction.* (emphasis added)

The UDHR then goes on to say in its Article 2 that

> [e]veryone is entitled to all the rights and freedoms set forth in this Declaration, without distinction of any kind, such as race, colour, sex, language, religion, political or other opinion, national or social origin, property, birth or other status.
>
> Furthermore, no distinction shall be made on the basis of the *political, jurisdictional or international status of the country or territory* to which a person belongs, whether it be independent, trust, non-self-governing or under any other limitation of sovereignty. (emphasis added)

The use of this concept of 'jurisdiction' in the UDHR was clearly meant to address the then still current problem of colonialism. Indeed, the second paragraph of Article 2 was inserted into the UDHR in lieu of a previous draft Article 3, which prescribed that the UDHR was to be applied equally in non-self-governing territories.[51] Concerns over colonialism also seem to motivate Article 14 ICESCR:

> Each State Party to the present Covenant which, at the time of becoming a Party, has not been able to secure in its metropolitan territory *or other territories under its jurisdiction* compulsory primary education, free of charge, undertakes, within two years, to work out and adopt a detailed plan of action for the progressive implementation, within a reasonable number of years, to be fixed in the plan, of the principle of compulsory education free of charge for all. (emphasis added)

[50] Thus, for example, the authoritative commentary on the CAT by Burgers and Danelius states that the reference to 'any territory under its jurisdiction' in the CAT was one to 'territories under the factual control of a State, including territories under military occupation.' J.H. Burgers and H. Danelius, *The United Nations Convention against Torture* (Nijhoff, 1988), at 133. See also ibid., at 131.

[51] See B. Simpson, *Human Rights and the End of Empire* (Oxford University Press, 2004), at 455.

The CERD 'also explicitly refers to territory in the jurisdiction clause in its Article 3: 'States Parties particularly condemn racial segregation and apartheid and undertake to prevent, prohibit and eradicate all practices of this nature in territories under their jurisdiction.' The link of jurisdiction to territory is made the clearest in the CAT, which speaks of 'any territory under its [the State's] jurisdiction' in seven separate clauses (Articles 2(1), 5(1)(a), 5(2), 11, 12, 13, and 16), and 'territory under whose jurisdiction' in one more clause (Article 7(1)).

The treaty practice of states thus shows that they employ two concepts of jurisdiction. The first, classical doctrine of jurisdiction in general international law refers to the state regulation of the conduct of persons, natural or legal, and the consequences of their actions under domestic law. In other words, it delimits the municipal legal orders of states, which can and do overlap. The second notion of jurisdiction, that is found, *inter alia*, in human rights treaties, denotes a certain kind of power that a state exercises over a territory and its inhabitants, i.e. it is *spatial* in nature. The two concepts may be related, but they cannot possibly be the same.[52]

One might object to this distinction by saying that the jurisdiction clauses in some human rights treaties do not refer to territories, but to persons within or subject to the state's jurisdiction—such as, for example, Article 1 ECHR and Article 2(1) ICCPR. These clauses can be interpreted—and indeed have been interpreted—as defining a particular kind of relationship between a state and an individual. Thus, human rights treaties should apply not (only) when a state exercises control over a *territory*, but (also) when it exercises authority and control over a *person*.[53] I will be examining the spatial and personal models of jurisdiction below in more detail.[54] Suffice it to say at this point that, textually, the jurisdiction clauses of most human rights treaties are 'primarily territorial' in nature—but not necessarily exclusively so. Even if they do, in fact, connote a relationship between a state and an individual, this would be one based on factual power, and would still not fall under the *other* meaning of the word 'jurisdiction' which denotes the state's extension of its *laws* to a particular individual.[55]

That the word 'jurisdiction' can mean several different things is apparent even from looking at dictionary definitions. Thus, among the general dictionaries, the *Oxford English Dictionary* attributes four meanings to the word 'jurisdiction', among which is the exercise of a *legal* authority or power, but *also* 'power or authority in general; administration, rule, control' and 'the extent or range of judicial or administrative power; the territory over which such power extends'.[56]

[52] The same conclusion was reached by Wilde in an excellent article—see R. Wilde, 'Triggering State Obligations Extraterritorially: The Spatial Test in Certain Human Rights Treaties', (2007) 40 *Israel L Rev* 503, especially at 508, 513–14. See also A. Ruth and M. Trilsch, 'Bankovic v. Belgium (Admissibility)', (2003) 97 *AJIL* 168, at 171; O. De Schutter, 'Globalization and Jurisdiction: Lessons from the European Convention on Human Rights', (2006) 6 *Baltic Yearbook of International Law* 183; A. Orakhelashvili, 'Restrictive Interpretation of Human Rights Treaties in the Recent Jurisprudence of the European Court of Human Rights', (2003) 14 *EJIL* 529, at 539 *et seq.*

[53] See, e.g., *Issa v. Turkey*, App. No. 31821/96, Judgment, 16 November 2004.

[54] See Chapter IV.

[55] See below, Chapter IV, Section 3.C.

[56] *OED Online*, available at <http://dictionary.oed.com/>.

Similarly, the *American Heritage Dictionary* defines 'jurisdiction' not only as '[t]he right and power to interpret and apply the law', but also as '[a]uthority or control' and '[t]he territorial range of authority or control'.[57] Among the specialized legal dictionaries, *Black's Law Dictionary* defines jurisdiction both as a 'government's general power to exercise authority over all persons and things within its territory' and a 'geographic area within which political or judicial authority may be exercised',[58] while 'territory' is defined as a 'geographical area included within a particular government's jurisdiction'.[59]

There is thus nothing extravagant in interpreting 'jurisdiction' to mean factual power. Likewise, the word 'jurisdiction' in its sense as control over territory and perhaps also individuals is in international law not used only in human rights treaties, but in a variety of other instruments. Let us now examine some of them, and delve a bit into the origins of jurisdiction clauses.

E. Origins of jurisdiction clauses

Where did the jurisdiction clauses in human rights treaties come from, and which was the first one? To answer that question, we first must take a brief look at the period following the end of the First World War and the establishment of the League of Nations. With limited exceptions, as with the rules on the treatment of aliens, classical international law did not protect individual rights. Individuals were, in the notorious words of Oppenheim, merely objects, not subjects of international law, which concerned itself solely with relations between nation-states. This classical position slowly began to erode in the first half of the twentieth century.

First, the peace treaties which ended the First World War established a legal regime for the protection of minorities in Europe. These minority treaties are in some ways the antecedents of the modern human rights regime created after the Second World War, both because of their limited successes and because of their much greater failures. As these treaties were the first systematic attempt to safeguard the rights of individuals in international law (sort of),[60] it is only natural to first look at them to see whether they contained anything like a jurisdiction clause. They do not. If we take as an example the 1919 Treaty of Saint Germain,[61] we will see that the state concerned, Austria, bound itself to guarantee certain rights only to its *inhabitants* or to its *nationals*. For instance, under Article 63(1) of the Treaty 'Austria undertakes to assure full and complete protection of life and liberty to all inhabitants of Austria without distinction of birth, nationality, language, race or religion' while according to Article 66(1) '[a]ll Austrian nationals shall be equal

[57] *American Heritage Dictionary* (3rd edn, 1992), at 978.

[58] *Black's Law Dictionary* (9th edn, 2009), at 927–8.

[59] Ibid., at 1611.

[60] See generally P. Thornberry, *International Law and the Rights of Minorities* (Oxford University Press, 1993); Simpson, above note 51, at 121 *et seq.*

[61] Treaty of Peace between the Allied and Associated Powers and Austria; Protocol, Declaration and Special Declaration, Austria; Protocol, Declaration and Special Declaration, St Germain-en-Laye, 10 September 1919, text available at <http://www.austlii.edu.au/au/other/dfat/treaties/1920/3.html>.

before the law and shall enjoy the same civil and political rights without distinction as to race, language or religion'.

However, a jurisdiction clause related to the protection of minorities can be gleamed in a failed American proposal to include, as Article 6, the following provision in the Covenant of the League of Nations:

The League of Nations shall require of all new States to bind themselves as a condition precedent to their recognition as independent or autonomous States, to accord to all racial or national minorities *within their several jurisdiction* exactly the same treatment and security, both in law and in fact, that is accorded to the racial or national majority of the people.[62]

The League of Nations Covenant as adopted only speaks of 'jurisdiction' when it provides in its Article 15 that the League Council will not intervene within matters which are solely within the domestic jurisdiction of a member state, which is yet another different meaning of the word 'jurisdiction'.[63] The Covenant does, however, contain one provision which is similar to a jurisdiction clause in Article 23(b), which stipulates that '[s]ubject to and in accordance with the provisions of international conventions existing or hereafter to be agreed upon, the Members of the League... undertake to secure just treatment of the native inhabitants of *territories under their control*' (emphasis added).

So, there are no jurisdiction clauses to be found either in the minority treaties, or in the League of Nations Covenant as adopted. Incidentally, they do not exist even in the modern treaties for the protection of minorities which were concluded under the auspices of the Council of Europe after the end of the Cold War, and are plainly territorially limited.[64] They are also only to be found in one of the numerous conventions adopted within the International Labour Organization (ILO), itself a child of the First World War, which are clearly intended to be applicable only within a state's territory and which moreover often contain elaborate colonial clauses.[65]

The word 'jurisdiction' in the sense of territory or territorial control is however used in bilateral treaties going back at least a hundred years. Thus, for example, the 1902 extradition treaty between the United States and Serbia provides in its Article I that

[t]he Government of the United States and the Government of Serbia mutually agree to

deliver up persons who, having been charged with or convicted of any of the crimes and offenses specified in the following article, committed *within the jurisdiction* of one of the high contracting parties, shall seek an asylum or be found *within the territories of the other*.[66] (emphasis added)

[62] Cited according to Simpson, above note 51, at 124 (emphasis added).

[63] A similar provision can be found in Art. 2(7) of the UN Charter.

[64] For instance, the preamble to the Framework Convention for the Protection of National Minorities, CETS No. 157, refers to the states parties' resolve 'to protect within their respective territories the existence of national minorities'.

[65] Furthermore, Art. 35 of the ILO Constitution sets out a general regime of applicability of ILO conventions to overseas territories, a sort of a super-colonial clause.

[66] Treaty on Extradition of 12 June 1902, 32 Stat. 1890, 12 Bevans 1238. Incidentally, this treaty still remains in force for the successor states of the former Yugoslavia.

Similarly, and quite appropriately, 'jurisdiction' as territorial control is also used in the bilateral treaty between the US and Cuba on the lease of Guantanamo Bay.[67] Under Article III of that treaty,

> While on the one hand the United States recognizes the continuance of the ultimate sovereignty of the Republic of Cuba over the above described areas of land and water, on the other hand the Republic of Cuba consents that during the period of the occupation by the United States of said areas under the terms of this agreement the United States *shall exercise complete jurisdiction and control over and within said areas*... (emphasis added)

This article thus clearly distinguished between sovereignty or title over territory on the one hand, which remained with Cuba, and factual jurisdiction and control over territory on the other, which was vested in the United States.

The earliest jurisdiction clause proper governing the scope of application of state obligations that I could find in a multilateral treaty is the one in Article 2 of the 1926 Slavery Convention,[68] which provides as follows:

> The High Contracting Parties undertake, each in respect of the *territories placed under its sovereignty, jurisdiction, protection, suzerainty or tutelage*, so far as they have not already taken the necessary steps:
>
> (a) To prevent and suppress the slave trade;
> (b) To bring about, progressively and as soon as possible, the complete abolition of slavery in all its forms. (emphasis added)

The formulation 'any territory placed under its sovereignty, jurisdiction, protection, suzerainty or tutelage' is repeated word for word in Articles 5 (forced labour), 9 (colonial clause), and 10 (denouncement) of the Convention. It shows very clearly that this 'jurisdiction' is a cognate of notions such as sovereignty and suzerainty, and moreover refers to a state's control over territory, not to the competence of the state to extend its domestic law to cover the conduct of a particular person.

To my knowledge, the only other multilateral treaty of the interwar period to have a jurisdiction clause was the 1930 ILO Forced Labor Convention,[69] which in its Article 26(1) provides that

> [e]ach Member of the International Labour Organisation which ratifies this Convention undertakes to apply it to the territories placed under its sovereignty, jurisdiction, protection, suzerainty, tutelage or authority, so far as it has the right to accept obligations affecting matters of internal jurisdiction; provided that, if such Member may desire to take advantage

[67] Agreement Between the United States and Cuba for the Lease of Lands for Coaling and Naval Stations, 23 February 1903, T.S. No. 418, available at <http://avalon.law.yale.edu/20th_century/dip_cuba002.asp>.

[68] Slavery, Servitude, Forced Labour and Similar Institutions and Practices Convention of 1926, 60 LNTS 253, entered into force 9 March 1927.

[69] Convention concerning Forced or Compulsory Labour (ILO No. 29), 39 UNTS 55, entered into force 1 May 1932.

of the provisions of article 35 of the Constitution of the International Labour Organisation, it shall append to its ratification a declaration stating:

(1) The territories to which it intends to apply the provisions of this Convention without modification;
(2) The territories to which it intends to apply the provisions of this Convention with modifications, together with details of the said modifications;
(3) The territories in respect of which it reserves its decision.

Bearing in mind the subject-matter of this treaty—prohibiting forced labour—it is apparent that its drafters modelled this provision on the jurisdiction clauses of the 1926 Slavery Convention. It is still fascinating, however, for several reasons. First, unlike with the several separate clauses in the Slavery Convention, this jurisdiction clause controls the scope of applicability of the entire treaty, much like Article 1 ECHR, not just that of a particular obligation, and appears to be the first of its kind. Secondly, to the list of terms in the clause found in the Slavery Convention one more is added after 'tutelage'—the word 'authority', which again demonstrates that the word 'jurisdiction' is used here with a similar meaning, as one on a spectrum of similar concepts. Thirdly, this is the only treaty provision that combines a jurisdiction clause with a colonial clause.

Now we must fast forward to the aftermath of the Second World War, when human rights law was born. The first two international human rights instruments adopted, both of which were hortatory instead of formally binding, were the American Declaration of the Rights and Duties of Man, in 1946, and the UDHR, in 1948. The former contains no provisions governing its territorial applicability,[70] and strictly speaking, neither does the latter. However, as we have seen, in its preamble the UDHR does indeed refer to territories under the states' jurisdiction, and in its Article 2 it again speaks of the jurisdictional status of territories.

The first true human rights jurisdiction clauses, however, did not appear in the usual suspects—the ECHR and the ICCPR—but in the peace treaties concluded after the war between the Allies and associated powers and the defeated powers of the Axis. For example, Article 15 of the 1947 Peace Treaty with Italy[71] provides as follows:

Italy shall take all measures necessary *to secure to all persons under Italian jurisdiction*, without distinction as to race, sex, language or religion, the enjoyment of human rights and of the fundamental freedoms, including freedom of expression, of press and publication, of religious worship, of political opinion and of public meeting.[72] (emphasis added)

[70] On the other hand, drafts of the Declaration did contain jurisdiction clauses—see Gondek, at 79–81.

[71] Available in (1948) 42 *AJIL Supplement* 47. See also Simpson, above note 51, at 328–31.

[72] An almost identical clause can be found in Article 4 of Annex VI to the Treaty, which served as the statute of the Free Territory of Trieste. In 1954, Italy, Yugoslavia, the United Kingdom, and the United States signed a Memorandum of Understanding which divided the Free State into Italian and Yugoslav areas—see UN Doc. S/3301. Annexed to this Memorandum of Understanding is a Special Statute which Italy and Yugoslavia agreed to enforce, and which provides for the 'common intention of the Italian and Yugoslav Governments to ensure human rights and fundamental freedoms without

This same mini-human rights convention containing a jurisdiction clause can be found in other peace treaties concluded after the Second World War.[73] A similar clause was inserted into Article 3 of the 1949 Statute of the Council of Europe, the treaty which created that international organization, which requires of a member state to 'accept the principles of the rule of law and of the enjoyment by all persons within its jurisdiction of human rights and fundamental freedoms'.

Therefore, jurisdiction clauses were not invented in the first human rights treaties to be drafted, the then single International Covenant and the ECHR, but have a considerably longer pedigree in the treaty practice of states, not to mention a meaning distinct from that concept of the jurisdiction of states which delimits the spheres of their municipal law. This is also shown by the preparatory work of the ECHR, which was cited by the European Court itself in *Bankovic*.[74]

As originally drafted by a committee of the Consultative Assembly of the Council of Europe, Article 1 ECHR extended its protections to all persons *residing* within the states parties' territories. This provision was later changed into the jurisdiction clause. As explained in the *travaux*:

> The Assembly draft had extended the benefits of the Convention to 'all persons residing within the territories of the signatory States.' It seemed to the [Expert Intergovernmental] Committee that the term 'residing' might be considered too restrictive. It was felt that there were good grounds for extending the benefits of the Convention to all persons in the territories of the signatory States, even those who could not be considered as residing there in the legal sense of the word. The Committee therefore replaced the term 'residing' by the words 'within their jurisdiction' which are also contained in Article 2 of the Draft Covenant of the United Nations Commission.[75]

This excerpt from the *travaux* makes several things clear. First, the reason for the insertion of the jurisdiction clause into Article 1 and the deletion of the reference to territory is only that the term 'residing' which was used in the previous draft could be interpreted to exclude persons temporarily (or illegally) present in the territory of the state. Secondly, the word 'jurisdiction' was used in a sense largely, though perhaps not entirely, synonymous with the word 'territory', as it is used, for example, in Article 12(1)(a) of the Inter-American Convention to Prevent and

discrimination of race, sex, language and religion *in the areas coming under their administration*' (emphasis added)—see E. Schwelb, 'The Trieste Settlement and Human Rights', (1955) 49 *AJIL* 240. It should be noted that the provisions of the Peace Treaty and of the Special Statute on human rights differ mainly in that the word 'jurisdiction' in the former was replaced by the word 'administration' in the latter, which again suggests that both of these words were used with the same or a very similar meaning.

[73] See Art. 2, Treaty of Peace with Bulgaria, (1948) 42 *AJIL Supplement* 179; Art. 6, Treaty of Peace with Finland, (1948) 42 *AJIL Supplement* 204; Art. 2, Treaty of Peace with Hungary, (1948) 42 *AJIL Supplement* 225; Art. 3, Treaty of Peace with Romania, (1948) 42 *AJIL Supplement* 252.

[74] *Bankovic*, para. 19.

[75] See A.H. Robertson (ed.), 3 *Collected Edition of the 'Travaux Préparatoires' of the European Convention on Human Rights* (Nijhoff, 1976), at 260.

Punish Torture cited above.[76] Thirdly, the reason why it was opted for by the drafters is because it was also contained in the draft UN Covenant on human rights, the provisions that were to become Article 2 ICCPR.[77] There is no indication in the *travaux* that the word 'jurisdiction' was used to connote the limits on the municipal legal orders of states imposed by international law.

F. Jurisdiction as power

The reader will hopefully not begrudge my spending so many pages on pointing out the obvious—that there is not *an* ordinary meaning, but *many* such meanings of the word 'jurisdiction' in international law. That word can be used to refer to the competence of a court or to that of any body which applies or interprets the law, to the jurisdiction of states to prescribe rules of their municipal law and to enforce them, or to the domestic jurisdiction of states, the domain in which they are to be free of outside interference. However, it can also be used more generally as a synonym for words such as power, authority, or control, either over people or over territory,[78] or as a synonym for the territory within which such power is exercised.[79]

It is this last meaning or group of meanings of the word 'jurisdiction' that is used in human rights treaties, as I hope to have proven in the preceding account. That notion of jurisdiction, however, is not an exclusive, autonomous concept that exists only in human rights treaties, which cannot be grasped by looking outside them. There is no fragmentation of general international law here, no self-contained regimes. The jurisdiction clauses in the Slavery Convention, the Forced Labour Convention, the peace treaties, and the UDHR all preceded those in the human rights treaties, and, as we have seen, one was almost included in the Covenant of the League of Nations. Such clauses also continue to be used to set a threshold for obligations in contemporary treaty practice, the most recent one being that in Article 9 of the Enforced Disappearances Convention, quoted above. They are also

[76] Of course, this extract from the *travaux* can be taken as a strong indication that the drafters did not envisage any extraterritorial application of the ECHR. Indeed, this is exactly how this passage from the *travaux* is interpreted by the Court. What it does not entail is that jurisdiction equals *title* over a territory. In other words, a state can have more (or less) territories under its jurisdiction (control) than it actually owns as a matter of law.

[77] Intermediate drafts of the Covenant, one produced by a drafting committee of the full Commission and another adopted at the fifth session of the Commission, which took place in May–June 1949, read 'all individuals within its jurisdiction'. See UN Doc. E/CN.4/95, at 16, UN Doc. E/800, at 15 and UN Doc. E/1371, also cited as UN Doc. E/CN.4/350, at 28. This is the formulation that was taken up by the drafters of the ECHR, but it was changed in the later drafts of the Covenant itself. On the drafting history of the ECHR and ICCPR with respect to jurisdiction clauses, see further Gondek, at 81 *et seq*.

[78] I will be exploring in more detail the nature of power or control over territory and over individuals in Chapter IV, Sections 2.C. and 3.C below.

[79] Thus, for example, Hersch Lauterpacht introduced the following provision (Art. 18) into his draft International Bill of Rights: 'The obligations of this Bill of Rights shall be binding upon States in relation both to their metropolitan territory and to any other territory under their control and jurisdiction.' See H. Lauterpacht, *International Law and Human Rights* (Archon Books, 1968; earlier editions 1930 and 1950), at 317 and 364.

to be found in instruments as varied as, *inter alia*, the Ottawa Convention on Landmines,[80] the Chemical Weapons Convention,[81] the Rio Declaration,[82] and the ENMOD Convention.[83] In all likelihood there are many more jurisdiction clauses in treaties both old and new that I have missed.

Of course, the reason for my belabouring the obvious is that the European Court in *Bankovic* simply assumed that the notion of 'jurisdiction' in Article 1 ECHR is the same as that concept of jurisdiction which determines when a state may apply rules of its domestic law, and relied on that assumption to restrict the extraterritorial application of the ECHR to some ill-defined exceptional circumstances.[84] All of *Bankovic* rests on that one, colossal *non sequitur*. To say that the scope of application of human rights treaties depends on the right of states to regulate certain types of conduct by their domestic law is nothing less than a category error.

The Court's own recounting of this position in a later case speaks for itself:

> In certain exceptional cases, jurisdiction is assumed on the basis of non-territorial factors, such as: acts of public authority performed abroad by diplomatic and consular representatives of the State; the criminal activities of individuals overseas against the interests of the State or its nationals; acts performed on board vessels flying the State flag or on aircraft or spacecraft registered there; and particularly serious international crimes (universal jurisdiction).[85]

Is the Court here saying that an individual who commits a crime against a state's interests or nationals is thereby within its jurisdiction for the purposes of Article 1 ECHR? That the UK would have the obligation of securing the human rights of a US national forging pounds sterling in the US, just because general international law gives the UK the right to criminalize such conduct? And then what are we to make of the reference to *universal* jurisdiction? Is it supposed to mean that any *génocidaire* on the planet is within the Article 1 jurisdiction of all ECHR states parties because they all have prescriptive jurisdiction over such acts?[86]

[80] Art. 5(2) of which reads: 'Each State Party undertakes to destroy or ensure the destruction of all anti-personnel mines in mined areas under its jurisdiction or control, as soon as possible but not later than ten years after the entry into force of this Convention for that State Party.'

[81] Art. 1(2) of which reads: 'Each State Party undertakes to destroy chemical weapons it owns or possesses, or that are located in any place under its jurisdiction or control, in accordance with the provisions of this Convention.' A similar clause is contained in numerous other articles of this Convention.

[82] Principle 2 of which reads: 'States have, in accordance with the Charter of the United Nations and the principles of international law, the sovereign right to exploit their own resources pursuant to their own environmental and developmental policies, and the responsibility to ensure that activities within their jurisdiction or control do not cause damage to the environment of other States or of areas beyond the limits of national jurisdiction.'

[83] Art. IV of which reads: 'Each State Party to this Convention undertakes to take any measures it considers necessary in accordance with its constitutional processes to prohibit and prevent any activity in violation of the provisions of the Convention anywhere under its jurisdiction or control.'

[84] This assumption has also been uncritically followed by some authors—see McGoldrick, above note 8; M. Pedersen, 'Territorial Jurisdiction in Article 1 of the European Convention on Human Rights', (2004) 73 *Nord J Int'l L* 279.

[85] *Assanidze v. Georgia* [GC], App. No. 71503/01, Judgment, 8 April 2004, para. 137 (recapitulating *Bankovic*).

[86] Cf. J.-P. Costa, 'L'Etat, le territoire et la Convention européenne des droits de l'homme', in M. Kohen (ed.), *Promoting Justice, Human Rights and Conflict Resolution through International Law:*

Asking these questions is to answer them. As I have demonstrated above, the notion of jurisdiction in human rights treaties relates essentially to a question of fact, of actual authority and control that a state has over a given territory or persons. 'Jurisdiction', in this context, simply means actual *power*, whether exercised lawfully or not—nothing more, and nothing less.[87]

I should note that this discussion is not meant to prejudge in any way whether the extraterritorial application of the ECHR or of any other human rights treaty should be extensive or exceptional. *Bankovic* may even be right when it comes to its ultimate result, if that result is based solely on the fact that the territory in which the applicants were located was not under the effective overall control of the defendant states[88]—though, as we shall see, it is questionable whether this requirement should apply when violations of a negative obligation are at stake. The *Bankovic* Court is most certainly not right, however, when it comes to the stated basis of its reasoning, as it curtailed the notion of jurisdiction in human rights treaties by reference to an entirely different concept.[89]

3. State Jurisdiction Is Not State Responsibility

A. *Loizidou*: a test of attribution?

Unfortunately, the confusion does not stop there, as it has proven necessary to distinguish the notion of state jurisdiction in human rights treaties from that of state responsibility, in particular from attribution or imputability of conduct. To explain this problem as clearly as possible, when the European Court found that Turkey was responsible for violations of the ECHR because it exercised 'effective overall control' over the area of northern Cyprus, did the Court establish that, as a matter of law, the acts of the so-called Turkish Republic of Northern Cyprus (TRNC) were *attributable* to Turkey, making all acts of the TRNC acts of the state of Turkey? In other words, is the 'effective overall control' test a test of attribution, and are jurisdiction and responsibility interchangeable concepts, or not? Terminological and conceptual inconsistencies in this regard have plagued the jurisprudence of the European Court,[90] and several judges have attempted to grapple with the distinction between jurisdiction and responsibility in their opinions.[91]

Liber Amicorum Lucius Caflisch (Nijhoff, 2007), 179, at 193 (citing *Assanidze*, and noting that the Court has never assumed a universality-based competence).

[87] It should be noted that one judge of the European Court—Loukis Loucaides—has expressly adopted this position, both in academic work and in several dissenting opinions. See L. Loucaides, 'Determining the Extra-territorial Effect of the European Convention: Facts, Jurisprudence and the *Bankovic* Case', (2006) 4 *EHRLR* 391.

[88] See Happold, above note 1.

[89] See also Wilde, above note 52, at 514; Gondek, at 375–7;A. Buyse, 'A Legal Minefield—The Territorial Scope of the European Convention', (2008) 1 *Inter-American and European Human Rights Journal* 269.

[90] See also Gondek, at 160 *et seq.*

[91] See, e.g., the opinions of Judges Loucaides and Kovler in *Ilascu*, and the joint dissenting opinion of Judges Golcuklu and Pettiti in *Loizidou* (preliminary objections).

This is again not an issue of purely academic interest, as the law of state responsibility is a body of secondary rules which is applicable in a wide variety of situations. A lot may depend on what the proper standard for attribution is. For example, the Appeals Chamber of the International Criminal Tribunal for the former Yugoslavia relied on *Loizidou* in support of its 'overall control' test of attribution in the *Tadic* case, which it used to determine whether the conflict in Bosnia was international or non-international in character. We will come to that issue in a moment; first, however, we must remind ourselves what the European Court actually held in *Loizidou.*

In its preliminary objections judgment the Court ruled as follows:

> Bearing in mind the object and purpose of the Convention, the responsibility of a Contracting Party may also arise when as a consequence of military action—whether lawful or unlawful—it exercises effective control of an area outside its national territory. The obligation to secure, in such an area, the rights and freedoms set out in the Convention, derives from the fact of such control whether it be exercised directly, through its armed forces, or through a subordinate local administration.[92]

At the merits stage, the Court affirmed its previous ruling, and held that

> [i]t is not necessary to determine whether, as the applicant and the Government of Cyprus have suggested, Turkey actually exercises detailed control over the policies and actions of the authorities of the 'TRNC'. It is obvious from the large number of troops engaged in active duties in northern Cyprus (see paragraph 16 above) that her army exercises effective overall control over that part of the island. Such control, according to the relevant test and in the circumstances of the case, entails her responsibility for the policies and actions of the 'TRNC' (see paragraph 52 above). Those affected by such policies or actions therefore come within the 'jurisdiction' of Turkey for the purposes of Article 1 of the Convention (art. 1). Her obligation to secure to the applicant the rights and freedoms set out in the Convention therefore extends to the northern part of Cyprus.[93]

As we have seen, the Court speaks of 'effective control of an area' in its preliminary objections judgment, and of 'effective overall control' of an area in its merits judgment, and generally prefers the latter formula in its subsequent case law.

B. The first possible reading of *Loizidou*

What are we to make of this test? Perhaps the most natural textual interpretation of this ruling is that the Court found all actions of the TRNC to be attributable to Turkey. This is indicated, first, by the Court labelling the TRNC as a local administration subordinate to Turkey, and, secondly, by the Court saying that Turkey's control over the area 'entails her responsibility' for the policies and actions of the TRNC. Indeed, this is how the *Loizidou* test was seen by the now defunct European Commission on Human Rights in the last of its reports on interstate

[92] *Loizidou v. Turkey*, App. No. 15318/89, Judgment (preliminary objections), 23 February 1995, para. 62.

[93] *Loizidou v. Turkey*, App. No. 15318/89, Judgment (merits), 28 November 1996, para. 56.

applications submitted by Cyprus against Turkey. In its reports on previous applications by Cyprus, as well as in its decisions regarding individual applications, the Commission distinguished between, on the one hand, acts of Turkish military forces and other Turkish authorities—for which Turkey was held responsible—and, on the other hand, acts of the Turkish Cypriot authorities, which were not attributable to Turkey. Accordingly, in those reports the question of attribution was examined separately in relation to each of the complaints submitted by Cyprus on the basis of the actual involvement of Turkish authorities or officers.[94] Now, however, after the merits judgment in *Loizidou*, the Commission felt that this distinction could no longer be maintained, since the Court found Turkey responsible for the actions of the TRNC.[95]

It is this reading of *Loizidou* that was picked up by the ICTY Appeals Chamber in *Tadic*,[96] in which it initiated a direct conflict of jurisprudence with the ICJ. As is well known, in *Tadic* the ICTY had to establish whether the conflict in Bosnia and Herzegovina could be qualified as international or as non-international, as that determined which rules of humanitarian law applied and which of the charges against the defendant could be sustained. The Appeals Chamber held that, in order to determine whether the conflict which was *prima facie* internal could be internationalized by the external involvement of Serbia and the control it had over the Bosnian Serbs, it had to have recourse to the general rules of state responsibility, since the rules of international humanitarian law did not provide an answer. In other words, because international armed conflicts are by definition interstate in nature, it had to answer the question whether the acts of the Bosnian Serbs were attributable to Serbia, and only if that answer was in the affirmative would the conflict have been one between two states, Bosnia and Serbia, and hence international in nature. To do that, the Appeals Chamber turned to the 'effective control' test developed by the ICJ in the *Nicaragua* case,[97] and found it lacking from the standpoint of customary international law. It therefore ruled that the proper standard of attribution to a state of acts committed by an organized armed group is 'overall control', which need not be exercised in a specific operation.

I have previously argued that *Tadic* was wrongly decided for several reasons.[98] First, the Appeals Chamber's resort to the law of state responsibility was inappropriate, since the qualification of an armed conflict is a matter exclusively for the primary rules of international humanitarian law, not for the secondary rules of state responsibility. The same legal test cannot logically be used to establish both what obligation a state has and whether a breach of that obligation is attributable to it.

Secondly, the Appeals Chamber's principal error in analysis is in its failure to distinguish between the two tests announced by the ICJ in *Nicaragua*. The first test, that of complete dependence or control, as set out in paragraph 109 of the

[94] *Cyprus v. Turkey*, App. No. 25781/94, Commission Report, 4 June 1999, para. 96.

[95] Ibid., paras 98–102.

[96] *Prosecutor v. Tadić*, IT-94-1, Appeals Chamber, Judgment, 15 July 1999 (hereinafter *Tadic*).

[97] *Military and Paramilitary Activities in and Against Nicaragua (Nicaragua v. United States)*, Judgment (merits), 27 June 1986, ICJ Reports 1986, at 14 (hereinafter *Nicaragua*).

[98] See M. Milanovic, 'State Responsibility for Genocide', (2006) 17 *EJIL* 553, at 576–81.

Nicaragua judgment,[99] operates at a general level by assessing whether the relationship between a state and non-state actor is so much of control on the one side and dependence on the other that the non-state actor has to be equated for legal purposes with an organ of the state. In other words, the non-state actor becomes a state organ *de facto*, i.e. an organ even though it is not designated as such by the state's own municipal law, and all or any of its acts become the acts of the state. The second test, that of effective control, as set out in paragraph 115 of the *Nicaragua* judgment,[100] kicks in only when the first test is not satisfied, and asks whether a *specific operation* of a non-state actor which is neither a *de jure* nor a *de facto* organ of the state was indeed conducted under the state's control. The Appeals Chamber unequivocally conflated the first test with the latter, contrary to the interpretations of *Nicaragua* professed both by the Prosecution and by Judge McDonald, who dissented from the Trial Chamber judgment in *Tadic*.[101]

That the Appeals Chamber's interpretation of *Nicaragua*, if nothing else, was fundamentally mistaken has been demonstrated by the ICJ's *Genocide* judgment,[102]

[99] *Nicaragua* , para. 109: 'What the Court has to determine at this point is whether or not the relationship of the contras to the United States Government was so much one of dependence on the one side and control on the other that it would be right to equate the contras, for legal purposes, with an organ of the United States Government, or as acting on behalf of that Government.' After assessing the available evidence, the Court concluded in para. 110 that it is 'unable to determine that the contra force may be equated for legal purposes with the forces of the United States'.

[100] Ibid., para. 115:

> The Court has taken the view (paragraph 110 above) that United States participation, even if preponderant or decisive in the financing, organizing, training, supplying and equipping of the contras, the selection of its military or paramilitary targets, and the planning of the whole of its operation, is still insufficient in itself, on the basis of the evidence in the possession of the Court, for the purpose of attributing to the United States the acts committed by the contras in the course of their military or paramilitary operations in Nicaragua. All the forms of United States participation mentioned above, and even the general control by the respondent State over a force with a high degree of dependency on it, would not in themselves mean, without further evidence, that the United States directed or enforced the perpetration of the acts contrary to human rights and humanitarian law alleged by the applicant State. Such acts could well be committed by members of the contras without the control of the United States. For this conduct to give rise to legal responsibility of the United States, it would in principle have to be proved that that State had effective control of the military or paramilitary operations in the course of which the alleged violations were committed.

[101] In the Appeals Chamber's own words (*Tadic*, para. 112):

> The Appeals Chamber does not subscribe to this interpretation [that two distinct tests were formulated by the ICJ in *Nicaragua*]. Admittedly, in paragraph 115 of the *Nicaragua* judgement, where 'effective control' is mentioned, it is unclear whether the Court is propounding 'effective control' as an alternative test to that of 'dependence and control' set out earlier in paragraph 109, or is instead spelling out the requirements of the same test. The Appeals Chamber believes that the latter is the correct interpretation. In *Nicaragua*, in addition to the 'agency' test (properly construed, as shall be seen in the next paragraph, as being designed to ascertain whether or not an individual has the formal status of a State official), the Court propounded only the 'effective control' test. This conclusion is supported by the evidently stringent application of the 'effective control' test which the Court used in finding that the acts of the contras were not imputable to the United States.

[102] *Application of the Convention on the Prevention and Punishment of the Crime of Genocide (Bosnia and Herzegovina v. Serbia and Montenegro)*, Judgment, 26 February 2007 (hereinafter the *Bosnian Genocide* merits judgment).

which affirmed both points of criticism raised above, by disputing the need to venture into state responsibility in order to determine the nature of an armed conflict[103] and by explicitly distinguishing between the test of complete control[104] and the test of effective control.[105] The *Tadic* overall control test was also previously rejected by the International Law Commission in its commentaries to the Articles on State Responsibility.[106]

Incidentally, the *Genocide* case is an excellent example of the immense practical importance of the various tests of attribution at hand. The ICJ found that Serbia did not exercise complete control over the Bosnian Serbs, which were therefore not its *de facto* organs, and it also found that there was no evidence that Serbia exercised effective control over the attack on Srebrenica in July 1995 during which the Bosnian Serbs committed genocide and killed thousands of Bosnian Muslims. Serbia consequently could not have been held responsible for the commission of genocide, though the ICJ found that it failed to prevent the Srebrenica genocide. Under the approach of the ICTY Appeals Chamber, however, Serbia would have been responsible for committing the Srebrenica genocide, as it undoubtedly participated in 'coordinating or helping in the general planning of the non-state actor's military activity'.[107] The obvious looseness of this test and the lack of connection between the actions of state organs and the act that is being imputed to the state are why the ICJ remarked that this test stretched the established paradigms of state responsibility to breaking point.[108]

My final argument against *Tadic* was that the authorities that it cited, chief among them being the *Yaeger* case[109] before the Iran-US Claims Tribunal and *Loizidou*, do not support its conclusion that they envisage an overall control test of attribution. The former case does not concern us at this time, so let us turn to the alternate, and in my view better, reading of *Loizidou*, which shows why viewing it as setting out a test of attribution is unwarranted.[110]

[103] Ibid., paras 404–05.

[104] Ibid., paras 391–3. [105] Ibid., paras 396–400.

[106] ILC Commentaries to the Draft Articles, extract from the Report of the International Law Commission on the work of its Fifty-third session, Official Records of the General Assembly, fifty-sixth session, Supplement No. 10 (A/56/10), chp.IV.E.2, available at <http://untreaty.un.org/ilc/texts/instruments/english/commentaries/9_6_2001.pdf> (ILC ASR), at 105–7.

[107] *Tadic*, para. 131.

[108] *Bosnian Genocide* merits judgment, para. 406.

[109] *Kenneth P. Yeager v. Islamic Republic of Iran*, (1987) 17 Iran-US Claims Tribunal Reports 92.

[110] For more on *Nicaragua* and *Tadic*, see A. de Hoogh, 'Articles 4 and 8 of the 2001 ILC Articles on State Responsibility, the Tadic Case and Attribution of Acts of Bosnian Serb Authorities to the Federal Republic of Yugoslavia', (2001) 72 *BYIL* 255; A. Cassese, 'The *Nicaragua* and *Tadic* Tests Revisited in Light of the ICJ Genocide Judgment in Bosnia', (2007) 18 *EJIL* 649; M. Spinedi, 'On the Non-Attribution of the Bosnian Serbs' Conduct to Serbia', (2007) 5 *JICJ* 829; R. Goldstone and R. Hamilton, '*Bosnia v. Serbia*: Lessons from the Encounter of the International Court of Justice with the International Criminal Tribunal for the Former Yugoslavia', (2008) 21 *Leiden JIL* 95, at 97–103; S. Talmon, 'The Responsibility of Outside Powers for Acts of Secessionist Entities', (2009) 58 *ICLQ* 493; K. Del Mar, 'The Requirement of "Belonging" under International Humanitarian Law', (2010) 21 *EJIL* 105; M. Milanovic, 'What Exactly Internationalizes an Internal Armed Conflict?,' *EJIL: Talk!*, 7 May 2010, available at <http://www.ejiltalk.org/what-exactly-internationalizes-an-internal-armed-conflict/>.

C. The second possible reading of *Loizidou*

In order to understand *Loizidou*, it is important to remind ourselves that in the jurisdiction clauses of all relevant treaties the notion of the state's jurisdiction is textually tied to the emergence of the state's obligation. It is a threshold criterion which determines whether the state incurs obligations under the treaty, and consequently whether any particular act of the state can be characterized as internationally wrongful. As set out in Article 2 of the ILC Articles on State Responsibility, '[t]here is an internationally wrongful act of a State when conduct consisting of an action or omission: (a) Is attributable to the State under international law; and (b) Constitutes a breach of an international obligation of the State.' The notion of state jurisdiction in human rights treaties is a question which falls under Article 2(b)—does an obligation exist—not under Article 2(a)—is a particular conduct attributable to a state.[111]

We must also remind ourselves of the basic point that human rights treaties generally contain (at least) two different types of obligations. The first are *negative* in nature—the obligation of the state to 'respect' human rights, i.e. not to have its organs, agents, or other persons whose acts are attributable to it commit human rights violations. The second are *positive* in nature—the obligation of the state to 'ensure' or 'secure' the human rights of persons within its jurisdiction, i.e. to prevent human rights violations committed by third states, private individuals, or non-state groups generally against other private individuals. Thus, for example, if the police arbitrarily shoot and kill someone, the state has failed to *respect* that person's right to life, as enshrined in Article 2 ECHR or Article 6 ICCPR. On the other hand, if the police have reasons to know that a murder is about to take place and they do nothing to prevent it, the state has failed to *secure* or *ensure* the right to life of the victim.[112] This distinction is explicit in the text of some human rights treaties, and implicit in others, most notably the ECHR.[113]

Distinguishing between the responsibility of a state for the acts of its own organs or forces and its responsibility for failing to prevent violations by forces not under its control has much support in the jurisprudence of the ICJ in particular. So, for example, in *Congo v. Uganda:*

> The Court, having concluded that Uganda was an occupying Power in Ituri at the relevant time, finds that Uganda's responsibility is engaged both for any acts of its military that violated its international obligations and for any lack of vigilance in preventing violations of

[111] See Lawson, above note 11, at 86.

[112] On the general importance of this distinction for the extraterritorial application of human rights treaties, see Chapter IV below, esp. Section 4.

[113] See generally A. Mowbray, *The Development of Positive Obligations under the European Convention on Human Rights by the European Court of Human Rights* (Hart Publishing, 2004); F. Sudre, 'Les obligations positives dans la jurisprudence européenne des droits de l'homme', (1995) *Revue trimestrielle des droits de l'homme* 363.

human rights and international humanitarian law by other actors present in the occupied territory, including rebel groups acting on their own account.[114]

Similarly, in the *Genocide* case the Court found the negative obligation not to commit genocide implicit in the positive state obligation to prevent genocide under Article 1 of the Genocide Convention.[115] Since it was impossible to prove to the required degree of certainty that the Srebrenica genocide was attributable to Serbia under either the test of complete control or the test of effective control,[116] the Court nonetheless found Serbia responsible for failing to prevent that genocide, i.e. for its own wrongful act of failing to exercise due diligence to prevent violations by third parties.[117]

Likewise, in the *Velasquez-Rodriguez* case, the Inter-American Court held that

> ... in principle, any violation of rights recognized by the Convention carried out by an act of public authority or by persons who use their position of authority is imputable to the State. However, this does not define all the circumstances in which a State is obligated to prevent, investigate and punish human rights violations, nor all the cases in which the State might be found responsible for an infringement of those rights. An illegal act which violates human rights and which is initially not directly imputable to a State (for example, because it is the act of a private person or because the person responsible has not been identified) can lead to international responsibility of the State, not because of the act itself, but because of the lack of due diligence to prevent the violation or to respond to it as required by the Convention.[118]

The same point has been made repeatedly by the Human Rights Committee in its interpretation of the ICCPR.[119]

In light of this distinction *Loizidou* becomes more comprehensible. Like the European Commission in the Cyprus cases which preceded *Loizidou*, the Court did not necessarily find that all of the acts of the TRNC were attributable to Turkey. What it did establish was that Turkey, by virtue of its effective overall control over northern Cyprus, had the *positive obligation* to prevent human rights violations, regardless of by whom they were committed. Indeed this is exactly what the Court itself says—'[t]he *obligation to secure*, in such an area, the rights and freedoms set out in the Convention, derives from the fact of such control whether it be exercised directly, through its armed forces, or through a subordinate local administration.'[120] Of course, bearing in mind that Article 1 ECHR does not make an explicit

[114] *Armed Activities on the Territory of the Congo (Congo v. Uganda)*, Judgment, 19 December 2005, para. 179.

[115] *Bosnian Genocide* merits judgment, paras 166–7.

[116] Ibid., paras 393–5, 408–15.

[117] Ibid., paras 428–38.

[118] *Velasquez Rodriguez Case*, Judgment of 29 July 1988, Inter-Am.Ct.H.R. (Ser. C) No. 4 (1988), para. 172.

[119] Human Rights Committee, General Comment No. 31, para. 8; see also Nowak, above note 6, at 37–41.

[120] *Loizidou v. Turkey*, App. No. 15318/89, Judgment (preliminary objections), 23 February 1995, para. 62 (emphasis added). See also *Bankovic*, para. 65: 'the scope of Article 1, at issue in the present case, is determinative of the very scope of the Contracting Parties' positive obligations'; *R. (on the application of Al-Skeini and Others) v. Secretary of State for Defence* [2007] UKHL 26, [2008] AC 153

distinction between positive and negative obligations this interpretation cannot be fully conclusive.

If we now take a look from this perspective at just how the ICTY Appeals Chamber in *Tadic* relied on *Loizidou*, we will see that its reliance was misplaced:

A similar approach [to the test of overall control] was adopted by the European Court of Human Rights in *Loizidou v. Turkey* (although in this case the question revolved around the possible control of a sovereign State over a State entity, rather than control by a State over armed forces operating in the territory of another State). The Court had to determine whether Turkey was responsible for the continuous denial to the applicant of access to her property in northern Cyprus and the ensuing loss of control over the property. The respondent State, Turkey, *denied that the Court had jurisdiction*, on the grounds that the act complained of was not committed by one of its authorities but, rather, was attributable to the authorities of the Turkish Republic of Northern Cyprus ('TRNC'). The Court dismissed these arguments and found that Turkey was responsible. In reaching the conclusion that the restrictions on the right to property complained of by the applicant were attributable to Turkey, the Court did not find it necessary to ascertain whether the Turkish authorities had exercised 'detailed' control over the specific 'policies and actions' of the authorities of the 'TRNC'. The Court was satisfied by the showing that *the local authorities were under the 'effective overall control' of Turkey.* [121]

First, the Appeals Chamber falls into a common trap when it speaks of Turkey denying that the European Court had jurisdiction. Turkey did indeed do so, but, as explained above, the issue under Article 1 ECHR and similar jurisdiction clauses that was addressed in *Loizidou* is whether a *state*, in this case Turkey, had jurisdiction, the issue of the Court's jurisdiction only being incidental.[122]

Moreover, the Appeals Chamber mistakenly refers to the European Court being satisfied that the TRNC was under the 'effective overall control' of Turkey. That is simply not the case. The Court's test is one of control *over an area or a territory*, not control over a *non-state actor*. For example, the Court says in *Loizidou* that '[i]t is obvious from the large number of troops engaged in active duties in northern Cyprus that her army exercises effective overall control over that part of the island'.[123] Similarly, in its practice under Article 2(1) ICCPR, the Human Rights Committee has also used an 'effective control' test, but also only in relation to control over territory rather than actors.[124]

In a response to my critique of *Tadic*, Professor Cassese, the principal author of the *Tadic* opinion, concedes as much:

(hereinafter *Al-Skeini HL*), para. 64 *in fine* (per Lord Rodger): '[T]he alleged conduct of the British forces [...] had no legal consequences under the Convention, unless there was that link and the deceased were within the jurisdiction of the United Kingdom at the time. For, only then would the United Kingdom have owed them any obligation in international law to secure their rights under article 2 of the Convention.'

[121] *Tadic*, para. 128 (emphasis added).

[122] See Section 1 above.

[123] *Loizidou* (merits), para. 56.

[124] See, e.g., Concluding Observations of the Human Rights Committee: Israel, UN Doc. CCPR/C/79/Add.93 (1998), para. 10. Generally on the spatial model of jurisdiction as control over territory, areas, or places, see Chapter IV, Section 2 below.

The only point that perhaps *Tadić* did not sufficiently clarify relates to *Loizidou*: there the ECtHR inferred the finding that control over the authorities that had breached the claimant's rights was in fact exercised by Turkey from the fact that Turkey had overall control *over the whole area of northern Cyprus* (the Court stated that 'it is not necessary to determine whether a Contracting Party actually exercises detailed control over the policies and actions of the authorities in the area situated outside its national territory, since even overall control of the area may engage the responsibility of the Contracting Party concerned:' *Loizidou* (merits), at para. 56. Thus, the Court preferred to refer to control over the area (from which it inferred control over the authorities operating there) rather than directly to control over the authorities that had violated Ms Loizidou's rights.[125]

But, of course, Professor Cassese's argument just begs the question whether the European Court did indeed infer Turkish control over an actor, the TRNC, from its control over a physical area of Cyprus. These are undoubtedly two different concepts. As stated above, and in all fairness both to the ICTY Appeals Chamber and to Professor Cassese, *Loizidou* could be reasonably interpreted as setting out a standard of attribution of all acts of the TRNC to Turkey, but that is not necessarily so. A wrongful act was certainly imputed to Turkey, but the question is what that act was: the conduct of the TRNC, or Turkey's own failure to prevent human rights violations in the territory over which it had jurisdiction. There are at least three more indicators that the latter interpretation is to be preferred.

First, though the European Court does frequently use the term 'imputability' in *Loizidou*, neither in *Loizidou* nor in any of its subsequent judgments does the Court use the words 'attribution' or 'imputability' in direct connection with its effective overall control of an area test. If it truly was a test of attribution, one could expect the European Court to have said so at least once. Certainly the Court sometimes seems to be applying its own rules of attribution, even though it never clearly says that it is doing so—for example, when it speaks in *Ilascu* of the Trasnistrian separatist regime in Moldova being under the 'decisive influence' of Russia[126]—but the Court always speaks of *state jurisdiction*, not attribution, as the consequence of satisfying the effective overall control of an area test.

Secondly, the Court in *Loizidou* never mentions *Nicaragua* when it first speaks of 'effective control', nor does it mention either *Nicaragua* or *Tadic* in any of its later

[125] Cassese, above note 110, at 658, n. 17 (emphasis in original).

[126] *Ilascu*, paras 392–4, esp. para. 392: 'All of the above proves that the "MRT", set up in 1991–1992 with the support of the Russian Federation, vested with organs of power and its own administration, remains under the effective authority, or at the very least under the decisive influence, of the Russian Federation, and in any event that it survives by virtue of the military, economic, financial and political support given to it by the Russian Federation.' Again, it is unclear whether the Court considers all of the actions of Transnistria (MRT) attributable to Russia or not. Compare this, however, to the ICJ's approach in *Nicaragua*, paras 109–15, later to be affirmed in the *Genocide* case. Cf. also the European Court's judgment in *Costello-Roberts v. United Kingdom*, App. No. 13134/87, Judgment, 25 March 1993, paras 25–8, where the Court found the UK's 'responsibility to be engaged' with regard to corporal punishment in private schools, without explaining whether this was because the school's acts were *attributable* to the UK and implicated its *negative* obligations, or rather whether the UK was responsible for violating its *positive* obligations under the ECHR for failing to prevent certain acts by entities that were not attributable to it.

judgments, nor does it invoke the work of the ILC on state responsibility.[127] Surely it is impossible that the European Court was unaware of the controversy regarding these two cases. After all, *Tadic* was a truly landmark, historic judgment which provoked a lot of debate on a number of issues, while the ICTY's disagreement with the ICJ was particularly commented on—one need only recall then ICJ President Guillaume lamenting before the UN General Assembly on the ICTY's disregard of the ICJ's case law.[128] If the European Court was truly adopting an interpretation of the law of state responsibility at odds with that of the pre-eminent international court, the ICJ, it would presumably have had the courtesy of saying so, as the ICTY itself did in *Tadic*. Yet, not a peep on this matter has ever been heard from the European Court. Moreover, when *Loizidou* was delivered, not a single author (and, one might add, no ICJ President) thought that the European Court was initiating a conflict of jurisprudence with the ICJ. A cynic might observe, though, that what the ICTY attempted to do openly, the European Court did by stealth.

Finally, and most importantly, this second reading should be preferred because it is the only way of reconciling *Loizidou* with the work of both the ILC and the ICJ on state responsibility, as well as with the subsequent case law of the European Court itself, *Bankovic* in particular. Now, the reader might wonder at my sudden desire for consistency with *Bankovic*, but if there is one thing that this case clearly shows, it is how state jurisdiction is different from state responsibility. The attribution question in *Bankovic* was fairly clear—it was undoubtedly the planes of the allied NATO countries which bombed the TV station in Belgrade, not the planes of some phantom actor. The difficult question of state responsibility that was not resolved by the Court in *Bankovic* since it declared the application inadmissible was whether the bombing was attributable to NATO as a separate international legal person, or to some of its member states, or to both. However, even though the bombing was most certainly attributable to someone, the Court still could not establish that the NATO states exercised effective overall control over Serbia, and therefore had obligations under the ECHR to the people of Serbia.

We can observe the exact same situation, though without the complexity added by the presence of NATO in the equation, in the *Al-Skeini* case before English courts, and ultimately the House of Lords. There was no problem of attribution in *Al-Skeini*, at least in principle—five of the applicants were shot by British troops while on patrol in Basra, and one was killed by British troops while he was in detention. However, their Lordships found that the first five applicants, though killed by British troops, were still not under British jurisdiction within the meaning of Article 1 ECHR. The primary ground for their Lordships' decision was based on a misinterpretation of *Bankovic* as holding that the effective control of an area test cannot apply at all to a territory outside the legal space (*espace juridique*) of the

[127] Indeed, in *Ilascu* the Court actually does invoke the work of the ILC on state responsibility in relation to the question of continuing violations—see *Ilascu*, paras 320–1. It does not, however, mention the ILC's work in regard to any question of attribution.

[128] See UN Doc. A/55/PV.41, at 7. See also Judge Guillaume's speech before the Sixth Committee, at <http://www.icj-cij.org/presscom/index.php?pr=85&p1=6&p2=1&search=%22tunisia%22&PHPSESSID=95d378896721ed931bcd585cf6290445>.

ECHR.[129] The alternative ground for disposing of the *Al-Skeini* claim, which interests us here, was that the United Kingdom did not exercise effective overall control over Basra, and that consequently the United Kingdom owed the applicants no obligations under the ECHR.[130] In other words, while it was beyond doubt that the killings in *Al-Skeini* were attributable to the UK, this still did not mean that the UK had jurisdiction over the victims, or over the area of Basra. Now, one can certainly claim that this approach is unsatisfactory, as I do, since there is no cogent reason to impose the jurisdiction threshold on a negative state obligation to refrain from doing harm.[131] What is clear, however, is that state jurisdiction and attribution are distinct concepts. Ultimately, the latter is an issue of state control over the *perpetrators* of human rights violations, while the former is a question of a state's control over the *victims* of such violations through its agents,[132] or, more generally, control over the territory in which they are located.

D. Attribution as a prerequisite for jurisdiction

As we have seen, within the conceptual framework of Article 2 ILC ASR, for a state to be responsible for a wrongful act the conduct in question must be attributable to it and amount to a breach of its international obligations—in our case a violation of a human rights treaty. The existence of state jurisdiction is a threshold question for the existence of (at least some) of the obligations under the relevant treaties. Likewise, both attribution and state jurisdiction are issues that go into the substance of a state's responsibility, and are properly not questions of a *court's* jurisdiction or the claim's admissibility, though the court's jurisdiction *ratione personae*, *materiae*, or *loci* depends on them.

As a matter of principle or logic, there is no hierarchical relationship between the issues of attribution and state jurisdiction—they are conceptually independent of each other. In some cases, however, attribution can actually be a prerequisite or a preliminary question for the existence of state jurisdiction. When a state exercises jurisdiction, i.e. control over a foreign territory or individuals, it by definition needs to do so through its own agents, i.e. persons whose acts are attributable to it. Turkey could not have had jurisdiction over northern Cyprus without having its soldiers there, nor could Russia have had jurisdiction over a part of Moldova without a military presence. For example, had it been disputed in *Loizidou* that Turkey had soldiers at all in northern Cyprus, the Court would first have had to

[129] *Al-Skeini HL*, paras 71–7 (per Lord Rodger), paras 109, 127 (per Lord Brown). See further below, Chapter III, Sections 4 and 6.

[130] Ibid., paras 81–4 (per Lord Rodger), para. 132 (per Lord Brown).

[131] See below, Chapter IV, Section 4. See also De Schutter, above note 52; J. Cerone, 'Jurisdiction and Power: The Intersection of Human Rights Law and the Law of Non-International Armed Conflict in an Extraterritorial Context', (2007) 40 *Israel L Rev* 72.

[132] See *Al-Skeini HL*, para. 64 (per Lord Rodger): 'It is important therefore to recognise that, when considering the question of jurisdiction under the Convention, the focus has shifted to the victim or, more precisely, to the link between the victim and the contracting state.'

establish whether this was the case before examining whether Turkey had control over that part of the island.[133]

In that sense, attribution of the conduct of Turkish troops in Cyprus to Turkey was a question that logically preceded the issue of Turkey's jurisdiction over northern Cyprus. Note, however, that the conduct attributed is not necessarily the act or omission that constitutes the alleged human rights violation, e.g. a taking of life or property. If the case is examined through the prism of the spatial model of jurisdiction as state control over an area, the conduct attributed is that through which the state exercises such control, e.g. through the establishment of a military government.[134] If, on the other hand, the case is examined under the personal model of jurisdiction as state authority and control over an individual, then there may well be significant, even total overlap between the state conduct through which it exercises such authority and control and the conduct which allegedly constitutes the human rights violation—e.g. an arbitrary deprivation of life or liberty.[135]

Likewise, once jurisdiction over an area is established, it does not imply attribution in the sense that anything that occurs within a state's jurisdiction is attributable to it. It would still be necessary to establish that the particular act that is alleged to be a human rights violation is attributable to the state. Or, even if the act in question is not attributable to the state, its responsibility may also arise for its failure to implement positive obligations under human rights treaties, e.g. to prevent human rights violations even by third parties.

4. Conclusion

Conceptual confusion frequently stems from the use of words which have several distinct meanings, often across different branches of the law. That is the case with the word 'jurisdiction' itself, which, as we have seen, has more meanings than one can count. 'Effective control' is also a homonym—there is the effective control test for the purposes of attribution, as developed by the ICJ in *Nicaragua*; there is

[133] See, in that regard, *Drozd and Janousek v. France and Spain* , App. No. 12747/87, Judgment, 26 June 1992, where the Court found, *inter alia*, that France and Spain could not be held responsible for the acts of their judges which they had seconded to Andorra. In effect, the Court considered that the acts of the judges were not attributable to the sending states. This is indeed the customary rule now codified in Art. 6 ILC ASR, stating that '[t]he conduct of an organ placed at the disposal of a State by another State shall be considered an act of the former State under international law if the organ is acting in the exercise of elements of the governmental authority of the State at whose disposal it is placed'. This is, in my view, how the case should be read (confusing though it is in some respects), and in fact the ILC cites *Drozd* in its commentary on Art. 6—see ILC ASR Commentary, at 98, n. 141.

[134] See also the *Behrami* case, where the issue of whether the conduct of the French contingent in KFOR in Kosovo was still attributable to France was logically prior to the issue of whether France exercised jurisdiction, i.e. effective overall control, over a part of Kosovo—see *Behrami and Behrami v. France, Saramati v. France, Germany and Norway* [GC] (dec.), App. Nos 71412/01 and 78166/01, 2 May 2007. See further M. Milanovic and T. Papic, 'As Bad As It Gets: The European Court of Human Rights' *Behrami and Saramati* Decision and General International Law', (2009) 58 *ICLQ* 267, at 271–4.

[135] See, e.g., *Gentilhomme, Schaff-Benhadji et Zerouki c. France*, App. Nos 48205/99, 48207/99, and 48209/99, Judgment, 23 April 2002.

'effective control' as sometimes used in humanitarian law to describe the threshold of the beginning of a belligerent occupation of a territory; there is effective (overall) control of an area as a test developed by the European Court for the purpose of determining a state's jurisdiction over territory; there is also effective control as used in international criminal law to describe the relationship a superior has to have over a subordinate so his command responsibility can be engaged.[136]

There is in principle nothing wrong with the *Bankovic* approach to interpreting the notion of state jurisdiction in Article 1 ECHR by reference to general international law. Again, I am not arguing that the word 'jurisdiction' should be given a special meaning autonomous to human rights law. Rather, the word has several different and equally ordinary meanings in general international law itself, and the question is hence which of these meanings—which of these *concepts*—the jurisdiction clauses of human rights treaties refer to.

My argument is simply that as a matter of semantic, textual interpretation the word 'jurisdiction' in various human rights treaties refers to a power that a state exercises over a territory, and perhaps also over individuals. When the state obtains this power it must, with due diligence, fulfil its obligation to secure or ensure the human rights of all persons within its jurisdiction. This power is a question of fact, of actual authority and control. Despite its name, it is not a legal competence, and it has absolutely nothing to do with that *other* notion of jurisdiction in international law which delimits the municipal legal systems of states. It is moreover not directly related to the concept of attribution in the law of state responsibility, even though both jurisdiction and attribution can be based on the same set of facts.

The first step in transforming a jurisprudence of compromise into a jurisprudence of principle is to agree on a common vocabulary. Only when it is clear *what* the courts are doing can we venture into the far more important issues of *why* and *whether* they should continue doing so.

[136] See, e.g., *Prosecutor v. Sefer Halilović* , Case No. IT-01-48-A, Appeals Chamber, Judgment, 16 October 2007, para. 59

III

Policy Behind the Rule

1. Introduction

The previous chapter of this study was dedicated mainly to establishing the semantic, ordinary meaning of the word 'jurisdiction' used in human rights treaties. That interpretative inquiry is not yet complete, as one major question still remains to be answered: whether the word 'jurisdiction' is used to connote solely a state's control over a territory, or also a state's authority and control over an individual, even when that individual is not located in a territory that is, as such, under the state's control. The relevance of that question, of course, is in the fact that it is easy to conceive of situations where a state exercises power over a person in a territory not under its control—say when Israeli agents abducted Adolf Eichmann from Argentina, when Russian agents allegedly poisoned Alexander Litvinenko in London, or when US drones engaged in targeted killings in Pakistan. As will be explained in the next chapter of this study, there are several possible models of extraterritorial application under the jurisdiction clauses of the various treaties, and textual interpretation alone does not suffice to opt for one of them.

This brings us from the realm of interpretation pure and simple, designed to establish the ordinary linguistic meaning of a legal text, to the realm of construction, the activity of translating that text, in light foremost of its object and purpose, into workable legal rules in cases where the text itself is vague and does not provide a clear answer.[1] That type of teleological inquiry into object and purpose, however, is impossible without first examining the policy considerations that favour or disfavour the extraterritorial application of human rights treaties, generally or in any given case, and either normatively or purely as a matter of fact. Indeed, I submit that it is the tension between these policy considerations that is the real cause of conflicts in the case law, particularly the case law of the European Court of Human Rights. It was not *really* the Court's failure to distinguish between the different

[1] On the interpretation/construction distinction, see L. Solum, *Semantic Originalism* (2008), available at <http://ssrn.com/abstract=1120244>, at 67–89. Of course, Art. 31(1) VCLT does not make a distinction between interpretation and construction, stating only that a 'treaty shall be interpreted in good faith in accordance with the ordinary meaning to be given to the terms of the treaty in their context and in the light of its object and purpose'. The distinction, however, is useful to highlight the difference between the nature of the inquiry in the previous chapter of this study and the one in those that follow. The VCLT does of course distinguish between interpretation and application—see further R. Gardiner, *Treaty Interpretation* (Oxford University Press, 2008), at 26–9.

meanings and concepts of 'jurisdiction' that produced *Bankovic*. Rather, *Bankovic* was the result of the Court's less than transparent weighing of different policy considerations, and its ultimate desire to come up with a superficial, legalistic rationale that would justify making the extraterritorial application of the ECHR *exceptional*. As we will see, many of these same policy considerations also inform the debates on the extraterritorial applicability of *domestic* human rights guarantees, which will to an extent also be examined in this part.

A legal analysis would be pointless if it did not examine the policy behind the rule, but it would also be pointless if it added nothing more. What I will attempt to establish is how these considerations of policy play out and influence the case law, and what lessons can be drawn from this. As a normative matter, the law certainly does not value each of these policy considerations equally, as, again, human rights treaties are not value-neutral instruments. As we will see, courts have developed their case law on the extraterritorial application of human rights treaties while acting out of universalist aspirations that are deeply embedded in the structure of human rights law. Yet these aspirations have almost always been tempered with considerations of effectiveness, since a law that would stray so far into pure normativity that it completely disregarded the realities of international relations would serve no useful purpose—especially so when it is courts, rather than states among themselves, which determine what a treaty really means. The law is thus constantly torn between universality and effectiveness, and niceties of legal principle are often lost, perhaps inevitably so, in the making of a judicial compromise between the two.[2]

2. Universality and Human Dignity

Every single applicant who demands protection against extraterritorial state action makes an appeal to universality. This, of course, makes perfect sense. If it is true, as the Universal Declaration of Human Rights exhorts us in its preamble, that 'the recognition of the inherent dignity and of the equal and inalienable rights of all members of the human family is the foundation of freedom, justice and peace in the world' and that states have pledged themselves to achieve 'the promotion of universal respect for and observance of human rights and fundamental freedoms', why should it then matter whether a state violates a person's rights within its territory or outside it? It is still exercising power over the individual, and there is no reason why that individual should be completely unprotected against the arbitrary exercise of that power, solely on the basis of his or her location.[3]

[2] Cf. M. Koskenniemi, *From Apology to Utopia: The Structure of International Legal Argument* (Cambridge University Press, 2005), at 16 *et seq.*

[3] See, e.g., R. Lawson, 'Life after Bankovic: On the Extraterritorial Application of the European Convention on Human Rights', in Coomans and Kamminga 83, at 86; O. Ben-Naftali and Y. Shany, 'Living in Denial: The Application of Human Rights Treaties in the Occupied Territories', (2003) 37 *Israel L Rev* 17, at 61–2; T. Meron, 'Extraterritoriality of Human Rights Treaties', (1995) 89 *AJIL* 78, at 80–1:

Universality and human dignity are undoubtedly the foundational principles of international human rights law, affirmed in binding treaties and non-binding declarations alike.[4] But universality and human dignity are concepts that are too amorphous,[5] in and of themselves, to overcome the threshold criterion of state jurisdiction that is found in most human rights treaties. The treaties say what they say, and the language of their jurisdiction clauses cannot be overridden merely by invoking universality. This is so not just because of the text, but also because these provisions have a purpose—they limit the obligations of contracting states to those situations where these obligations could realistically be met and human rights effectively protected.

Universality in its purest form—that all states have direct and enforceable human rights obligations vis-à-vis all individuals in the world—has the drawback of appearing hopelessly utopian. Whether and how universality can be reconciled with effectiveness is an issue that we will return to time and again. At the very least, however, universality, so clearly embraced by all the major treaties, creates a baseline, a default position: it is the wholesale *denial* of any rights to an individual affected by state action, not the *extension* of those rights, that must be justified. This is not to say that there is in international law a formal presumption of extraterritorial applicability of human right treaties.[6] The relevance of this baseline becomes apparent once rights are recognized or denied on the basis of criteria such as citizenship or regionalism, when these criteria themselves have to be tested and justified.

In the case law universality generally emerges as a response to attempts to curtail the scope of application of human rights treaties. Although, as we will see, courts will invoke other considerations, such as sovereignty or citizenship, when *denying* that individual rights guarantees apply extraterritorially, they will invoke universality when *accepting* that such guarantees do indeed apply. It is, in other words, a moral appeal, an appeal to conscience that courts will always turn to in justifying an expansive approach to rights protection.[7]

> In view of the purposes and objects of human rights treaties, there is no a priori reason to limit a state's obligation to respect human rights to its national territory. Where agents of the state, whether military or civilian, exercise power and authority (jurisdiction, or *de facto* jurisdiction) over persons outside national territory, the presumption should be that the state's obligation to respect the pertinent human rights continues. That presumption could be rebutted only when the nature and the content of a particular right or treaty language suggest otherwise.

[4] Similarly, see C. Keitner, 'Rights Beyond Borders', (2011) 36 *Yale J. Int'l L.* (forthcoming), draft available on SSRN at <http://ssrn.com/abstract=1480886>, esp. at 7 *et seq.*

[5] On the largely nebulous—and instrumentalist—role of 'dignity' in human rights adjudication, see especially C. McCrudden, 'Human Dignity and Judicial Interpretation of Human Rights', (2008) 19 *EJIL* 655.

[6] See above Chapter I, Section 3, esp. note 38.

[7] This is apparent, for example, from the Human Rights Committee's jurisprudence, starting from *Lopez Burgos v. Uruguay* (1981) 68 ILR 29, para. 12.3: 'it would be unconscionable to so interpret the responsibility under article 2 of the Covenant as to permit a State party to perpetrate violations of the Covenant on the territory of another State, which violations it could not perpetrate on its own territory'.

Perhaps somewhat surprisingly, universality also has a darker, less human rights friendly side, at least from the standpoint of the individual applicants. It is commonly said that a corollary of the universality of human rights is their indivisibility and interdependence,[8] and the European Court deftly used these aspects of universality against the applicants in the *Bankovic* case.

The applicants argued that the notion of jurisdiction in Article 1 ECHR was flexible, in that it did not require in all circumstances a level of state control necessary to secure the respect of all rights guaranteed by the ECHR. In the applicants' view, the fact that the respondent states could not possibly be expected to, say, secure the freedom of religion or the right to free and fair elections to the inhabitants of Serbia that they were bombing did not mean that the respondent states were absolved from *any* human rights obligations, most notably their obligation to respect the applicants' right to life. The Court disagreed, holding that

> ... the applicants' submission is tantamount to arguing that anyone adversely affected by an act imputable to a Contracting State, wherever in the world that act may have been committed or its consequences felt, is thereby brought within the jurisdiction of that State for the purpose of Article 1 of the Convention. The Court is inclined to agree with the Governments' submission that the text of Article 1 does not accommodate such an approach to "jurisdiction". Admittedly, the applicants accept that jurisdiction, and any consequent State Convention responsibility, would be limited in the circumstances to the commission and consequences of that particular act. However, the Court is of the view that the wording of Article 1 does not provide any support for the applicants' suggestion that the positive obligation in Article 1 to secure "the rights and freedoms defined in Section I of this Convention" can be divided and tailored in accordance with the particular circumstances of the extra-territorial act in question and, it considers its view in this respect supported by the text of Article 19 of the Convention. Indeed the applicants' approach does not explain the application of the words "within their jurisdiction" in Article 1 and it even goes so far as to render those words superfluous and devoid of any purpose.[9]

The Court in *Bankovic* thus adopted an all or nothing conception of the extraterritorial application of the ECHR—though, as we will see, the Court's case law is actually neither very strict nor consistent in applying this approach. The idea apparently behind it is that the indivisibility of human rights, as an aspect of their universality, is essential for protecting the *integrity* of the human rights regime. In other words, the Court is concerned that dividing and tailoring human rights in an extraterritorial context would serve to impermissibly dilute them, and diminish the regime as a whole. To what extent that concern is justified remains to be seen, but first we need to examine some of the responses to universality.

[8] See, e.g., para. 5 of the Vienna Declaration of the World Conference on Human Rights, which proclaims that '[a]ll human rights are universal, indivisible and interdependent and interrelated'. Vienna Declaration and Programme of Action, UN Doc. A/CONF.157/23, 12 July 1993. Note that the indivisibility and interdependence of human rights are not a *necessary* corollary of universality, but these are ideas which are frequently associated.

[9] *Bankovic and Others v. Belgium and Others* [GC] (dec.), App. No. 52207/99, 12 December 2001, para. 75.

3. Sovereignty and Territory

The most common argument against extending fundamental rights guarantees to persons outside a state's borders is that these rights were meant to operate between a state and the inhabitants of its own territory. This argument has figured most prominently in the post-9/11 debates and litigation on the treatment and process that is due to suspected terrorists detained in Guantanamo, in Afghanistan, or in Iraq. Indeed, the motivation of the Bush administration behind the creation of a detention facility in Guantanamo, as opposed to somewhere in the United States proper, was precisely that formal categories such as sovereignty and territory would diminish or extinguish the government's obligations towards persons detained there, both under international and under domestic law.[10]

The United States has in particular relied on a restrictive interpretation of Article 2(1) ICCPR, which provides that each state party 'undertakes to respect and to ensure to all individuals within its territory and subject to its jurisdiction the rights recognized in the present Covenant'. According to the United States, the language of this provision is clear, as it requires that an individual be *both* within a state's territory *and* subject to its jurisdiction in order to have rights vis-à-vis that state—an interpretation that is in the US view supported by the ICCPR's negotiating history. Since Guantanamo is a territory under the sovereignty of Cuba that has only been leased by Cuba to the United States, it is not US territory for the purposes of Article 2(1) ICCPR, thus rendering the treaty inapplicable.[11]

The US government made the same argument with regard to the CAT, though here its textual argument was significantly weakened by the fact that the CAT's many jurisdiction clauses (for example, in Articles 2, 5, and 16) designate the scope of each party's obligations to 'any territory under its jurisdiction'. Unlike with the ICCPR, the word 'jurisdiction' in the CAT refers to territory itself, and as explained above, the ordinary meaning of this word is broader than legal title or sovereignty and relates to factual control over territory. In respect of the CAT, therefore, the US

[10] See Memorandum from Patrick F. Philbin and John C. Yoo, Deputy Assistant Attorneys General, Office of Legal Counsel, to William J. Haynes II, General Counsel, Department of Defense, 28 December 2001, available in K. Greenberg et al. (eds), *The Torture Papers* (Cambridge University Press, 2005), at 29 *et seq.*

[11] See, e.g., the Opening Statement of Matthew Waxman, Head of US Delegation before the UN Human Rights Committee, 17 July 2006, available at <http://2001-2009.state.gov/g/drl/rls/70392.htm>; Reply of the Government of the United States of America to the Report of the Five UN Special Rapporteurs on Detainees in Guantanamo Bay, Cuba (2006), available at <http://www.asil.org/pdfs/ilib0603212.pdf>, at 25 *et seq.* For a detailed defence of the US argument, see M. Dennis, 'Application of Human Rights Treaties Extraterritorially in Times of Armed Conflict and Military Occupation', (2005) 99 *AJIL* 119, and M. Dennis and A. Surena, 'Application of the International Covenant on Civil and Political Rights in Times of Armed Conflict and Military Occupation: The Gap Between Legal Theory and State Practice', (2008) *EHRLR* 714; for a rebuttal see N. Rodley, 'The Extraterritorial Reach and Applicability in Armed Conflict of the International Covenant on Civil and Political Rights: a Rejoinder to Dennis and Surena', (2009) *EHRLR* 628; N. Lubell, *Extraterritorial Use of Force Against Non-State Actors* (Oxford University Press, 2010), at 195 *et seq.*

government relied primarily on its argument that the law of armed conflict as *lex specialis* displaced any norms of human rights law.[12]

The position that human rights treaties generally have no scope of extraterritorial application was rejected both by the Human Rights Committee[13] and by the Committee Against Torture,[14] while the International Court of Justice also affirmed that the ICCPR may apply extraterritorially, particularly in territories under belligerent occupation.[15] Of course, authority is not a substitute for argument, and the fact that the Human Rights Committee and the ICJ adopted a position does not necessarily make that position correct. The US textual argument in respect of the ICCPR is a serious one and needs to be examined more fully, as it will be in Chapter IV of this study.[16]

As far as the US argument pertains to the CAT or human rights treaties generally it is untenable, not just on the text of the various jurisdiction clauses, but more importantly in the policy behind the argument—or the lack of any policy recognized by international law as legitimate. It is one thing to say that, universality and human dignity notwithstanding, states should not have obligations under human rights law that they could not realistically be expected to keep. But when such obligations *can* be realistically kept, i.e. when the state exercises control over a territory even if the sovereignty over the territory nominally belongs to someone else, why should the *title* over the territory matter?[17] This is especially so, of course, for situations that were manifestly created with the purpose of evading otherwise applicable legal rules.[18]

Certainly, sovereignty and territory have always had a central place in international law. As noted by Oppenheim, 'a State without a territory is not possible ... it is the space within which the State exercises its supreme, and normally exclusive, authority'.[19] But sovereignty has equally never been a matter of naked title or pure form. Above all else, its source is in the *effectiveness* of state power over a territory and its inhabitants. At the same time, as explained by Judge Huber serving as

[12] See, e.g., the Opening Remarks by John Bellinger, Legal Adviser, US Department of State, before the UN Committee Against Torture, 5 May 2006, available at <http://www.state.gov/g/drl/rls/68557.htm>.

[13] Concluding Observations of the Human Rights Committee: United States of America, UN Doc. CCPR/C/USA/CO/3/Rev.1, 18 December 2006, para. 10.

[14] Conclusions and Recommendations of the Committee against Torture: United States of America, UN Doc. CAT/C/USA/CO/2, 25 July 2006, paras 14 and 15.

[15] *Legal Consequences of the Construction of a Wall in the Occupied Palestinian Territory,* Advisory Opinion, 9 July 2004, ICJ Reports 2004, at 136, paras 109–11; *Armed Activities on the Territory of the Congo (Congo v. Uganda)*, Judgment, 19 December 2005, ICJ Reports 2005, at 168, paras 179, 216–17.

[16] See Chapter IV, Section 5 below.

[17] The term 'sovereignty' can be used to connote either the actual authority and control that a state possesses over a territory, or its mere title or ownership over the territory, and it is the latter that should in my view be perfectly irrelevant for the application of human rights treaties. On these two meanings of 'sovereignty', see generally R. Wilde, *International Territorial Administration: How Trusteeship and the Civilizing Mission Never Went Away* (Oxford University Press, 2008), at 99–101 and the sources cited therein.

[18] See also Section 7 below.

[19] R.Y. Jennings and A.D. Watts (eds), *Oppenheim's International Law* (9th edn, 1992), at 563–4.

arbitrator in the *Island of Palmas* case, international law does not protect sovereignty for its own sake:

> Territorial sovereignty, as has already been said, involves the exclusive right to display the activities of a State. This right has as corollary a duty: the obligation to protect within the territory the rights of other States, in particular their right to integrity and inviolability in peace and in war, together with the rights which each State may claim for its nationals in foreign territory. Without manifesting its territorial sovereignty in a manner corresponding to circumstances, the State cannot fulfil this duty. Territorial sovereignty cannot limit itself to its negative side, i.e. to excluding the activities of other States; for it serves to divide between nations the space upon which human activities are employed, in order to assure them at all points the minimum of protection of which international law is the guardian.[20]

In its *Namibia* advisory opinion, the ICJ likewise emphasized that

> The fact that South Africa no longer has any title to administer the Territory does not release it from its obligations and responsibilities under international law towards other States in respect of the exercise of its powers in relation to this Territory. Physical control of a territory, and not sovereignty or legitimacy of title, is the basis of State liability for acts affecting other States.[21]

When a state hides behind its lack of sovereignty over a territory that it nonetheless controls to deny the applicability of human rights treaties to the inhabitants of this territory it is actually turning the notion of sovereignty on its head, by assigning legal significance to a sovereignty deprived of all effectiveness. The idea that, rather than the United States, it is Cuba that has to guarantee the rights of persons detained at Guantanamo Bay, does not pass muster even on its own merits, let alone when confronted with the universality imperative behind all human rights treaties.

International law is a territorial system, as its principal subjects, states, are by their very definition territorial entities. But there is nothing magical about state sovereignty. Title or no title, its foundation is power, 'physical control over territory'. When a state exercises such power outside its borders, something that states have done throughout history, the mere lack of sovereignty on its part is not an excuse for denying the application of a human rights treaty, particularly because the text of the treaties (with the possible exception of the ICCPR) does *not* refer to title, but to jurisdiction over territory. Again, while it is perfectly true that the Westphalian international order rests on territoriality, and that international law provides for the authority of states to administer their territory and be free of outside interference,[22] it makes no sense to condition respect for human rights on the *lawfulness* of a state's control over territory, rather than on the *fact* of such control, and even less so to actually require *title* over the territory in question.

[20] *Island of Palmas Arbitration*, (1949) 2 *UNRIAA* 829, 839.

[21] *Legal Consequences for States of the Continued Presence of South Africa in Namibia (South West Africa) notwithstanding Security Council Resolution 276 (1970)*, Advisory Opinion, ICJ Reports 1971, 16, at 54, para. 118.

[22] See, e.g., K. Raustiala, *Does the Constitution Follow the Flag? The Evolution of Territoriality in American Law* (Oxford University Press, 2009), at 5–12.

Imagine, moreover, if the applicability of a human rights treaty was raised with regard to a territory whose title is disputed between two states or is even truly uncertain, as is frequently the case. Would a court examining the issue actually have addressed as a preliminary question the merits of the territorial dispute, rather than focus on the state which actually controls the territory, as a matter of fact? Surely not.

In short, those who argue that title over territory should matter for determining the scope of human rights need to give a good *normative* reason as to why this should be the case. Invoking Westphalia just does not cut it. To my mind at least, it is unimaginable that in the wake of the Second World War states could have created a regime in which Nazi Germany would have had no human rights obligations towards the subjugated millions of Poland, Russia, or the rest of occupied Europe, merely because it did not have legal title over the territories that it controlled, when it is the slaughter of those same millions which gave birth to the human rights movement in the first place. Hence, though this study is devoted to the *extraterritorial* application of human rights treaties, this is to an extent a misnomer: the application of these treaties should *never* depend on naked title over a territory, but on actual power exercised over it.

4. Sovereignty and Comity to the Territorial State

A. The sovereignty of Iraq in *Al-Skeini*

Sovereignty has given rise to one more argument for curtailing the extraterritorial scope of human rights treaties: that requiring state agents to respect their human rights obligations while acting in the territory of another state would amount to an infringement on the sovereignty of that state. This argument is at least partly based on the fallacy, discussed above,[23] that the notion of state jurisdiction in human rights treaties is the same concept of jurisdiction that delimits the municipal legal systems of states, rather than on factual power over territory. As we have seen, the greatest exponent of this fallacy is the European Court in *Bankovic*, where it stressed that the extraterritorial application of the ECHR can only be exceptional, and purportedly needs to be justified on the grounds of general international law. But the European Court is not alone. In the House of Lords' decision in *Al-Skeini*, the question presented was whether the United Kingdom exercised jurisdiction over Basra in the sense of Article 1 ECHR. Lord Carswell considered as follows:

> I would only observe that any extraterritorial jurisdiction of one state is *pro tanto* a diminution or invasion of the territorial jurisdiction of another, which must lead one to the conclusion that such extra-territorial jurisdiction should be closely confined. It clearly exists by international customary law in respect of embassies and consulates. It has been conceded by the Secretary of State that it extends to a military prison in Iraq occupied and controlled by agents of the United Kingdom. Once one goes past these categories, it would

[23] See Chapter II, Section 2.

in my opinion require a high degree of control by the agents of the state of an area in another state before it could be said that that area was within the jurisdiction of the former. The test for establishing that is and should be stringent, and in my judgment the British presence in Iraq falls well short of that degree of control.[24]

In other words, according to his Lordship (and pursuant to his reading of *Bankovic*), it would be an *intrusion upon Iraq's sovereignty* to apply the ECHR to the British troops stationed there. With respect, this position shows just how misguided it is to treat the concept of 'jurisdiction' in the ECHR as being identical to the notion of jurisdiction in general international law. As if Iraq's sovereignty was not already infringed by an invasion and occupation of its territory by hundreds of thousands of foreign troops—it would actually be the imposition of human rights obligations on the occupiers that would violate it. This is not a position that is in my view tenable, though, as we will see, this is not to say that it is not inspired by some very real fears and concerns.

B. Canadian case law on sovereignty and comity

Similar considerations have informed some municipal case law on the extraterritorial application of domestic human rights guarantees. The most instructive example in that regard is the Canadian jurisprudence on the extraterritorial application of the Canadian Charter of Rights and Freedoms.[25] Like most constitutions, and some human rights treaties, the Charter contains no provision dealing with its territorial scope of application. All it says in s. 32(1) is that it applies to governmental action, be it the action of the federal government, that of the provinces, or of the local authorities.[26] Rules on the Charter's extraterritorial application thus have to be constructed by Canadian courts. The first case that I will examine is *Hape* before the Supreme Court of Canada,[27] in importance (and internal consistency) arguably the Canadian *Bankovic*.

The facts of the case were as follows. The Royal Canadian Mounted Police (RCMP) were conducting a money laundering investigation against the appellant. The appellant ran a company in the Turks and Caicos Islands, the offices of which the RCMP wished to search as part of their investigation. They obtained permission to do so from the Turks and Caicos official in charge of all police investigations. They conducted three separate searches, and were at all times accompanied by the said Turks and Caicos official. However, the RCMP did not have a judicial warrant for these searches, either from a Canadian court or from a court in the Turks and Caicos. When documents seized in these searches were used against him

[24] *R. (on the application of Al-Skeini and Others) v. Secretary of State for Defence* [2007] UKHL 26 [2008] AC 153 (hereinafter *Al-Skeini HL*), para. 97.

[25] For a general overview, see Keitner, above note 4, at 26 *et seq.*

[26] S. 32(1) reads as follows: 'This Charter applies (a) to the Parliament and government of Canada in respect of all matters within the authority of Parliament including all matters relating to the Yukon Territory and Northwest Territories; and (b) to the legislature and government of each province in respect of all matters within the authority of the legislature in each province.'

[27] *R. v. Hape*, 2007 SCC 26 (hereinafter *Hape*).

at his trial in Canada, the appellant argued that the searches violated his Charter right to be free from unreasonable search and seizure, and that the documents should be excluded. The question thus presented to the Supreme Court was whether the Charter applied to the acts of Canadian organs or agents when they were acting extraterritorially.

Justice LeBel began the discussion for the majority as follows:

> Section 32 does not expressly impose any territorial limits on the application of the *Charter*. By virtue of state sovereignty, it was open to the framers to establish the jurisdictional scope of the *Charter*. Had they done so, the courts of this country would have had to give effect to a clear expression of that scope. However, the framers chose to make no such statement. Consequently, as with the substantive provisions of the *Charter*, it falls upon the courts to interpret the jurisdictional reach and limits of the *Charter*. Where the question of application involves issues of extraterritoriality, and thereby necessarily implicates interstate relations, the tools that assist in the interpretation exercise include Canada's obligations under international law and the principle of the comity of nations. As I will explain, the issue of applying the *Charter* to activities that take place abroad implicates the extraterritorial enforcement of Canadian law. The principles of state jurisdiction are carefully spelled out under international law and must guide the inquiry in this appeal.[28]

Emphasizing the importance of the principle of sovereign equality in international law,[29] of comity as an interpretative principle,[30] and of the need for domestic law to conform to international law,[31] Justice LeBel held that the extraterritorial reach of the Charter must be informed by these considerations, most notably by international legal rules on jurisdiction.[32] After reviewing earlier case law,[33] Justice LeBel disapproved of the Court's previous position that the Charter will apply to acts of Canadian law enforcement authorities engaged in governmental action where the application of Charter standards will not conflict with the concurrent territorial jurisdiction of the foreign state,[34] and concluded that

> [t]he issue in these cases is the applicability of the *Charter* to the activities of Canadian officers conducting investigations abroad. The powers of prescription and enforcement are both necessary to application of the *Charter*. The *Charter* is prescriptive in that it sets out what the state and its agents may and may not do in exercising the state's powers. Prescription is not in issue in the case at bar, but even so, the *Charter* cannot be applied if compliance with its legal requirements cannot be enforced. Enforcement of compliance with the *Charter* means that when state agents act, they must do so in accordance with the requirements of the *Charter* so as to give effect to Canadian law as it applies to the exercise of the state power at issue. However, as has already been discussed, Canadian law cannot be enforced in another state's territory without that state's consent. Since extraterritorial enforcement is not possible, and enforcement is necessary for the *Charter* to apply, extraterritorial application of the *Charter* is impossible.[35]

[28] Ibid., para. 33 (per LeBel J).
[29] Ibid., paras 40–6 (per LeBel J).
[30] Ibid., paras 47–52 (per LeBel J).
[31] Ibid., paras 53–6 (per LeBel J).
[32] Ibid., paras 57–65 (per LeBel J).
[33] Ibid., paras 70–82 (per LeBel J).
[34] Ibid., paras 83–4 (per LeBel J), citing *R. v. Cook* [1998] 2 SCR 597.
[35] Ibid., para. 85 (per LeBel J).

As noted by a commentator, if you are struggling to digest the logic of this passage, you are not alone.[36] Recall the fundamental distinction between the jurisdiction to enforce, which can extraterritorially only be exercised lawfully with the territorial state's consent, and the jurisdiction to prescribe, the extraterritorial exercise of which is in no way exceptional.[37] Contrary to what the learned Justice says, it is precisely *prescription* that is at issue in the case at bar, if the Charter's scope of application is to be determined by reference to international rules on the scope of domestic law generally. For a state to prescribe that its organs and agents must respect the human rights of persons outside the state's borders is no more an exorbitant exercise of the jurisdiction to prescribe than, say, if the state's prohibition of murder extended to the state's nationals or agents abroad. The *enforcement* of this prescription, on the other hand, is limited to the state's own territory and its own courts and organs. Such were the circumstances of the present case, as the appellant sought to enforce the provisions of the Charter in Canada before a Canadian court, not a court of the Turks and Caicos.

The mere application of a law, the Charter, to an extraterritorial situation is not enforcement, but prescription. The sovereignty of the Turks and Caicos was only implicated in the extraterritorial act of the search itself, to which the officials of the Turks and Caicos consented.[38] Had Canadian law actually commanded the police or other officials of the Turks and Caicos to act in a certain way, that indeed would have been an exorbitant exercise of the jurisdiction to prescribe, and a violation of the sovereignty of the territorial state. But, again, international law simply does not require the consent of the territorial state when another state is merely applying its law to its own officials or nationals who are acting extraterritorially.[39]

Justice LeBel added yet another twist to the territorial scope of the Charter, by leaving open the possibility of its application if Canada's agents acting abroad violated a human rights treaty to which Canada is a party.[40] But of course it is the extraterritorial applicability of these very human rights treaties that is so hotly contested—not to mention the fact that a situation where the territorial scope of a state's constitution is defined by the scope of treaties to which the state is a party, and not by the constitution itself, can only be qualified as bizarre. However, the Court's purpose in adopting this approach seems to be obvious—it wanted to keep its options open in cases in which Canadian agents operating abroad did something more morally repugnant than a mere warrantless search and seizure.[41]

This methodologically unhappy state of affairs has persisted in subsequent Canadian case law. In the *Afghan Detainees* case, two non-governmental organizations brought an application for judicial review on behalf of non-Canadian

[36] J. Stribopoulos, 'The Charter's Unstated Territorial Limits: *R. v. Hape*', *The Court*, 8 June 2007, available at <http://www.thecourt.ca/2007/06/08/the-charters-unstated-territorial-limits-r-v-hape/>.

[37] See above, Chapter II, Section 2.B.

[38] See also Stribopoulos, above note 36; P.H. Verdier, '*R. v. Hape*', (2008) 102 *AJIL* 143, 147; Keitner, above note 4, at 30–1.

[39] Cf. *Hape*, paras 106–7 (per LeBel J).

[40] Ibid., para. 101 (per LeBel J).

[41] In that regard, see the discussion of the *Khadr* case at Section 5.C. below.

nationals detained by Canadian forces in Afghanistan, claiming that these detainees were at significant risk of being subjected to torture if transferred by the Canadian forces to Afghan authorities. In its decision,[42] the Federal Court applied the *Hape* judgment of the Supreme Court, and asked the following question: 'whether the Government of Afghanistan has consented to the application of Canadian law, including the Charter, to the conduct of Canadian Forces personnel in relation to the detention of individuals on Afghan soil'.[43] It was indisputable that the Afghan government had consented to the presence and operation of Canadian troops within its territory. However, according the Court, there was no evidence that the Afghan government had agreed to the 'wholesale forfeiture of its sovereignty', that the application of Canadian law would presumably entail, by consenting to the presence of foreign troops.[44] In particular, 'there has been no consent by the Government of Afghanistan to having Canadian Charter rights conferred on its citizens, within its territory.'[45] The Court thus concluded that Afghan detainees in Canadian custody were protected solely by international humanitarian law, not Canadian law. The decision of the Federal Court was affirmed by the Federal Court of Appeal on identical grounds.[46]

C. Comity as a distraction from effectiveness

The reasoning behind this Canadian line of cases proceeds from a faulty premise. Again, international law—the law of jurisdiction in particular—in no way prohibits states from regulating through their domestic law the actions of their agents when they are acting abroad. States do this all the time, and they do not need the consent of the territorial state to extend the applicability of their laws. In fact, in the *Afghan Detainees* case the appellants rightly pointed out that Canadian criminal and military legislation extended to the acts of Canadian soldiers abroad.[47] If it was in conformity with international law, as the Canadian courts appeared to have thought it was, to extend the criminal laws of Canada to Canadian soldiers in Afghanistan, why would extending the norms of the Charter of Rights and Freedoms to those same soldiers be a violation of Afghan sovereignty? That this regulation may also involve creating rights in domestic law for foreigners affected by the extraterritorial acts of state agents is immaterial. The exercise of prescriptive jurisdiction is limited to the state organs or agents in question, as domestic law is commanding them—not the people of the territorial state whose lives they might affect—how they may or may not behave.

[42] *Amnesty International Canada & British Columbia Civil Liberties Association v. Chief of the Defence Staff for the Canadian Forces et al.*, 2008 FC 336 (hereinafter *Afghan Detainees FC*).
[43] Ibid, paras 145, 151 *et seq.*
[44] Ibid., para. 157.
[45] Ibid., para. 172.
[46] *Amnesty International Canada & British Columbia Civil Liberties Association v. Chief of the Defence Staff for the Canadian Forces et al.*, 2008 FCA 401.
[47] *Afghan Detainees FC*, paras 269–72.

The sovereignty-as-comity rationale for denying the extraterritorial applicability either of a human rights treaty, as in Lord Carswell's judgment in *Al-Skeini*, or of a domestic human rights instrument, as in *Hape* and *Afghan Detainees*, is nothing more than a distraction, the reddest of herrings. The organs or agents of a state can be present in the territory of another state lawfully or unlawfully. The sovereignty of the territorial state has either already been infringed, or that state has given its consent to the presence of foreign agents on its soil. Its sovereignty is therefore a non-issue. Where the true policy behind these cases lies is not in comity towards the sovereignty of the territorial state, but in the fear that an extraterritorial legal regime would prove ineffective, impracticable, or unworkable.

Al-Skeini, for instance, is imbued with the concern that the ECHR, adapted to European mores and sensibilities, is unsuitable for application in a country like Iraq, as we will soon see.[48] In *Hape*, on the other hand, Justice LeBel makes an enormous effort to outline what he calls 'the theoretical and practical difficulties arising out of an attempt to apply *Charter* standards outside Canada'.[49] He thus argues, for example, that extending Charter guarantees extraterritorially would hamper cooperative investigations, precisely at a time when criminal activity has become globalized.[50] It is above all effectiveness, or the lack thereof, that is the major theme of Justice LeBel's opinion.[51] Likewise, in the *Afghan Detainees* case the Federal Court expressed its concern that if the Charter were to be applied extraterritorially

> . . . an Afghan insurgent detained by members of the Canadian Forces in Kandahar province could end up having entirely different rights than would Afghan insurgents detained by soldiers from other NATO partner countries, in other parts of Afghanistan. The result would be a hodgepodge of different foreign legal systems being imposed within the territory of a state whose sovereignty the international community has pledged to uphold.
>
> This would be a most unsatisfactory result, in the context of a United Nations-sanctioned multinational military effort, further suggesting that the appropriate legal regime to govern the military activities currently underway in Afghanistan is the law governing armed conflict—namely international humanitarian law.[52]

Of course, there is a policy response to each of these policy objections. For example, rather than denying the applicability of the Charter altogether, the *Hape* Court could simply have said that Canadian agents engaging in searches abroad nonetheless have to obtain warrants from Canadian courts, and that Parliament should accommodate that requirement by passing the necessary legislation.[53] Or, the Court could have interpreted the reasonableness requirement of the substantive Charter rule as saying that an extraterritorial search would be reasonable under the Charter if it conformed to the domestic law of the state where it was conducted, subject perhaps to some minimum standard. That this latter approach would

48 See Section 6 below.
49 *Hape*, para. 86 (per LeBel J).
50 Ibid., paras 96 *et seq* (per LeBel J).
51 See also Stribopoulos, above note 36; Keitner, above note 4, at 31.
52 *Afghan Detainees FC*, paras 275–6.
53 See Stribopoulos, above note 36.

significantly reduce Charter guarantees in some cases of searches and seizures, as Justice LeBell pointed out in his judgment,[54] is to an extent a valid concern for the integrity of the Charter regime. Still, however, this is not a sufficient justification for his all or nothing approach, as other rights, such as the right to life or the prohibition of torture, would not and could not be interpreted downwards in a similar fashion in an extraterritorial context. When it comes to the *Afghan Detainees* case, a 'hodgepodge' of legal norms is unavoidable in situations involving a plurality of actors, not just in relation of their own municipal legal systems, but also in respect of the many different treaties to which these states may or may not be parties. That IHL applies, moreover, is in and of itself not reason enough for human rights law not to apply, particularly when that law—unlike IHL—can be judicially enforced.[55]

Whether these policy responses are considered to be persuasive is for the time being beside the point. The point that I do wish to make is that it is these policy debates that truly matter, while comity and the concern for the territorial state's sovereignty are an entirely fictitious objection to the extraterritoriality of human rights guarantees, be they domestic or international. This is not to say that serious legal problems could not arise from such extraterritorial application, even when the application itself or its specifics are uncontested—an issue that I will address in the section on norm conflicts and effectiveness below.[56] Now, however, I will turn to the much more central place that sovereignty and citizenship play when the extraterritorial application of domestic human rights guarantees is at stake.

5. Sovereignty, Citizenship, and the Social Contract

A. US case law on citizenship and extraterritoriality

The extraterritorial application of domestic human rights norms exposes a set of considerations that is largely missing, at least superficially, from the judicial discourse on the extraterritorial application of international human rights treaties. While it is impossible to read human rights treaties without acknowledging their universalist aspirations, domestic constitutions, including their bills of rights, have an alternative reading—that of a social contract operating between the members of the polity that adopted it. From this standpoint, it is *citizenship*, above all else, that determines the constitutional protections that individuals are entitled to. In the constitution-as-social-contract scheme of things, foreigners, aliens, or *others* generally cannot expect to be entitled to an equal standard of treatment as those belonging to the same community.

Some jurisdictions explicitly give central importance to citizenship, while in others the reliance on citizenship tends to be more implicit. The best example of the

[54] *Hape*, para. 88 (per LeBel J).
[55] On the co-application of human rights and IHL, see further Chapter V below.
[56] See Section 9.

former is the United States.[57] Consider, for instance, the *Verdugo-Urquidez* case,[58] whose facts are, but for one important point, identical to the Canadian *Hape* case that we just examined. A US agent conducted the search of the premises in Mexico of a Mexican national suspected of drug trafficking. He had no warrant to do so from a US court. At his trial in the United States, the defendant challenged the admissibility of the evidence obtained in the search, on the basis of the Fourth Amendment to the US Constitution.[59]

The trial court agreed that the evidence should be dismissed, and so did the majority of the Court of Appeals. The dissenting judge, however, 'viewed the Constitution as a "compact" among the people of the United States, and the protections of the Fourth Amendment were expressly limited to "the people"'.[60] This was the theme taken up by Chief Justice Rehnquist, writing for the plurality of the Supreme Court. In his view, the available historical data showed that 'the purpose of the Fourth Amendment was to protect the people of the United States against arbitrary action by their own Government; it was never suggested that the provision was intended to restrain the actions of the Federal Government against aliens outside of the United States territory'.[61]

Chief Justice Rehnquist distinguished the Supreme Court's earlier decision in *Reid v. Covert*,[62] which involved an attempt by Congress to subject the wives of American servicemen to trial by military tribunals without the protection of the Fifth and Sixth Amendments to the US Constitution, which the Court found to be unconstitutional. In Chief Justice Rehnquist's view, the basis of that decision was not some broader notion of extraterritoriality, but the fact that the individuals concerned were US citizens.[63] He also distinguished previous cases that recognized the constitutional rights of aliens (lawfully) present on US territory,[64] as well as the *Lopez-Mendoza* case,[65] where a majority of the Supreme Court assumed that illegal aliens present on US soil also possessed constitutional rights, stating that these illegal aliens 'were in the United States voluntarily and presumably had accepted some societal obligations; but respondent had no voluntary connection with this country that might place him among "the people" of the United States'.[66] He concluded by emphasizing the fact that, for better or for worse, we live in a world of nation-states in which the US government must be able to function effectively in the company of other sovereign nations. Applying the Constitution extraterritorially would have significant and deleterious consequences for the United States in conducting activities beyond its boundaries, particularly when it came to the use of armed force.[67]

[57] For a comprehensive overview of the extraterritoriality of the US Constitution, see Raustiala, above note 22.

[58] *United States v. Verdugo-Urquidez*, 494 U.S. 259 (1990).

[59] Which reads: '[t]he right of the people to be secure in their persons, houses, papers, and effects, against unreasonable searches and seizures, shall not be violated, and no Warrants shall issue, but upon probable cause, supported by Oath or affirmation, and particularly describing the place to be searched, and the persons or things to be seized.'

[60] *Verdugo-Urquidez*, at 264.

[61] Ibid., at 266.

[62] *Reid v. Covert*, 354 U.S. 1 (1957).

[63] *Verdugo-Urquidez*, at 270.

[64] Ibid., at 271.

[65] INS v. Lopez-Mendoza, 468 U.S. 1032 (1984).

[66] *Verdugo-Urquidez*, at 273.

[67] Ibid., at 273–5.

Chief Justice Rehnquist's reasoning thus turned around two principal strands of policy: citizenship and effectiveness.[68] Justices Kennedy and Stevens filed separate opinions concurring in the judgment, but downplaying the importance of citizenship. Justice Kennedy in particular emphasized the role of effectiveness, by arguing that what governs the disposition of the case is that the application of the Fourth Amendment requirements for a reasonable search and seizure in an extraterritorial context would have been anomalous and impracticable.[69] Justices Brennan and Marshall dissented in a universalist vein, arguing that by exercising its power and enforcing its law against an individual outside its territory, the United States is precluded from arguing that the Constitution does not travel with its agents.[70]

Citizenship played an even greater role in the now extensive litigation on the applicability of the writ of habeas corpus to persons detained by the US military outside the United States, particularly in Guantanamo Bay, Cuba. For instance, in *Al-Odah*,[71] the US Court of Appeals for the DC Circuit found that aliens detained outside US sovereign territory had no constitutional rights, and that having no such rights these could not be vindicated through habeas corpus as protected by the US Constitution and the applicable statute.

In its ruling, the Court of Appeals relied on *Eisentrager*, a post-Second World War case involving the detention of German nationals by US forces in a US military prison in Germany.[72] Indeed, *Eisentrager* proved to be absolutely central for the Bush administration's strategy in its 'global war on terror', and is deserving of discussion at some length.

Writing for the Court in *Eisentrager*, Justice Jackson considered that

> [m]odern American law has come a long way since the time when outbreak of war made every enemy national an outlaw, subject to both public and private slaughter, cruelty, and plunder. But even by the most magnanimous view, our law does not abolish inherent distinctions recognized throughout the civilized world between citizens and aliens, nor between aliens of friendly and of enemy allegiance, nor between resident enemy aliens who have submitted themselves to our laws and nonresident enemy aliens who at all times have remained with, and adhered to, enemy governments.
>
> With the citizen we are now little concerned, except to set his case apart as untouched by this decision and to take measure of the difference between his status and that of all categories of aliens. Citizenship as a head of jurisdiction and a ground of protection was old when Paul invoked it in his appeal to Caesar. The years have not destroyed nor diminished the importance of citizenship, nor have they sapped the vitality of a citizen's claims upon his government for protection.[73]

Justice Jackson continued by saying that

> [t]he alien, to whom the United States has been traditionally hospitable, has been accorded a generous and ascending scale of rights as he increases his identity with our society. Mere lawful presence in the country creates an implied assurance of safe conduct and gives him

[68] Similarly, see Raustiala, above note 22, at 172–7.
[69] *Verdugo-Urquidez*, at 275–9.
[70] Ibid., at 279 *et seq.*
[71] *Al Odah v. United States*, 321 F.3d 1134 (D.C. Cir. 2003), esp. at 1141.
[72] *Johnson v. Eisentrager*, 339 U.S. 763 (1950).
[73] Ibid., at 768–9.

certain rights; they become more extensive and secure when he makes preliminary declaration of intention to become a citizen, and they expand to those of full citizenship upon naturalization. . . . [I]n extending constitutional protections beyond the citizenry, the Court has been at pains to point out that it was the alien's presence within its territorial jurisdiction that gave the Judiciary power to act.[74]

He further held that

[i]t is war that exposes the relative vulnerability of the alien's status. The security and protection enjoyed while the nation of his allegiance remains in amity with the United States are greatly impaired when his nation takes up arms against us. While his lot is far more humane and endurable than the experience of our citizens in some enemy lands, it is still not a happy one. But disabilities this country lays upon the alien who becomes also an enemy are imposed temporarily as an incident of war, and not as an incident of alienage.[75]

Justice Jackson then turned to effectiveness, arguing that extending the writ of habeas corpus to enemy aliens detained outside US sovereign territory would pose severe practical difficulties and would seriously hamper the ability of the Executive to wage war, especially in battlefield conditions.[76] He further delved into the impossibility of extending constitutional rights such as the freedom of speech, trial by jury, or the right to bear arms in an extraterritorial context, especially during a military occupation.[77] The petition for habeas corpus was thus dismissed:

Justice Black, whom Justices Douglas and Burton joined, dissented. In his view,

not only is United States citizenship a 'high privilege,' it is a priceless treasure. For that citizenship is enriched beyond price by our goal of equal justice under law—equal justice not for citizens alone, but for all persons coming within the ambit of our power. This ideal gave birth to the constitutional provision for an independent judiciary with authority to check abuses of executive power and to issue writs of habeas corpus liberating persons illegally imprisoned.[78]

He further argued that

[s]ince the Court expressly disavows conflict with the *Quirin* or *Yamashita* decisions, it must be relying not on the status of these petitioners as alien enemy belligerents, but rather on the fact that they were captured, tried, and imprisoned outside our territory. The Court cannot, and, despite its rhetoric on the point, does not, deny that, if they were imprisoned in the United States, our courts would clearly have jurisdiction to hear their habeas corpus complaints. Does a prisoner's right to test legality of a sentence then depend on where the Government chooses to imprison him? Certainly the *Quirin* and *Yamashita* opinions lend no support to that conclusion, for, in upholding jurisdiction, they place no reliance whatever on territorial location. The Court is fashioning wholly indefensible doctrine if it permits the executive branch, by deciding where its prisoners will be tried and imprisoned, to deprive all federal courts of their power to protect against a federal executive's illegal incarcerations.

If the opinion thus means, and it apparently does, that these petitioners are deprived of the privilege of habeas corpus solely because they were convicted and imprisoned overseas, the Court is adopting a broad and dangerous principle. The range of that principle is

[74] Ibid., at 769–71. [75] Ibid., at 771–2. [76] Ibid., at 779.
[77] Ibid., at 782–3. [78] Ibid., at 791.

underlined by the argument of the Government brief that habeas corpus is not even available for American citizens convicted and imprisoned in Germany by American military tribunals. While the Court wisely disclaims any such necessary effect for its holding, rejection of the Government's argument is certainly made difficult by the logic of today's opinion. Conceivably, a majority may hereafter find citizenship a sufficient substitute for territorial jurisdiction, and thus permit courts to protect Americans from illegal sentences. But the Court's opinion inescapably denies courts power to afford the least bit of protection for any alien who is subject to our occupation government abroad, even if he is neither enemy nor belligerent, and even after peace is officially declared.[79]

Justice Black then attacked Justice Jackson's effectiveness rationale:

It has always been recognized that actual warfare can be conducted successfully only if those in command are left the most ample independence in the theater of operations. Our Constitution is not so impractical or inflexible that it unduly restricts such necessary independence. It would be fantastic to suggest that alien enemies could haul our military leaders into judicial tribunals to account for their day to day activities on the battlefront. Active fighting forces must be free to fight while hostilities are in progress. But that undisputable axiom has no bearing on this case or the general problem from which it arises.

When a foreign enemy surrenders, the situation changes markedly. If our country decides to occupy conquered territory either temporarily or permanently, it assumes the problem of deciding how the subjugated people will be ruled, what laws will govern, who will promulgate them, and what governmental agency of ours will see that they are properly administered. This responsibility immediately raises questions concerning the extent to which our domestic laws, constitutional and statutory, are transplanted abroad. Probably no one would suggest, and certainly I would not, that this nation either must or should attempt to apply every constitutional provision of the Bill of Rights in controlling temporarily occupied countries. But that does not mean that the Constitution is wholly inapplicable in foreign territories that we occupy and govern.[80]

He concluded with powerful rhetoric:

Perhaps, as some nations believe, there is merit in leaving the administration of criminal laws to executive and military agencies completely free from judicial scrutiny. Our Constitution has emphatically expressed a contrary policy.

As the Court points out, Paul was fortunate enough to be a Roman citizen when he was made the victim of prejudicial charges; that privileged status afforded him an appeal to Rome, with a right to meet his 'accusers face to face.' Acts 25:16. But other martyrized disciples were not so fortunate. Our Constitution has led people everywhere to hope and believe that, wherever our laws control, all people, whether our citizens or not, would have an equal chance before the bar of criminal justice.

Conquest by the United States, unlike conquest by many other nations, does not mean tyranny. For our people 'choose to maintain their greatness by justice, rather than violence.' Our constitutional principles are such that their mandate of equal justice under law should be applied as well when we occupy lands across the sea as when our flag flew only over thirteen colonies. Our nation proclaims a belief in the dignity of human beings as such, no matter what their nationality or where they happen to live. Habeas corpus, as an instrument to protect against illegal imprisonment, is written into the Constitution. Its use by courts

[79] Ibid., at 795–6. [80] Ibid., at 796–7.

cannot, in my judgment, be constitutionally abridged by Executive or by Congress. I would hold that our courts can exercise it whenever any United States official illegally imprisons any person in any land we govern. Courts should not for any reason abdicate this, the loftiest power with which the Constitution has endowed them.[81]

The tension in *Eisentrager* between universality and the 'dignity of human beings as such' on the one hand, and citizenship and effectiveness on the other, is palpable. For the majority of the justices, citizenship and effectiveness carried the day. The Court was understandably reluctant to accept that the Constitution protected thousands upon thousands of aliens detained in the wake of the Second World War, but its reluctance then is still felt today. The Supreme Court's *Eisentrager* holding was the principal basis for the view of the Bush administration lawyers some fifty years later that non-US nationals detained in Guantanamo would not be entitled to any constitutional protections.[82] As a matter of policy, lack of judicial supervision was precisely the reason why the administration chose to turn Guantanamo into a terrorist detention centre. As we have seen from *Al-Odah*, that strategy was initially successful before lower US courts. Before the Supreme Court, however, the story turned out differently—but that journey was neither quick nor inevitable.

B. The Guantanamo cases

The first Guantanamo habeas corpus case to be heard by the Supreme Court was *Rasul v. Bush*,[83] and it split the Court badly along ideological lines. In a 5 to 4 ruling, Justice Stevens delivered the opinion of the Court.

In its initial foray into addressing the Bush administration's legal strategy in the 'war on terror', the Court exhibited quite a bit of caution. Justice Stevens approached the case solely on statutory grounds, without reaching the question whether persons detained in Guantanamo had a constitutional right to habeas corpus, or indeed any constitutional rights at all. He also did his best not to overturn, but (rather unpersuasively) distinguish *Eisentrager*, by asserting that its statutory predicate was overruled in another case.[84] He thus ultimately held that statutory habeas corpus attaches to all persons within the 'territorial jurisdiction' of US courts; since Guantanamo was, pursuant to the treaty between the United States and Cuba, under US 'complete jurisdiction and control', even if Cuba retained sovereignty,[85] habeas corpus was available to persons detained there.[86] Regarding citizenship, Justice Stevens stated that

[81] Ibid., at 797–8.
[82] See above note 10. See also John C. Yoo, 'Memorandum for William J. Haynes II, Re: Military Interrogation of Alien Unlawful Combatants Held Outside the United States', US Department of Justice, Office of the Legal Counsel, 14 March 2003, available at <http://www.aclu.org/pdfs/safefree/yoo_army_torture_memo.pdf>.
[83] *Rasul v. Bush*, 542 U.S. 466 (2004).
[84] Ibid., at 475–9.
[85] See above, Chapter II, note 67 and accompanying text.
[86] *Rasul*, at 480–4.

[c]onsidering that the [habeas] statute draws no distinction between Americans and aliens held in federal custody, there is little reason to think that Congress intended the geographical coverage of the statute to vary depending on the detainee's citizenship. Aliens held at the base, no less than American citizens, are entitled to invoke the federal courts' authority under § 2241 [the habeas statute].[87]

In his concurring opinion, Justice Kennedy disagreed with the Court that *Eisentrager* could be distinguished on statutory grounds, but thought that it could be distinguished on the facts, namely because '[f]rom a practical perspective, the indefinite lease of Guantanamo Bay has produced a place that belongs to the United States, extending the "implied protection" of the United States to it', Cuba's formal sovereignty notwithstanding.[88]

In his dissent, Justice Scalia rather thoroughly debunked the Court's distinguishing of *Eisentrager*.[89] As for the relevance of citizenship, he considered that

[n]either party to the present case challenges the atextual extension of the habeas statute to United States citizens held beyond the territorial jurisdictions of the United States courts; but the possibility of one atextual exception thought to be required by the Constitution is no justification for abandoning the clear application of the text to a situation in which it raises no constitutional doubt.[90]

In other words, in Justice Scalia's view, discrimination between citizens and non-citizens should be read into the habeas statute, so it could indeed apply extraterritorially to the former.[91] Justice Scalia then turned to an effectiveness argument:

The consequence of this holding, as applied to aliens outside the country, is breathtaking. It permits an alien captured in a foreign theater of active combat to bring a § 2241 petition against the Secretary of Defense. Over the course of the last century, the United States has held millions of alien prisoners abroad.... A great many of these prisoners would no doubt have complained about the circumstances of their capture and the terms of their confinement. The military is currently detaining over 600 prisoners at Guantanamo Bay alone; each detainee undoubtedly has complaints—real or contrived—about those terms and circumstances. The Court's unheralded expansion of federal-court jurisdiction is not even mitigated by a comforting assurance that the legion of ensuing claims will be easily resolved on the merits.... From this point forward, federal courts will entertain petitions from these prisoners, and others like them around the world, challenging actions and events far away, and forcing the courts to oversee one aspect of the Executive's conduct of a foreign war.[92]

Finally, Justice Scalia disputed the Court's discussion of the unique status of Guantanamo:

The Court does not explain how 'complete jurisdiction and control' without sovereignty causes an enclave to be part of the United States for purposes of its domestic laws. Since 'jurisdiction and control' obtained through a lease is no different in effect from 'jurisdiction and control' acquired by lawful force of arms, parts of Afghanistan and Iraq should logically be regarded as subject to our domestic laws. Indeed, if 'jurisdiction and control' rather than

[87] Ibid., at 481. [88] Ibid., at 487. [89] Ibid., at 489–98. [90] Ibid., at 497.
[91] See also Ibid., at 501–2. [92] Ibid., at 498–9 (internal citations omitted).

sovereignty were the test, so should the Landsberg Prison in Germany, where the United States held the *Eisentrager* detainees.[93]

On this point, Justice Scalia has it exactly right. There is indeed no reason of principle why jurisdiction and control over a territory obtained through a treaty should be treated differently from jurisdiction and control obtained through the use of force, lawful and even unlawful, as *Loizidou* attests. But then again, lack of *sovereignty* or title over territory is also not a principled reason for denying any extraterritorial protection of individual rights. Justice Scalia never explains why sovereignty alone should matter. Moreover, if considerations of effectiveness should preclude aliens from invoking habeas corpus, why do they not equally affect citizens as well?

On the same day as *Rasul*, the Court decided *Hamdi v. Rumsfeld*,[94] holding that a US citizen detained in the United States as an enemy combatant had a constitutional right for the legality of his detention to be reviewed by an independent court. Though no single opinion commanded the majority of the Court, eight of the nine justices thought that US citizens detained in the United States were entitled to constitutional due process, and it was indeed Justice Scalia who was most fiercely opposed to the Executive's asserted broad power of detention.

Congress acted swiftly to overrule *Rasul* by passing the Detainee Treatment Act 2005 (DTA) which stripped US courts of jurisdiction to entertain petitions by non-US nationals detained abroad. In *Hamdan v. Rumsfeld*, the Court's next encounter with Guantanamo, the Court held that the jurisdiction-stripping provisions of the DTA did not apply retrospectively to petitions that were pending at the time the DTA was passed, and that all detainees in US custody were entitled to the minimum fair trial protections of Common Article 3 of the 1949 Geneva Conventions.[95] In response to *Hamdan*, Congress passed the Military Commissions Act 2006 (MCA) which quite explicitly stripped US courts of jurisdiction even with respect to pending petitions.

The stage was thus set for the *Boumediene* case,[96] where the Court considered whether aliens detained in Guantanamo had a *constitutional*, as opposed to statutory, right to habeas corpus, and answered that question in the affirmative in yet another 5 to 4 decision. After reviewing earlier case law, Justice Kennedy, writing for the Court, could not accept the government's argument that 'at least as applied to noncitizens, the Constitution necessarily stops where de jure sovereignty ends'.[97] In his view, it is 'practical considerations' above all else that govern the extraterritorial

[93] Ibid., at 501.

[94] *Hamdi v. Rumsfeld*, 542 U.S. 507 (2004).

[95] *Hamdan v. Rumsfeld*, 126 S. Ct. 2749 (2006).

[96] *Boumediene v. Bush*, 128 S. Ct. 2229 (2008). For general comments on the decision, see, e.g., G. Neuman, 'The Extraterritorial Constitution after *Boumediene v. Bush*', (2009) 82 *S. Cal. L. Rev.* 259; D. Jenkins, 'Habeas Corpus and Extraterritorial Jurisdiction after *Boumediene*: Towards a Doctrine of "Effective Control" in the United States', (2009) 9 *HRLR* 306; R. Chesney, 'Case Note: *Boumediene v. Bush*', (2008) 102 *AJIL* 848.

[97] *Boumediene*, at 2253.

application of the Constitution,[98] and these same practical considerations were also the primary basis for the *Eisentrager* precedent, which he unsurprisingly, but not altogether candidly, chose to distinguish rather than overrule.[99] Though, as I have explained above, practical considerations were absolutely crucial in *Eisentrager*, Justice Jackson also clearly based his ruling on sovereignty and citizenship. However, according to Justice Kennedy,

> Even if we assume the *Eisentrager* Court considered the United States' lack of formal legal sovereignty over Landsberg Prison as the decisive factor in that case, its holding is not inconsistent with a functional approach to questions of extraterritoriality. The formal legal status of a given territory affects, at least to some extent, the political branches' control over that territory. *De jure* sovereignty is a factor that bears upon which constitutional guarantees apply there.[100]

He was of the opinion that earlier case law supported 'the idea that questions of extraterritoriality turn on objective factors and practical concerns, not formalism.'[101] He thus concluded that

> . . . at least three factors are relevant in determining the reach of the Suspension Clause: (1) the citizenship and status of the detainee and the adequacy of the process through which that status determination was made; (2) the nature of the sites where apprehension and then detention took place; and (3) the practical obstacles inherent in resolving the prisoner's entitlement to the writ.[102]

Applying this functional framework, Justice Kennedy considered that: (1) though they were not citizens, it was in doubt whether the detainees were *enemy* aliens, and the process created by the Executive to establish their status fell well short of mechanisms that would obviate the need for habeas corpus review; (2) in 'every practical sense Guantanamo is not abroad; it is within the constant jurisdiction of the United States', and is thus distinguishable from the military prison in *Eisentrager* that was under the joint jurisdiction of the four Allied Powers that were occupying Germany (this of course left open the question whether habeas corpus would be available to detainees in other US facilities, such as Bagram); and (3) there were no practical obstacles in conducting habeas review.[103] Justice Kennedy also dismissed the lack of a precedent exactly on point in support of his ruling:

> It is true that before today the Court has never held that noncitizens detained by our Government in territory over which another country maintains *de jure* sovereignty have any rights under our Constitution. But the cases before us lack any precise historical parallel. They involve individuals detained by executive order for the duration of a conflict that, if measured from September 11, 2001, to the present, is already among the longest wars in American history. . . . The detainees, moreover, are held in a territory that, while technically not part of the United States, is under the complete and total control of our Government. Under these circumstances the lack of a precedent on point is no barrier to our holding.[104]

[98] Ibid., at 2257.
[99] Ibid., at 2257 *et seq.*
[100] Ibid., at 2258.
[101] Ibid., at 2258.
[102] Ibid., at 2259.
[103] Ibid., at 2259–62.
[104] Ibid., at 2262.

The dissenting opinions of both Chief Justice Roberts and Justice Scalia put a heavy emphasis on effectiveness arguments, particularly on the deference that the political branches of government were due from the courts in matters of war and foreign policy. Justice Scalia rehearsed his arguments from *Rasul*:

> The writ of habeas corpus does not, and never has, run in favor of aliens abroad; the Suspension Clause thus has no application, and the Court's intervention in this military matter is entirely *ultra vires*. . . . The game of bait-and-switch that today's opinion plays upon the Nation's Commander in Chief will make the war harder on us. It will almost certainly cause more Americans to be killed. That consequence would be tolerable if necessary to preserve a time-honored legal principle vital to our constitutional Republic. But it is this Court's blatant *abandonment* of such a principle that produces the decision today.[105]

Dismissing the majority's distinguishing of *Eisentrager* as unpersuasive (which it was), Justice Scalia summed up his stance on the relevance of citizenship to the extraterritorial application of the Constitution: 'merely because citizenship is not a *sufficient* factor to extend constitutional rights abroad does not mean that it is not a *necessary* one'.[106]

C. Evaluating citizenship as a ground for extraterritoriality of individual rights

Though questions of effectiveness are ubiquitous in all cases on the extraterritoriality of individual rights, be they domestic or international, it is the reliance on citizenship in US case law that is particularly striking. Taken as a value in and of itself, citizenship stands in manifest opposition to the principle of universality.[107] When it comes to protecting an individual from the arbitrary exercise of state power, why should a formal tie between the individual and the state make him worthy of protection, and not his inherent human dignity? If, moreover, citizenship is not the determining consideration for the protection of individual rights when a state acts *within* its borders, why should it matter when a state acts *outside* its borders?[108]

In the United States at least, citizenship matters because of the prevalence of social contract conceptions of the US Constitution, including its guarantees of individual rights, as opposed to more universalist viewpoints. As explained by Neuman,

> Social contract rhetoric has played a significant role in American constitutionalism. Social contract theory seeks to legitimate government through the idea of an actual or hypothetical agreement embodying the consent of the governed who have established the state and empowered it to govern. Some accounts of social contract theory identify a limited class of

105 Ibid., at 2294. 106 Ibid., at 2301.
107 See Ben-Naftali and Shany, above note 3, at 61–3.
108 See also Raustiala, above note 22, at 170–2.

'members' as the proper beneficiaries of the contract. The beneficiaries have rights based in the contract; nonbeneficiaries are relegated to whatever rights they may have independent of the contract. A skeptic who did not ascribe normative force to social contract arguments still could invoke the idea of a social contract as a historically-grounded tool for interpreting American constitutionalism. This sort of reasoning is evident in Chief Justice Rehnquist's opinion in *Verdugo-Urquidez.*[109]

The question is thus whether the purpose of habeas corpus specifically or the Constitution in general is to protect *American* rights and *citizens'* rights, or conversely, whether they are meant to protect *human* rights against an irresponsible government.[110] We have already seen this tension in *Eisentrager,* with Justice Jackson focusing on the importance of citizenship, and Justice Black arguing that it is the nation's 'belief in the dignity of human beings as such, no matter what their nationality or where they happen to live'[111] that should determine the scope of the Constitution's guarantees of individual rights. Ultimately, the only real answer that can be given to this question is ideological.[112] It is certainly possible, as Neuman noted in the quotation above, to attempt an historical inquiry into how that question was answered in the past. This was done, as we have seen, by both the majority and the minority in *Rasul* and *Boumediene*. An intellectually honest approach to this historical inquiry would reach the unsurprising conclusion that the same debates we wage today were waged before, and that sometimes universality (or, at least, non-discrimination between citizens and aliens) and sometimes a social contract approach had the upper hand.[113] We should also not forget that it was America's founders who declared: '[w]e hold these truths to be self-evident, that all men are created equal, that they are endowed by their Creator with certain unalienable Rights, that among these are Life, Liberty and the pursuit of Happiness.' Citizenship hardly fits well with this universalist sentiment.[114] Be that as it may, it is in any case for American lawyers and the American polity to determine what conception of individual rights their Constitution enshrines in the twenty-first century.[115] If they do wish to continue assigning such crucial importance to citizenship—as is evident not just from the dissents in the Guantanamo cases, but

[109] G. Neuman, 'Whose Constitution?', (1991) 100 *Yale L.J.* 909, 917.
[110] See, in that vein, T. Endicott, '*Habeas Corpus* and Guantánamo Bay: A View from Abroad', *American Journal of Jurisprudence* (forthcoming), draft available on SSRN at <http://ssrn.com/abstract=982412>.
[111] *Eisentrager*, at 797–8.
[112] See also Raustiala, above note 22, at 246–7.
[113] See generally Neuman, above note 109. See also Raustiala, above note 22, at 54–7.
[114] See, e.g., L. Henkin, 'The Constitution as Compact and as Conscience: Individual Rights abroad and at Our Gates', (1985–86) 27 *Wm. & Mary L. Rev.* 11, esp. at 32: 'Our federal government must not invade the individual rights of any human being. The choice in the Bill of Rights of the word "person" rather than "citizen" was not fortuitous; nor was the absence of a geographical limitation. Both reflect a commitment to respect the individual rights of all human beings.'
[115] For a general account of the evolution of the importance of citizenship in US law, see P. Spiro, *Beyond Citizenship: American Identity After Globalization* (Oxford University Press, 2008).

also from the discussions of the majority of the Court as well as from other cases[116]—they are free to do so.[117]

What is undeniably true, however, is that some legal systems have already decided that citizenship should *not* be the basis for the protection of individual rights, whether extraterritorially or not. It is above all international law that has unambiguously adopted the principle of universality of human rights.[118] In that international law is by no means alone. Compare for a moment the Canadian with the US case law, with best comparison points being *Hape* and *Verdugo-Urquidez*, since they both concerned extraterritorial searches.

In *Verdugo-Urquidez*, the petitioner lost because he was not a US citizen, and because the Court could not see how the Fourth Amendment could effectively apply to an extraterritorial search. In *Hape*, on the other hand, the petitioner actually *was* a Canadian citizen, but he still lost on grounds of effectiveness. What is so striking is precisely that the Canadian Supreme Court, unlike its US counterpart, gave no significance to citizenship.[119] Indeed, at least one reason that it did not do so is that a citizenship-based argument in favour of the extraterritorial application of the Canadian Charter would have undermined the effectiveness argument against extraterritorial application. In other words, if the provisions of the Charter on unreasonable searches could effectively apply to the premises of a Canadian citizen outside Canada on the basis of citizenship alone, then one could not honestly say that these same provisions could not be applied effectively vis-à-vis non-citizens. This in fact also undermines the dissenting justices' effectiveness-based arguments in the Guantanamo cases that granting non-citizens habeas corpus review would undermine the ability of the Executive to successfully prosecute a

[116] Thus, for example, on the same day that it decided *Boumediene*, the US Supreme Court also delivered its opinion in *Munaf v. Geren*, 128 S.Ct. 2207 (2008), where it unanimously held that US citizens detained abroad by US forces acting as a part of a military coalition would have a statutory right to habeas corpus, though their petition failed on the merits of the case. See also *Hamdi v. Rumsfeld*, 542 U.S. 507 (2004). The idea that citizens are somehow by definition entitled to more rights than aliens, including in matters such as executive detention, is prevalent even in the literature which broadly endorses the US Supreme Court's judgments on Guantanamo. See, e.g., R. Fallon and D. Meltzer, 'Habeas Corpus Jurisdiction, Substantive Rights, and the War on Terror', (2007) 120 *Harv. L. Rev.* 2029, 2083 (stating that such a conclusion is based on the 'assumption that the Constitution is a continuing compact among the American people, established and accepted principally for the benefit of Americans. A constitution under which the government owes special obligations to its citizens can of course grant rights to noncitizens, as ours does and should. But the moral foundations for the rights of aliens are different in kind from the moral foundations of citizens' rights, because aliens are by definition outsiders to the fair scheme of social cooperation and mutual advantage that the Constitution aims to establish among the American people. The rights of aliens, resting as they do on a different basis, can accordingly be narrower in scope.')

[117] See further in that regard the discussions in Neuman, above note 109; S. Cleveland, 'Embedded International Law and the Constitution Abroad', (2010) 110 *Colum. L. Rev.* 225; J.A. Kent, 'A Textual and Historical Case Against a Global Constitution', (2007) 95 *Geo. L. J.* 463; Fallon and Meltzer, above note 116; Raustiala, above note 22.

[118] Thus, for example, the ADHR emphatically declares that 'the essential rights of man are not derived from the fact that he is a national of a certain state, but are based upon attributes of his human personality'.

[119] See Verdier, above note 38, at 148. Citizenship was only mentioned by Binnie J. in *Hape*, para. 187, as he admonished the majority of the Court not to be too rigid in its approach, as it might have serious repercussions for future cases.

war—when that exact same rationale applies equally to citizens detained by the military, in respect of whom it was undisputed that they have a right to habeas corpus.[120]

I am not saying that citizenship plays no role whatsoever in the legal discourse on the extraterritoriality of individual rights protection outside the United States. Take Canada again as an example, this time the *Khadr* case[121] before the Canadian Supreme Court. Omar Ahmed Khadr is a Canadian national, since 2002 detained by US forces in Guantanamo as an enemy combatant. He was originally detained in Afghanistan for taking part in hostilities against US forces there, but has since been charged with murder and other crimes before a military commission. At the time of his capture he was only fifteen years of age, which made him the youngest prisoner in Guantanamo. In 2003, he was questioned in Guantanamo by Canadian intelligence agents, who shared the product of this interview with US authorities. Before Canadian courts, he sought an order that would require Canadian authorities to disclose all documents relevant to the charges against him, including the records of the interview.

The Court held that Khadr was indeed entitled to such disclosure under the Charter of Rights and Freedoms. To do that, however, it had to distinguish its ruling in *Hape* that the Charter cannot apply on foreign soil due to comity and respect for the sovereignty of the territorial state. It did so by relying on Justice LeBel's dictum in *Hape* that comity would end if Canada's agents were involved in violations of Canada's international human rights obligations.[122] Because the US Supreme Court itself held in *Hamdan* that the United States government had breached its obligations toward the detainees under Common Article 3 of the Geneva Conventions, the United States was owed no deference by Canada and the Charter applied to Canadian agents in Guantanamo.[123]

Though the Court in *Khadr* quite neatly managed to avoid assessing on its own whether the United States had violated Khadr's human rights by referring to a decision of the US Supreme Court,[124] its analysis of Canada's complicity in that human rights violation—in fact not a human rights violation proper, but a violation of international humanitarian law—leaves something to be desired. I have already explained above why I believe that the whole comity rationale for denying extraterritorial application of human rights guarantees does not withstand scrutiny. In any event, it is hard to read *Khadr* without having the impression that the fact that Khadr was a Canadian national had an impact on the Court, even though the Court does not actually base its holding on Khadr's citizenship. I cannot help but wonder what the Court's holding would have been if Khadr had had Afghan

[120] Cf. *A and Others v. Secretary of State for the Home Department* [2004] UKHL 56, [2005] 2 AC 68 (the *Belmarsh* case), where the House of Lords held that a derogation under Art. 15 ECHR to allow for preventive detention under Art. 5 ECHR is impermissible if it is limited only to non-citizens.

[121] *Canada (Justice) v. Khadr*, 2008 SCC 28. See also its companion case *Canada (Prime Minister) v. Khadr*, 2010 SCC 3, in which the Court dealt with the remedy for the violation of Khadr's rights.

[122] *Hape*, para. 101 (per LeBel J).

[123] *Khadr*, paras 2–3, 21, 25–6.

[124] See also Keitner, above note 4, at 34–5.

nationality, and had been interrogated by Canadian agents because he was a subject of interest.[125]

Though citizenship as a policy consideration is thus not completely irrelevant in jurisdictions other than the United States, its relevance nonetheless seems rather marginal when compared to considerations of effectiveness. Indeed, invocations of citizenship are often treated as pernicious.[126]

Whatever the relevance of citizenship in respect of the extraterritorial application of domestic human rights guarantees, it is clear that it should have no relevance to the extraterritorial application of international human rights treaties, which are not social compacts, but compacts between nations. Again, this is not to say that, as a matter of fact, considerations of citizenship do not influence judges when they interpret the relevant treaties. What I am saying is that *normatively* citizenship is not a legitimate consideration on which the application of a human rights treaty is to be based, as it directly contradicts the principle of universality.[127] While some national legal systems have as yet not decided whether their constitutional instruments have their basis in a social compact or in a universality theory, that value or ideological choice has already been made in the international system. In the aftermath of the Second World War, states have jointly decided—in word, if not in deed—that the moral foundation of individual rights cannot be citizenship or membership in a polity, but can only be found in the dignity inherent in every human being. From the standpoint of the international legal system, universality is no longer just one of many competing ideological viewpoints—*universality is the law*. This decision is of course no accident, but is born of bitter experience. We need little reminding how the Nazi persecution of German Jews escalated by the stripping of their German citizenship and the enactment of the Nuremberg Laws,[128] or how apartheid South Africa transferred the citizenship of a large part of its black population to supposedly independent Bantustans, so that it could continue discriminating against them, but now on the superficially more palatable basis of citizenship rather than race.[129]

[125] Indeed, a case practically identical to *Khadr* arose in the lower Canadian courts, with one important difference—that the applicants were not Canadian citizens. For that reason, the Federal Court found that they were not entitled under the Charter to the disclosure of information gathered by Canadian agents who interrogated them in Guantanamo. See *Slahi v. Canada*, 2009 FC 160, esp. paras 39–48. The judgment was rather tersely affirmed by the Federal Court of Appeal—see *Slahi v. Canada*, 2009 FCA 259. For its part, the Supreme Court chose to remain silent on the issue, and refused to grant leave to appeal—see *Slahi v. Canada*, 2010 CarswellNat 297.

[126] Of interest in that regard is ongoing debate in the UK government proposals to enact a British Bill of Rights, which would supplement or supplant the Human Rights Act and the ECHR—see, e.g., Joint Committee on Human Rights, 'Twenty-Ninth Report,' 21 July 2008, available at <http://www.publications.parliament.uk/pa/jt200708/jtselect/jtrights/165/16506.htm>, esp. para. 84: 'We call on the Government to decouple the debate about a Bill of Rights from the debate about citizenship and the rights and duties of the citizen, and to ensure that in future the universality of fundamental human rights is explicitly recognised in documents and speeches relating to a Bill of Rights.'

[127] See also J.M. Piret, '*Boumediene v. Bush* and the Extraterritorial Reach of the US Constitution: A Step Towards Judicial Cosmopolitanism?', (2008) 4 *Utrecht Law Review* 81.

[128] See, e.g., I. Kershaw, *Hitler 1889–1936: Hubris* (Norton, 1999), at 560–70.

[129] See, e.g., J. Crawford, *The Creation of States in International Law* (Oxford University Press, 2nd edn, 2006), at 338–48.

Citizenship has retained only relatively minor importance in human rights law. It is solely political rights—the rights to vote in democratic elections and to stand for office—that have remained tied to citizenship, because these rights are ultimately tied to membership in the polity. Immigration is of course another field where citizenship has remained broadly relevant. But generally speaking, when citizenship has been used to create privileges or deny rights, it has been treated as a ground of discrimination that is prohibited unless reasonably and objectively justified.[130] The European Court even considers it a particularly suspect classification, saying that 'very weighty reasons would have to be put forward before the Court could regard a difference of treatment based exclusively on the ground of nationality as compatible with the Convention'.[131] It certainly may not be used as the sole basis for the extraterritorial application of a human rights treaty. For example, it would be normatively unacceptable to argue that a British national in the custody of British troops in Iraq would be within the jurisdiction of the United Kingdom and that the ECHR would thus apply, while an Iraqi national in the same situation would not be within the United Kingdom's jurisdiction. That international law recognizes nationality as a basis for prescriptive jurisdiction, i.e. as a basis for the extension of domestic law to a person, is irrelevant, since, as explained above, the concept of state jurisdiction in human rights treaties is grounded in factual power, not in the extension of domestic law.[132]

To put this another way, as I am writing this in Cambridge, I am within the prescriptive jurisdiction of two states: the United Kingdom, in whose territory I am residing, and Serbia, my state of nationality. If I were, say, to commit murder, I could be punished under both UK law, on the basis of the territoriality principle, and under Serbian law, pursuant to the active personality principle. But even though both states have extended their prescriptive or legislative jurisdiction to me, i.e. both of them have laws commanding me not to kill other people on pain of criminal punishment, I am still 'within the jurisdiction' in the sense of Article 1 ECHR or the relevant jurisdiction clauses of other human rights treaties of only one state—the United Kingdom. Thus, if I was the victim of murder by an unknown assailant, it would be the United Kingdom, not Serbia, who would have the obligation to conduct an effective investigation into my murder pursuant to Article 2 ECHR. This would be the case irrespective of whether the murderer himself was also a Serbian national.[133] What counts is that I am located in a *territory* which is under the United Kingdom's jurisdiction, i.e. under its effective overall control.

Case law on the extraterritoriality of human rights treaties has thankfully been free, as a general matter, of the notion that extraterritoriality might depend on the nationality of either the victim or the perpetrator of a human rights violation. There are some older cases, however, where this fallacy is either present or is hinted

[130] See, e.g., CERD General Recommendation No. 30 UN Doc. HRI/GEN/1/Rev.7/Add.1 (2004); M. Nowak, *CCPR Commentary* (Engel, 2nd revised edn, 2005), at 54–5, 618–23.

[131] *Gaygusuz v. Austria*, App. No. 17371/90, Judgment, 16 September 1996, para. 42.

[132] See above, Chapter II. See also below, Chapter IV, Section 3.C.3.

[133] The situation might be different if I was killed by a Serbian *agent*, i.e. a person whose acts are attributable to Serbia, whether of Serbian nationality or not. See Chapter IV, Section 4.C. below.

at—most notably cases of the now defunct European Commission on Human Rights. Thus, for example, in *X v. Federal Republic of Germany*,[134] a case with a rather colourful set of facts where the applicant alleged that his ECHR rights were violated by German consular officials *inter alia* because they denied that he was of noble birth, the Commission remarked that

> ... in certain respects, the nationals of a Contracting State are within its 'jurisdiction' even when domiciled or resident abroad; [...] in particular, the diplomatic and consular representatives of their country of origin perform certain duties with regard to them which may, in certain circumstances, make that country liable in respect of the Convention.[135]

However, this was not one such case, and the application was inadmissible. In its first decision on northern Cyprus,[136] the Commission likewise remarked that

> [i]n Art. 1 of the Convention, the High Contracting Parties undertake to secure the rights and freedoms defined in Section 1 to everyone 'within their jurisdiction' (in the French text: 'relevant de leur juridiction'). The Commission finds that this term is not, as submitted by the respondent Government, equivalent to or limited to the national territory of the High Contracting Party concerned. It is clear from the language, in particular of the French text, and the object of this Article, and from the purpose of the Convention as a whole, that the High Contracting Parties are bound to secure the said rights and freedoms to all persons under their actual authority and responsibility, whether that authority is exercised within their own territory or abroad. The Commission refers in this respect to its decision on the admissibility of Application No. 1611/62—X. v/Federal Republic of Germany—Yearbook of the European Convention on Human Rights, Vol. 8, pp. 158–169 (at pp. 168–169).
>
> The Commission further observes that nationals of a State, including registered ships and aircrafts, are partly within its jurisdiction wherever they may be, and that authorised agents of a State, including diplomatic and consular agents and armed forces, not only remain under its jurisdiction when abroad but bring any other persons or property 'within the jurisdiction' of that State, to the extent that they exercise authority over such persons or property. Insofar as, by their acts or omissions, they affect such persons or property, the responsibility of the State is engaged.[137]

To the extent that the Commission says or wants to say that nationals of a state can be within its jurisdiction wherever they may be it is guilty of a category error, as I have explained above. That a state may legislate for its nationals does not mean that it *ipso facto* has jurisdiction over them in the purely factual Article 1 sense. The extraterritorial application of the ECHR simply cannot depend on the nationality of the alleged victim of a human rights violation. But as we can see from the Commission's reasoning quoted above, this was a fairly marginal point, nothing more than an *obiter dictum*. The Commission's actual holding is that state organs or agents through their own actions bring persons under their authority and control,

[134] *X v. Federal Republic of Germany*, App. No. 1611/62, 25 September 1965, 8 *Yearbook of the European Convention on Human Rights* 158.
[135] Ibid., at 168.
[136] *Cyprus v. Turkey* (dec.), App. Nos 6780/74 and 6950/75, 26 May 1975.
[137] Ibid., at 136, para. 8.

and that this equals jurisdiction in the sense of Article 1.[138] I will turn to this question in the next chapter of this study.[139] For now, however, assuming that the Commission's views are correct, it will suffice to observe that it is not the nationality of the perpetrators of the violation that matters, but their status as organs or agents of a state.

In conclusion, citizenship or nationality alone is not a legitimate basis for the extraterritorial applicability of a human rights treaty. While it is possible for a right to be directly tied to nationality—for example, Article 12(4) ICCPR provides that '[n]o one shall be arbitrarily deprived of the right to enter his own country'—linking the application of a human rights treaty as such to nationality would not only have no textual basis, but would also be contrary to the treaty's object and purpose. It is in particular when universality, rather than social contract, is seen as the underpinning of individual rights, that it is not form, but the substance of state power and control over individuals that matter. 'The domain of human rights has no place for passports.'[140]

6. Relativism and Regionalism

A. Relativism and regionalism before the European Court

That universality is entrenched in international law does not mean that it has remained unchallenged. Culturally relativist responses to universality are as old as universality itself. For all the lofty affirmations of the universality of human rights at UN conferences, vast differences between, say, Europe and Iraq undeniably remain as a matter of fact. It is likewise undeniable that the strength of the European human rights regime has historically been contingent on a large number of shared values and morals. For the relativist, saying that *European* human rights should apply in Iraq in just the same way as they apply in Europe is as hopelessly utopian as universality itself.

The universalist–relativist debate is again essentially ideological, and it will not be rehearsed here.[141] What interests me here is how relativist considerations influence cases on the extraterritorial application of human rights guarantees. The short answer is that relativist considerations do matter, explicitly or (more often) implicitly. As we will see, a relativist argument can be made in two basic ways. The first is that some societies are simply not developed enough, economically or politically, to allow for the application of human rights standards that have evolved in more advanced countries. Because it positively reeks of the old imperialist distinction between civilized and uncivilized nations and directly contradicts the

[138] The Commission's reasoning (like that of the Court) also exhibits a degree of confusion between state jurisdiction and state responsibility, on which see Chapter II, Section 3 above.

[139] See below, Chapter IV, Sections 3 and 4.

[140] R. Dworkin, *Is Democracy Possible Here?* (Princeton University Press, 2006), at 48.

[141] See generally J. Donnelly, *Universal Human Rights in Theory and Practice* (Cornell University Press, 2nd edn, 2003), at 89 *et seq*, and the authorities cited therein.

principle of universality, such unabashed relativism is a bit too hard to swallow for modern sensibilities. It is thus often disguised in more palatable form. For instance, it can be argued that it is not the difference in the level of development that should militate against the universal application of the same standards, but difference pure and simple in a multicultural world. Far from assuming an imperial civilizing mission for human rights, or indeed Kipling's 'white man's burden',[142] it is (supposedly) the *respect* for the different cultures of other societies that requires that foreign standards not be imposed upon them.

Relativist considerations are especially pronounced in the European context.[143] Not only does relativism have quite a pedigree in Europe and its now largely defunct imperialist projects, but emphasizing the *regional* nature of the European human rights regime can be yet one more way of disguising relativist considerations that would otherwise not be welcome in today's polite society. That these considerations are at play is however evident even from the very text of the ECHR. In its preamble, for example, the states parties considering the adoption of the UDHR state that 'this Declaration aims at securing the *universal* and effective recognition and observance of the Rights therein declared' (emphasis added), and reaffirm 'their profound belief in those fundamental freedoms which are the foundation of justice and peace in the world and are best maintained on the one hand by an effective political democracy and on the other by a common understanding and observance of the human rights upon which they depend'. But they also emphasize that they are 'the governments of *European* countries which are like-minded and have a *common heritage* of political traditions, ideals, freedom and the rule of law' (emphasis added). And then there is the colonial clause in Article 56 of the treaty, which allows for its extension to territories for whose international relations a state party is responsible, but provides in its paragraph 3 that '[t]he provisions of this Convention shall be applied in such territories with due regard, however, to local requirements'.[144]

The European Court case that most stressed the *regional* nature of the ECHR was of course *Bankovic*. There the Court was faced with the applicants' argument that 'any failure to accept that they fell within the jurisdiction of the respondent States would defeat the *ordre public* mission of the Convention and leave a regrettable vacuum in the Convention system of human rights' protection'.[145] The Court's response was intriguing, and it bears quoting in full:

> The Court's obligation, in this respect, is to have regard to the special character of the Convention as a constitutional instrument of *European* public order for the protection of individual human beings and its role, as set out in Article 19 of the Convention, is to ensure

142 See Raustiala, above note 22, at 3 *et seq.*

143 They are also particularly prominent in American case law dealing with the United States' colonial domains in the late nineteenth and early twentieth century—see generally Raustiala, above note 22, at 59 *et seq.*

144 For an erudite, entertaining, and far more insightful account of the interplay between the end of colonialism and the drafting of the ECHR than any I could ever give, see B. Simpson, *Human Rights and the End of Empire* (Oxford University Press, 2004).

145 *Bankovic*, para. 79.

the observance of *the engagements undertaken* by the Contracting Parties (the above-cited Loizidou judgment (*preliminary objections*), at § 93). It is therefore difficult to contend that a failure to accept the extra-territorial jurisdiction of the respondent States would fall foul of the Convention's *ordre public* objective, which itself underlines the essentially regional vocation of the Convention system, or of Article 19 of the Convention which does not shed any particular light on the territorial ambit of that system.

It is true that, in its above-cited *Cyprus v. Turkey* judgment (at § 78), the Court was conscious of the need to avoid 'a regrettable vacuum in the system of human-rights protection' in northern Cyprus. However, and as noted by the Governments, that comment related to an entirely different situation to the present: the inhabitants of northern Cyprus would have found themselves excluded from the benefits of the Convention safeguards and system which they had previously enjoyed, by Turkey's 'effective control' of the territory and by the accompanying inability of the Cypriot Government, as a Contracting State, to fulfil the obligations it had undertaken under the Convention.

In short, the Convention is a multi-lateral treaty operating, subject to Article 56 of the Convention, in an essentially regional context and notably in the legal space (*espace juridique*) of the Contracting States. The FRY clearly does not fall within this legal space. The Convention was not designed to be applied throughout the world, even in respect of the conduct of Contracting States. Accordingly, the desirability of avoiding a gap or vacuum in human rights' protection has so far been relied on by the Court in favour of establishing jurisdiction only when the territory in question was one that, but for the specific circumstances, would normally be covered by the Convention.[146]

These dicta from *Bankovic* have provoked much debate and argument in recent years. Their significance depends on how broadly one reads them. At their narrowest, the Court here does nothing more than reject the applicants' argument that there would exist a vacuum in the ECHR system of protection, on the grounds that the applicants were never protected by the ECHR in the first place. This narrowest reading is in my view correct, as was the Court's rejection of the applicants' vacuous vacuum argument, at least in those terms.[147]

A more expansive reading would be an *a contrario* one—what if, say, the bombing of Belgrade had happened not in 1999, but in 2009, when Serbia was a party to the ECHR? Would the Court's entire analysis in *Bankovic* collapse at that point, and would the applicants then actually be 'within the jurisdiction' of the respondent states? Such an interpretation is not only possible, but is also inspired, one might say, by a whiff of hypocrisy and double standards about the Court's decision.[148] It is hard to escape the impression that the outcome of *Bankovic* would not have been the same had it been Berlin or Brussels that were bombed, rather than Belgrade.[149]

[146] Ibid., para. 80 (explanatory footnote omitted, emphasis in original).

[147] See, e.g., R. Wilde, 'The "Legal Space" or "Espace Juridique" of the European Convention on Human Rights: Is It Relevant to Extraterritorial State Action', (2005) 2 *EHRLR* 115; Gondek, at 174 *et seq*.

[148] See, in that regard, E. Roxstrom, M. Gibney, and T. Einarsen, 'The NATO Bombing Case (*Bankovic* et al. *v. Belgium* et al.) and the Limits of Western Human Rights Protection', (2005) 23 *B.U. Int'l L.J.* 55.

[149] See also Lawson, above note 3, at 114–15.

This reading of course just begs the question whether the vacuum or gap argument was actually all that important in *Bankovic* and in *Cyprus v. Turkey*. To my mind at least, the argument was both marginal and unpersuasive. Yes, it would indeed have been regrettable for Cypriots who possessed rights under the ECHR to lose them because their country was invaded by another sovereign. But they undoubtedly would have lost them vis-à-vis the invader had the invader been a state that was *not* a party to the ECHR. The vacuum was avoided solely because Turkey *was* a party to the ECHR, and because it exercised jurisdiction, i.e. effective overall control, over an area that it invaded. In short, the desirability *vel non* of a vacuum in protection does not and cannot affect the operation of Article 1 ECHR, since a state party either has jurisdiction over a territory or it does not.

But this brings me to another possible reading of *Bankovic*. Yes, it is true that, technically speaking, all the Court does is dismiss a generally meritless argument. But the Court nonetheless says so much more. It speaks of the ECHR as the 'constitutional instrument of *European* public order' (emphasis in original) and the 'essentially regional vocation' of the ECHR system. The ECHR is, according to the Court, a treaty operating in an 'essentially regional context and notably in the legal space (*espace juridique*) of the Contracting States', that 'was not designed to be applied throughout the world, even in respect of the conduct of Contracting States'.

All this talk about the *European* nature of the ECHR and its *espace juridique* would be entirely irrelevant if all the Court was doing was dismissing the vacuum in coverage argument on the grounds that there was no coverage in the first place. What the Court really does here is to send a message: the European system of human rights protection is successful precisely because it is a *European* system of human rights protection, and we do not want to jeopardize it by opening it up to others more than is strictly necessary. The thought of the floodgates opening to litigation by 'aliens' affected by a European state's overseas military adventure, and of the consequences this might have for the ECHR regime already tottering under the pressure of thousands of applications against human rights paragons such as Russia and Turkey, was to the Court understandably unappealing. Think of it what you will, but this fear of a deluge of litigation shaped the Court's decision and the message that it wanted to send. And that message was well heard, as we will now see from the *Al-Skeini* case.

B. Relativism and regionalism in *Al-Skeini*

Recall that there were six applicants in *Al-Skeini*, all of them Iraqis. Five were shot by British troops on patrol in Basra, or died in the course of British military action. The sixth, Baha Mousa, was actually apprehended by British troops, and beaten to death while in British custody. All six applicants claimed that the ECHR and the HRA applied extraterritorially to the acts of British organs or agents in Iraq, and that they were protected by Article 2 ECHR, which enshrines the right to life. The applicants relied on the positive obligation of states to conduct an investigation into a killing, and were not merely contending that the killings themselves were unlawful under Article 2. In the first instance, the case was decided by the Divisional Court;[150] it was then considered by the Court of Appeal,[151] and finally by the House of Lords.[152] As of the time of writing, it was pending before the Grand Chamber of the European Court, which held oral hearings in the case in June 2010.[153]

The applicants' argument was fairly straightforward. The United Kingdom was the occupying power in Basra. It had troops on the ground, and effective overall control over Basra. Therefore, per *Loizidou*, it exercised jurisdiction over the Basra area within the meaning of Article 1 ECHR, and the ECHR applied extraterritorially under a *spatial* conception of jurisdiction. Alternatively, the applicants submitted that they were as individuals under the authority and control of British soldiers, and that this brought them within the United Kingdom's jurisdiction, jurisdiction this time conceived of *personally*. For its part, the UK government contended, relying on *Bankovic*, that the 'effective overall control of an area' notion of jurisdiction applies only to areas that already fall within the *espace juridique* of the ECHR, i.e. to the territories of states parties, such as Cyprus, but not to third states such as Iraq. Even if did, the government submitted that its control over Southern Iraq was not effective for the purposes of Article 1 ECHR, despite it being an occupying power. It further argued that the 'authority and control over a person' notion of jurisdiction was insufficiently supported by earlier case law, and that it was at any rate manifestly contrary to *Bankovic*.

The Divisional Court agreed with the government on both points. It considered that the European Court assigned critical significance to the ECHR as the constitutional instrument of *European* public order, and to the vacuum argument generally.[154] Its holding that the effective overall control of an area test cannot apply to an area that is itself not a part of an ECHR contracting state was only confirmed by the apparent contradiction between an expansive interpretation of Article 1 ECHR and the colonial clause, namely that the United Kingdom could choose whether or not to apply the ECHR to its overseas dependencies, but that it had no choice whether to apply it in Iraq.[155] But the Divisional Court nonetheless took great care to reject the government's more overtly relativist argumentation:

> We prefer to put the matter in this way, rather than on the basis which seemed at one point to be urged on us by Mr Greenwood [counsel for the government], that there were territories in the world, such as Iraq, for which the Convention was not designed and for which they might not be ready. That seemed to us an unhappy submission to have to make

[150] *R. (on the application of Al-Skeini and Others) v. Secretary of State for Defence* [2004] EWHC 2911 (Admin), [2004] All ER (D) 197 (Dec) (hereinafter *Al-Skeini DC*).

[151] *R. (on the application of Al-Skeini and Others) v. Secretary of State for Defence* [2005] EWCA Civ 1609, [2005] All ER (D) 337 (Dec) (hereinafter *Al-Skeini CA*).

[152] *R. (on the application of Al-Skeini and Others) v. Secretary of State for Defence* [2007] UKHL 26, [2008] AC 153 (hereinafter *Al-Skeini HL*).

[153] See M. Milanovic, 'Grand Chamber Hearings and Preview of Al-Skeini and Al-Jedda', *EJIL: Talk!*, 9 June 2010, available at <http://www.ejiltalk.org/grand-chamber-hearings-and-preview-of-al-skeini-and-al-jedda/>.

[154] *Al-Skeini DC*, paras 274–7.

[155] Ibid., para. 278; see also above, Chapter I, Section 4.B.

about a country which was one of the cradles of civilisation. No one knows to whom the baton or batons of the human race will be handed. The Convention was not created because of the humanity of Europe, but because of its failures.[156]

Well put. But for all its admirable sentiment, the Divisional Court did not explain why the regional nature of the ECHR, that was so stressed by the European Court in *Bankovic*, actually *mattered*. If all the Convention did was to create a regional enforcement mechanism for norms and values that are truly *universal*, why then should the application of this treaty be confined solely to the territories of the contracting states?

The Divisional Court thus found that the five applicants killed by British troops while on patrol did not fall within the UK's jurisdiction. However, the Court found that the sixth applicant, who was killed while in British custody, was within the UK's jurisdiction, on the grounds that 'a British military prison, operating in Iraq with the consent of the Iraqi sovereign authorities, and containing arrested suspects, falls within even a narrowly limited exception exemplified by embassies, consulates, vessels and aircraft, and in the case of *Hess v. United Kingdom*, a prison'.[157]

What is so remarkable about this conclusion is that it is unsupported by any broader consideration of principle. It in fact exemplifies the casuistry that is so prevalent in the current case law on the extraterritorial application of the ECHR—for which most of the blame, of course, must be laid at the feet of Strasbourg. It is fair to say that, as a substantive matter, the (perhaps mistaken) killing of civilians by patrolling troops in a combat area is less repugnant, morally or legally, then the beating to death of a defenceless prisoner. But that does not mean that the two situations are at all distinct as to whether the ECHR applies in the first place. If the *espace juridique* doctrine promulgated by the European Court in *Bankovic* really meant that the effective overall control notion of jurisdiction cannot apply beyond the borders of European states, how could the ECHR then not apply in Basra, but apply to a *prison* in Basra?

The applicants appealed the Divisional Court's ruling. On appeal, the government strategically conceded that the ECHR applied to Baha Mousa, but it maintained its objections regarding the extraterritorial applicability of the ECHR to the other five applicants. The Court of Appeal affirmed the Divisional Court's judgment, though its reasons were somewhat different. With regard to Mr Mousa, Lord Justice Brooke thought that he was within the UK's jurisdiction because he 'came within the control and authority of the UK from the time he was arrested at the hotel and thereby lost his freedom at the hands of British troops', not on the basis of an analogy between a UK prison and an embassy.[158] In other words, he applied a personal, rather than spatial, conception of jurisdiction. With respect to the other five applicants, he also disagreed with the Divisional Court's ruling that the effective overall control of an area test can apply only within the *espace juridique* of the ECHR. In his view, it was the fact that the UK actually did not have such control over Basra, because of the intensity of the fighting and its lack of adequate

[156] Ibid., para. 279. [157] Ibid., para. 287. [158] *Al-Skeini CA*, para. 108.

troops, that meant that the UK did not have jurisdiction over Basra within the meaning of Article 1 ECHR, even though it was an occupying power.[159] If the UK *did* have effective control over Basra, per *Bankovic* it would have had to secure to everyone in Basra all rights and freedoms guaranteed by the ECHR, and '[o]ne only has to state that proposition to see how utterly unreal it is'.[160] Though he dismissed the government's *espace juridique* argument, he nonetheless remarked that

> . . . it is in any event very much open to question whether an effort by an occupying power in a predominantly Muslim country to inculcate what the ECtHR has described (in Golder v UK (1975) 1 EHRR 524 at para 34) as 'the common spiritual heritage of the member states of the country [*sic*] of Europe' during its temporary sojourn in that country would have been consistent with the Coalition's goal, which was to transfer responsibility to representative Iraqi authorities as early as possible.[161]

The learned judge was so taken with relativism that he apparently forgot not only about the member states within the Council of Europe which are indeed 'predominantly Muslim,' e.g. Turkey, Albania, and Azerbaijan, but also the growing Muslim population even in the countries of 'old Europe'.[162]

The case then went to the House of Lords, where *espace juridique* made a grand comeback. Five law lords heard the case. Lord Bingham held that, whatever the extraterritorial applicability of the ECHR, the HRA had no extraterritorial application, but the other four law lords disagreed—I will turn to his opinion shortly. For his part, Lord Rodger agreed with the Divisional Court that the effective overall control of an area test can only apply to territories that are already a part of the *espace juridique* of the ECHR. In his view,

> The essentially regional nature of the Convention is relevant to the way that the [European] court operates. It has judges elected from all the contracting states, not from anywhere else. The judges purport to interpret and apply the various rights in the Convention in accordance with what they conceive to be developments in prevailing attitudes in the contracting states. This is obvious from the court's jurisprudence on such matters as the death penalty, sex discrimination, homosexuality and transsexuals. The result is a body of law which may reflect the values of the contracting states, but which most certainly does not reflect those in many other parts of the world. So the idea that the United Kingdom was obliged to secure observance of all the rights and freedoms as interpreted by the European Court in the utterly different society of southern Iraq is manifestly absurd. Hence, as noted in *Bankovic*, 11 BHRC 435, 453–454, para 80, the court had 'so far' recognised jurisdiction based on effective control only in the case of territory which would normally be covered by the Convention. If it went further, the court would run the risk not only of colliding with the jurisdiction of other human rights bodies but of being accused of human rights imperialism.[163]

[159] Ibid., paras 119–25. [160] Ibid., para. 124. [161] Ibid., para. 126.

[162] See R. Wilde, 'Compliance with human rights norms extraterritorially: "human rights imperialism"?', in L. Boisson de Chazournes and M. Kohen (eds), *International Law and the Quest for its Implementation/Le droit international et la quête de sa mise en œuvre, Liber Amicorum Vera Gowlland-Debbas* (Brill/Martinus Nijhoff, 2010), 319, at 329.

[163] *Al-Skeini HL*, para. 78.

According to Lord Rodger and his reading of *Bankovic*, the European Convention on Human Rights should not apply to Iraq precisely because it is the *European* Convention on Human Rights. Indeed, applying it would amount to 'human rights imperialism'—in no small part because of the *Bankovic* holding that the Convention is an all or nothing package.[164] Because it is 'manifestly absurd' to require the UK to secure, say, the right to equality of women, homosexuals, and transsexuals in Iraq, it should also not be expected to respect the right to life of Iraqis when it acts through its own troops.

Lord Rodger further argued that even if the *espace juridique* doctrine did not preclude the effective overall control of area notion of state jurisdiction, the UK could still not be said to have been in effective control of Basra.[165] Baroness Hale agreed with both points of Lord Rodger's reasoning,[166] and so did Lord Carswell[167] and Lord Brown.[168] Lord Brown in particular, like the Divisional Court, considered that the interaction between the jurisdiction clause in Article 1 ECHR and the colonial clause in Article 56 ECHR mandated only one reasonable interpretation of the northern Cyprus cases: that the sole rationale for the effective overall control of an area test is the avoidance of a vacuum in coverage.[169] As he put it,

How then could that principle [of effective overall control] logically apply to any other territory outside the area of the Council of Europe? As the respondent submits, it would be a remarkable thing if, by the exercise of effective control, for however short a time, over non-Council of Europe territory, a state could be fixed with the article 1 obligation to secure within that territory, without regard to local requirements, all Convention rights and freedoms whereas, despite its exercise of effective control over a dependent territory, perhaps for centuries past, the state will not be obliged to secure any Convention rights there unless it has made an article 56 declaration and even then it would be able to rely on local requirements.[170]

Lord Brown further opined that

. . . except where a state really does have effective control of territory, it cannot hope to secure Convention rights within that territory and, unless it is within the area of the Council of Europe, it is unlikely in any event to find certain of the Convention rights it is bound to secure reconcilable with the customs of the resident population. Indeed it goes further than that. During the period in question here it is common ground that the UK was an occupying power in Southern Iraq and bound as such by Geneva IV and by the Hague Regulations. Article 43 of the Hague Regulations provides that the occupant 'shall take all the measures in his power to restore and ensure, as far as possible, public order and safety, while respecting, unless absolutely prevented, the laws in force in the country.' The appellants argue that occupation within the meaning of the Hague Regulations necessarily involves the occupant having effective control of the area and so being responsible for securing there all Convention rights and freedoms. So far as this being the case, however, the occupants' obligation is to respect 'the laws in force', not to introduce laws and the means to enforce them (for example, courts and a justice system) such as to satisfy the requirements of

164 *Al-Skeini HL*, para. 79.
165 Ibid., para. 83.
166 Ibid., paras 85–92.
167 Ibid., paras 96–7.
168 Ibid., para. 109.
169 Ibid., paras 111–14.
170 Ibid., para. 113.

the Convention. Often (for example where Sharia law is in force) Convention rights would clearly be incompatible with the laws of the territory occupied.[171]

As for the sixth applicant, Mr Mousa, Lord Brown recognized UK jurisdiction over him 'only on the narrow basis found established by the Divisional Court, essentially by analogy with the extra-territorial exception made for embassies'.[172]

C. Relativism and regionalism evaluated: another face of effectiveness

What are we thus to make of this jurisprudence on the ECHR's *espace juridique*, and how does *Al-Skeini* fit the case law of the European Court? Can the effective overall control of an area test truly apply only to territories that were already covered by the ECHR, as the House of Lords held?

As indicated above, the answer to the latter question must be no. First, the argument that the ECHR cannot apply outside the territories of its contracting parties taken collectively is textually equally as untenable as the argument that the ECHR cannot apply outside the territory of the specific contracting party whose conduct is being assessed. Nothing in Article 1 supports such an interpretation. The word 'jurisdiction' in Article 1 either means 'effective overall control of an area', or it does not. If it does, the legal status of the territory over which such control is exercised is irrelevant. In fact, the only textual support for limiting the extraterritorial application of the ECHR to the contracting states' *espace juridique* is the colonial clause in Article 56, in other words the clear contradiction between requiring the UK, for example, to abide by the ECHR in Iraq, but giving it the choice whether to apply the ECHR or not to its own dependent territories.[173] But that contradiction is equally manifest if the UK had to abide by the ECHR while acting in the territory of another member state, such as Cyprus. The only cure for this contradiction is to deny the possibility of *any* application of the ECHR outside a state's (metropolitan) territory, an option that the European Court itself finds unpalatable.[174]

Secondly, the view that the *espace juridique* is an indispensable condition for the extraterritorial application of the ECHR is not supported by the case law on which the House of Lords relies. The European Court did not invoke the vacuum in coverage argument in *Loizidou*, the case in which it devised the effective control of an area test, but only in *Cyprus v. Turkey*, where it was of marginal importance.[175] The same was true of *Bankovic*, where the vacuum argument was only the *fifth* of the applicants' alternative arguments, and was treated by the Court as such.[176] If the Court in *Bankovic* was truly announcing a bright-line rule that the ECHR cannot apply outside the territories of states parties to the treaty, as the House of Lords thought it did, it could have easily dispensed with the case in two paragraphs,

[171] Ibid., para. 129. [172] Ibid., para. 132.
[173] On the colonial clause generally, see Chapter I, Section 4.B.
[174] See also *Loizidou* (preliminary objections), paras 86–8.
[175] *Cyprus v. Turkey* [GC], App. No. 25781/94, Judgment, 10 May 2001, para. 78.
[176] *Bankovic*, paras 79–80.

and would not have needed to engage in the extensive and detailed (if fatally flawed) discussion of extraterritoriality that it did.[177]

Thirdly, not only is the House of Lords' judgment in *Al-Skeini* not in accordance with *Loizidou*, *Cyprus v. Turkey*, and *Bankovic*, but it is also contrary to the European Court's post-*Bankovic* case law. Thus, in *Issa v. Turkey*, the applicants were persons living in a part of northern Iraq that was invaded by Turkish forces, who claimed that the ECHR applied to Turkish actions in Iraq. A Chamber of the European Court held as follows:

> The Court does not exclude the possibility that, as a consequence of the military action, the respondent State could be considered to have exercised, temporarily, effective overall control of a particular portion of the territory of northern Iraq. Accordingly, if there is a sufficient factual basis for holding that, at the relevant time, the victims were within that specific area, it would follow logically that they were within the jurisdiction of Turkey (and not that of Iraq, which is not a Contracting State and clearly does not fall within the legal space (*espace juridique*) of the Contracting States (see the above-cited *Bankovic* decision, § 80).
>
> However, notwithstanding the large number of troops involved in the aforementioned military operations, it does not appear that Turkey exercised effective overall control of the entire area of northern Iraq. This situation is therefore in contrast to the one which obtained in northern Cyprus in the *Loizidou v. Turkey* and *Cyprus v. Turkey* cases (both cited above). In the latter cases, the Court found that the respondent Government's armed forces totalled more than 30,000 personnel (which is, admittedly, no less than the number alleged by the applicants in the instant case—see § 63 above—but with the difference that the troops in northern Cyprus were present over a much longer period of time) and were stationed throughout the whole of the territory of northern Cyprus. Moreover, that area was constantly patrolled and had check points on all the main lines of communication between the northern and southern parts of the island.[178]

Thus, the Court in *Issa* declared the application inadmissible, but on the factual ground of lack of effective control by Turkey over northern Iraq. This of course presumes that the ECHR *could* apply in Iraq, even though Iraq is not a part of its *espace juridique*—a holding totally at odds with *Al-Skeini*. The law lords were well aware of this, but chose to disregard *Issa*, considering it to be in conflict with *Bankovic*, a 'watershed' decision by the Grand Chamber of the European Court.[179] But at least on this (admittedly very superficial) level, *Issa* is *not* at odds with *Bankovic*, as I have explained above. In disregarding *Issa*, their Lordships emphasized the fact that there were no other cases in which the European Court used the effective overall control of an area test for a territory outside the ECHR's *espace juridique*. But that point entirely misses its mark, as no application asserting the application of this test to a non-European territory was actually filed before *Issa*.[180]

[177] See also Wilde, above note 147, at 118 *et seq*; T. Thienel, 'The ECHR in Iraq', (2008) 6 *JICJ* 115, at 119 *et seq*.

[178] *Issa v. Turkey*, App. No. 31821/96, Judgment, 16 November 2004, paras 74 and 75.

[179] *Al-Skeini HL*, para. 108.

[180] The Court *did*, however, examine a number of cases dealing with extraterritorial situations outside the Council of Europe, but under the personal conception of 'jurisdiction' as authority and control over individuals—see below, Chapter IV, Section 3. See also Wilde, above note 147, at 120 *et seq*.

To sum up, the *espace juridique* doctrine, developed by the English courts in *Al-Skeini* as a limitation on the extraterritorial application of the ECHR, has no support either in the text of the treaty or in the case law of the European Court.[181] Now that *Al-Skeini* is before the European Court, it will hopefully take the opportunity to make this clear. On the other hand, the alternative holding of *Al-Skeini*, that British troops actually lacked effective overall control over Basra, is far from implausible, and will be addressed in more detail in the next chapter of this study.[182]

But having said that, although their Lordships' decision in *Al-Skeini* does not rest well with the text and holding in *Bankovic*, it nonetheless sits very well with *Bankovic*'s overall spirit and tenor. Though the judges of the European Court were considerably more coy than their English counterparts in expressing relativist sentiment, it is again hard to escape the impression that they to a great extent shared it. All that talk about the essentially regional and European nature of the ECHR was not there for nothing. Whether they say it or not, the judges in Strasbourg are fully aware that the ECHR's strength—its *European-ness*—is also its weakness, if the ECHR is to be applied extraterritorially. In other words, relativist considerations plainly matter for judges in such cases, be it in the less (Iraq is 'not ready' for the application of the ECHR) or the more politically correct variant (applying the ECHR in Iraq would amount to 'human rights imperialism').[183]

The question, therefore, is not *whether*, but *should* such considerations matter. To that normative question one could give an almost reflexive universalist answer—much as I have already given one with regard to citizenship. When Lord Carswell states that it would be 'manifestly absurd' for the UK to have to safeguard the ECHR rights of homosexuals and transsexuals in Iraq, and that this would moreover amount to 'human rights imperialism', a universalist response would be that yes, the UK *does* have an obligation to protect the rights of sexual minorities, and no, there is nothing imperialist or absurd about it. These rights are not just European rights; universality of human rights is the law, and that is that.[184] When Lord Brown argues that occupying states have an obligation under Article 43 of the 1907 Hague Regulations to '[respect], unless absolutely prevented, the laws in force in the [occupied] country', and that the ECHR would often be incompatible with such laws, especially Sharia, the universalist response would be—so what! To the

[181] See Wilde, above note 147; Thienel, above note 177; Lawson, above note 3, at 114–15; P. Leach, 'The British Military in Iraq—the Applicability of the *Espace Juridique* Doctrine under the European Convention on Human Rights', (2005) *PL* 448.

[182] See Chapter IV, Section 2.C.3.

[183] Relativism as grounds for denying individual rights protection is of course itself a theme with a distinctly imperialist pedigree, as evident, for example, from the US Supreme Court's early twentieth century *Insular Cases*, in which it held that the protections of the US Constitution did not fully extend to non-metropolitan territories over which the US had title, such as Puerto Rico. See generally Raustiala, above note 22, at 79 *et seq*. See also *Reid v. Covert*, at 14 (per Black J.) (limiting the applicability of *Insular Cases* and limiting them to 'rules and regulations to govern temporarily territories with wholly dissimilar traditions and institutions'.)

[184] Not to mention that Iraq is a party to the ICCPR, which contains much the same rights as the ECHR, and has been interpreted in much the same way by the Human Rights Committee as the ECHR by the European Court.

extent that these laws, Sharia included, are incompatible with international human rights, these laws—say, for example, the punishment of adultery by death by stoning—would have to go. An occupying power would indeed be 'absolutely prevented' from allowing the continuing existence of laws that are contrary to its human rights obligations. Article 43 of the Hague Regulations needs to be read flexibly, in light of the numerous developments in international law during the past century, especially the advent of human rights.[185] Or, perhaps, ultimately there would be a norm conflict between the UK's obligations under the ECHR and its obligations under IHL, since states are perfectly capable of entering into contradictory commitments.[186]

Regardless of whether one finds the universalist response to be persuasive, it is beyond doubt that a gap opens up between the factual and the normative. On the one hand, the differences, cultural or otherwise, between Europe and Iraq are clearly vast. On the other, the single defining characteristic of human rights *qua human* rights is that they are immanent in the human condition alone. 'Human rights imperialism' is an oxymoron if the rights in question really are *human* rights. But the gap still remains. Perhaps the wisest way of bridging it would be to dispense for a time with the canards of the old universalist–relativist debate. The issue is not so much whether the ECHR and its case law are contrary to the cultural mores of the territory where it is to be applied, but whether the obligations arising from this treaty can be realistically complied with.

If observed solely from this standpoint of effectiveness, the concerns expressed by most of the British judges in *Al-Skeini* become clearer, but are also exposed as somewhat exaggerated. The positive obligation to secure or ensure the human rights of persons within a state's jurisdiction, that arises once a state obtains effective overall control over an area, is not an absolute one. It is an obligation of due diligence, an obligation for a state to do all that can reasonably be expected of it, taking into account all of the relevant circumstances.[187] There is absolutely no reason why this obligation should be read in exactly the same way when a state applies the ECHR in its own territory in peacetime conditions and when a state invades and occupies a foreign land. As explained by Lord Justice Sedley in his opinion in *Al-Skeini*:

> No doubt it is absurd to expect occupying forces in the near-chaos of Iraq to enforce the right to marry vouchsafed by Art. 12 or the equality guarantees vouchsafed by Art. 14. But I do not think effective control involves this. If effective control in the jurisprudence of the ECtHR marches with international humanitarian law and the law of armed conflict, as it clearly seeks to do, it involves two key things: the de facto assumption of civil power by an occupying state and a concomitant obligation to do all that is possible to keep order and protect essential civil rights. It does not make the occupying power the guarantor of rights;

[185] See Thienel, above note 177, at 124–7.

[186] See Section 9, as well as Chapter V below.

[187] See also *Velasquez Rodriguez Case*, Judgment of 29 July 1988, Inter-Am.Ct.H.R. (Ser. C) No. 4 (1988), para. 172; *Application of the Convention on the Prevention and Punishment of the Crime of Genocide*, Judgment, 27 February 2007, para. 430.

nor therefore does it demand sufficient control for all such purposes. What it does is place an obligation on the occupier to do all it can.[188]

Beside the considerable degree of flexibility that is inherent in the very concept of due diligence obligations, it should also be borne in mind that the universality of human rights does not demand total uniformity. Thus, for example, the assembled heads of state and government affirmed in the 2005 World Summit Outcome document that

> ... all human rights are universal, indivisible, interrelated, interdependent and mutually reinforcing and that all human rights must be treated in a fair and equal manner, on the same footing and with the same emphasis. *While the significance of national and regional particularities and various historical, cultural and religious backgrounds must be borne in mind,* all States, regardless of their political, economic and cultural systems, have the duty to promote and protect all human rights and fundamental freedoms.[189]

The ECHR system itself allows its member states a significant 'margin of appreciation', i.e. the liberty to adopt different positions and solutions for some of the same problems, that is especially pertinent in respect of moral controversies. As noted by Lord Rodger in *Al-Skeini*,[190] the European Court develops the ECHR as a living instrument by reference to a growing consensus among European states on a given issue, which thus serves as a limiting factor on the states' margin of appreciation. And so, for example, the Court found decades ago that there was a European consensus against the criminalization of homosexuality, and that therefore the few states that were outliers on the issue had to conform to the opinion of the majority, which demonstrated that such criminalization was not necessary in a democratic society.[191] On the other hand, there is still no European consensus on the question when life that is worthy of legal protection truly begins,[192] and states are hence free either to prohibit or to liberalize abortion within certain limits. Similarly, there is no consensus on whether states should recognize same-sex marriages, and they accordingly still enjoy a wide margin of appreciation.[193]

To the extent that some of the state obligations that can arise under the ECHR are subject to an evolving interpretation dependent on a European consensus, there is no bar to varying the content of such obligations in an extraterritorial context.[194] Again, the positive obligation to secure or ensure human rights requires a state to do all that it reasonably can, not to do the impossible. In determining the content of this obligation, it is entirely feasible to give 'due regard for local requirements', even without an explicit provision to that effect in the treaty.[195] How that content is to

[188] *Al-Skeini CA*, para. 196.
[189] World Summit Outcome Document 2005, UN Doc. A/RES/60/1, para. 121.
[190] *Al-Skeini HL*, para. 78.
[191] *Dudgeon v. United Kingdom,* App. No. 7525/76, Judgment, 22 October 1981.
[192] *Vo v. France*, App. No. 53924/00, Judgment, 8 July 2004.
[193] *Schalk and Kopf v. Austria,* App. No. 30141/04, Judgment, 24 June 2010.
[194] See also Wilde, above note 162, at 332 *et seq*; Thienel, above note 177, at 122–4.
[195] One could say that this would even be *required* by the principle of self-determination, which courts could fully take into account when interpreting the ECHR—see Wilde, above note 162, at 339

be determined, and how that content is to be made effective, are questions requiring policy and value judgments, and are best answered in each given case. That in *some* cases this inquiry will be difficult is not reason enough to deny the possibility of the extraterritorial applicability of the ECHR or some other human rights treaty altogether, only to then invent an unprincipled 'embassy' or 'prison' exception to this supposed rule when our moral intuitions do not permit the result that our soldiers outside our borders are free to torture and kill at will.

If, therefore, relativist considerations are put in the proper perspective, that of effectiveness, the European Court's exegesis in *Bankovic* on the ECHR's 'essentially regional' character becomes more or less meaningless. Without relativism, 'Europe' is just an arbitrary geographical label, devoid of any substance. That a human rights treaty is a regional one is irrelevant to the question of its (extra-)territorial application. It is the whole of Russia, from Kaliningrad to Vladivostok, that is a party to the ECHR; the territorial application of the Convention does not end with the Urals. Nor does the Convention's writ in Turkey stop at the Bosphorus, nor are Georgia, Armenia, and Azerbaijan somehow by definition not susceptible to its coverage, merely because they are not in 'Europe,' even though they are parties to the ECHR. There is, in other words, no principled reason for the ECHR's territorial scope of application to be any different in a given situation than that of the ICCPR or the CAT merely on the basis that the former is a regional, and the latter are universal instruments. To the extent that common heritage, cultural or otherwise, can influence the application of a human rights treaty, considerations of effectiveness are far from trivial and should be addressed on their own merits.

7. Preventing Arbitrary Distinctions and Results and the Abuse of Law

Preventing arbitrariness is an instrumental rule of law consideration that frequently comes out in favour of the extraterritorial application of a human rights instrument. Consider, for example, the position of Justice Kennedy in *Boumediene*:

> Yet the Government's view is that the Constitution had no effect there [in Guantanamo], at least as to noncitizens, because the United States disclaimed sovereignty in the formal sense of the term. The necessary implication of the argument is that by surrendering formal sovereignty over any unincorporated territory to a third party, while at the same time entering into a lease that grants total control over the territory back to the United States, it would be possible for the political branches to govern without legal constraint.
>
> Our basic charter cannot be contracted away like this. The Constitution grants Congress and the President the power to acquire, dispose of, and govern territory, not the power to decide when and where its terms apply. Even when the United States acts outside its borders, its powers are not 'absolute and unlimited' but are subject 'to such restrictions as are expressed in the Constitution.' *Murphy v. Ramsey*, 114 U.S. 15, 44, 5 S.Ct. 747, 29 L.

et seq. See also R. Wilde, 'Complementing Occupation Law? Selective Judicial Treatment of the Suitability of Human Rights Norms', (2009) 42 *Israel L Rev* 80.

Ed. 47 (1885). Abstaining from questions involving formal sovereignty and territorial governance is one thing. To hold the political branches have the power to switch the Constitution on or off at will is quite another. The former position reflects this Court's recognition that certain matters requiring political judgments are best left to the political branches. The latter would permit a striking anomaly in our tripartite system of government, leading to a regime in which Congress and the President, not this Court, say 'what the law is.' *Marbury v. Madison*, 1 Cranch 137, 177, 2 L.Ed. 60 (1803).[196]

Note how Justice Kennedy considers deferring to the judgment of the political branches, an issue that I will come to shortly. However, he refuses to do so, emphasizing the Supreme Court's own institutional role in preserving the rule of law, and does so on the grounds that it would be unacceptable to 'hold the political branches have the power to switch the Constitution on or off at will'. It is the *arbitrary* nature of the government's theory that, by choosing where to imprison people it can thereby choose to operate without legal constraint, made it unpalatable to an institution, the Supreme Court, which sees itself as the final guardian against such arbitrariness. The Court was even more reluctant to accept the government's approach because it was not the product of historical circumstance but was quite deliberate—on the advice of its lawyers, the government chose Guantanamo as a detention facility *precisely* in order to be free of any legal constraint.[197]

Similarly, consider the Human Rights Committee's universality-inspired holding in *Lopez Burgos* that 'it would be unconscionable to so interpret the responsibility under article 2 of the Covenant as to permit a State party to perpetrate violations of the Covenant on the territory of another State, which violations it could not perpetrate on its own territory'.[198] For his part, in his individual opinion in *Lopez Burgos* Christian Tomuschat considered that to 'construe the words "within its territory" [in Article 2(1) ICCPR] pursuant to their strict literal meaning as excluding any responsibility for conduct occurring beyond the national boundaries would, however, lead to utterly absurd results'.

Tomuschat appears to be invoking the idea—perhaps even a full-fledged canon of interpretation—that legal norms should always be interpreted so as to avoid

[196] *Boumediene*, at 2258–9.

[197] See, e.g., Memorandum from Patrick F. Philbin and John C. Yoo, Deputy Assistant Attorneys General, Office of Legal Counsel, to William J. Haynes II, General Counsel, Department of Defense, 28 December 2001, available in K. Greenberg et al. (eds), *The Torture Papers* (Cambridge University Press, 2005), at 29 *et seq*.. The OLC's advice was rightly criticized for

> treat[ing] the Constitution the way aggressive lawyers treat the tax code, as an annoyance to be evaded. It must be followed, if you cannot figure out how to get around it, but it is not of any normative significance. And Guantanamo is in fact much like a tax shelter, something created for no purpose but circumvention of the law . . . Treating the Constitution like the tax code disrespects it, and to the extent that the Constitution is a charter of American values, it disrespects those values.

K. Roosevelt, 'Detention and Interrogation in the Post-9/11 World', (2008) 42 *Suffolk U. L. Rev.* 1, at 29.

[198] *Lopez Burgos v. Uruguay* (1981) 68 ILR 29, para. 12.3. See also *Celiberti de Casariego v. Uruguay* (1981) 68 ILR 41.

absurd results. However, whether courts or quasi-judicial bodies actually have the authority to disregard or 'fix' treaty or statutory language merely because, in their view, it leads to absurd results is at the very least non-obvious. In the international law-specific context, Article 32 VCLT merely allows for recourse to supplementary means of interpretation, such as the *travaux*, if interpretation under Article 31 leads to manifestly absurd or unreasonable results. It does not, however, exclude the possibility that the absurd or unreasonable result might, in fact, be confirmed by having recourse to the supplementary means.[199] That result may be exactly what states wanted.

This is not to say that an assessment of arbitrariness or absurdity is irrelevant when it comes to the extraterritorial application of human rights instruments. Far from it. My point is that this assessment requires a *policy and value judgment*, and that this judgment depends on some sort of moral or ideological baseline. When it comes to human rights, that baseline is universality. It is *because* we normatively regard human rights as rooted in a universally valid notion of human dignity that it is absurd or arbitrary to make these rights dependent on naked title to territory.

However, because an assessment of arbitrariness is linked to some moral foundation as a reference point, it may not on its own contribute much if that foundation itself is contested. For example, for most European lawyers it would be absurd and arbitrary to say that a UK national held in Iraq by the UK would be protected from torture under the ECHR, but that an alien would not be so protected. In mainstream American legal thought, however, it is not regarded as absurd that a US national held in Iraq would have constitutional rights, while an alien would not, because universality is largely (still) not seen as the foundation of individual rights under the US Constitution.[200] Recall that even with regard to territorial title, it was only the barest 5 to 4 majority of the Supreme Court in *Boumediene* that thought that it would be arbitrary to make individual rights exclusively contingent upon it, but even the majority thought it to be relevant.

8. Political Questions, Deference, and Institutional Incompetence

Because of the political implications arising from judicial intervention in any given case, deference is a major theme in the case law on the extraterritorial application of human rights guarantees.[201] In the domestic context, particularly the US one, the

[199] See also Gardiner, above note 1, at 328–30.

[200] See above, Section 5.

[201] By 'deference' I mean the conscious choice of one decision-maker to withhold making its own judgment on a matter—at least to an extent—opting rather to respect the judgment of another decision-maker that appears to be more competent to make this judgment according to some predefined criterion, such as expertise or democratic legitimacy. The term seems appropriate enough to me in this context, although one should note the criticism it has attracted from high quarters:

> ...although the word "deference" is now very popular in describing the relationship between the judicial and the other branches of government, I do not think that its overtones of servility, or perhaps gracious concession, are appropriate to describe what is happening. In a society based upon the rule of law and the separation of powers, it is necessary to decide

issue may be one of the separation of powers, and the proper degree of deference that courts should accord to the political branches. In the UK context, with domestic courts applying an international human rights treaty, we also have deference to Strasbourg. For its part, the Strasbourg Court may also feel reluctant to absorb the political fallout that a controversial decision in favour of terrorists or what have you would bring both from the general public in European states and from the states themselves, thus tempting it to produce various rationales for dismissing applications on jurisdictional or admissibility grounds, rather than examining them on the merits. In all of these cases, courts are motivated by fear of their own institutional incompetence, perceived or real, and/or lack of a democratic, political legitimacy.

Consider, for example, how Justice Scalia concludes his opinion in *Rasul*:

> Departure from our rule of *stare decisis* in statutory cases is always extraordinary; it ought to be unthinkable when the departure has a potentially harmful effect upon the Nation's conduct of a war. The Commander in Chief and his subordinates had every reason to expect that the internment of combatants at Guantanamo Bay would not have the consequence of bringing the cumbersome machinery of our domestic courts into military affairs. Congress is in session. If it wished to change federal judges' habeas jurisdiction from what this Court had previously held that to be, it could have done so.... For this Court to create such a monstrous scheme in time of war, and in frustration of our military commanders' reliance upon clearly stated prior law, is judicial adventurism of the worst sort.[202]

Faced with such venomous criticism from the bench itself, it is little wonder that the intervention of the Supreme Court into Guantanamo came slowly and half-heartedly. It ultimately did come, however, with the majority of the Supreme Court refusing to defer any longer to the judgment of the political branches. In the words of Justice Kennedy in *Boumediene*:

> Our opinion does not undermine the Executive's powers as Commander in Chief. On the contrary, the exercise of those powers is vindicated, not eroded, when confirmed by the Judicial Branch. Within the Constitution's separation-of-powers structure, few exercises of judicial power are as legitimate or as necessary as the responsibility to hear challenges to the authority of the Executive to imprison a person. Some of these petitioners have been in custody for six years with no definitive judicial determination as to the legality of their detention. Their access to the writ is a necessity to determine the lawfulness of their status, even if, in the end, they do not obtain the relief they seek.
>
> Because our Nation's past military conflicts have been of limited duration, it has been possible to leave the outer boundaries of war powers undefined. If, as some fear, terrorism continues to pose dangerous threats to us for years to come, the Court might not have this luxury. This result is not inevitable, however. The political branches, consistent with their

> which branch of government has in any particular instance the decision-making power and what the legal limits of that power are. That is a question of law and must therefore be decided by the courts.

BBC v. ProLife Alliance [2003] UKHL 23, [2004] 1 AC 185, para. 75 (per Lord Hoffmann). To this I would only add that in time of war or other great public emergency servility and concession are not uncommon attitudes for courts to take.

202 *Rasul*, at 506.

independent obligations to interpret and uphold the Constitution, can engage in a genuine debate about how best to preserve constitutional values while protecting the Nation from terrorism.... The laws and Constitution are designed to survive, and remain in force, in extraordinary times. Liberty and security can be reconciled; and in our system they are reconciled within the framework of the law. The Framers decided that habeas corpus, a right of first importance, must be a part of that framework, a part of that law.[203]

The split within the Court remained as large as ever, with Chief Justice Roberts responding in the following terms:

So who has won? Not the detainees. The Court's analysis leaves them with only the prospect of further litigation to determine the content of their new habeas right, followed by further litigation to resolve their particular cases, followed by further litigation before the D.C. Circuit—where they could have started had they invoked the DTA procedure. Not Congress, whose attempt to 'determine—through democratic means—how best' to balance the security of the American people with the detainees' liberty interests, has been unceremoniously brushed aside. Not the Great Writ, whose majesty is hardly enhanced by its extension to a jurisdictionally quirky outpost, with no tangible benefit to anyone. Not the rule of law, unless by that is meant the rule of lawyers, who will now arguably have a greater role than military and intelligence officials in shaping policy for alien enemy combatants. And certainly not the American people, who today lose a bit more control over the conduct of this Nation's foreign policy to unelected, politically unaccountable judges.[204]

It is hard to escape the impression that had there been another major terrorist attack on US soil since 9/11, it would have been highly unlikely that the Supreme Court would have garnered enough courage to intervene with regard to the administration's indefinite detention policy. The Chief Justice's reference to the 'rule of lawyers', i.e. 'unelected, politically unaccountable judges', is of course overly dramatic. It is precisely because judges are less amenable to political pressure and control that they are a safeguard against individual rights being swept away at a majority's whim. On the other hand, judges are not, and should not be, completely insulated from political accountability for their actions, and it would be fanciful to suggest that judicial intervention in the face of overwhelming public opinion can be strong, or can persist indefinitely.

Chief Justice Roberts is right to point out, however, that it would be nonsensical to confine the Court's approach to the extraterritoriality of the Constitution to the facts of Guantanamo, 'a jurisdictionally quirky outpost'. Yet it is precisely in cases of great political controversy that courts tend to issue narrow rulings that are confined to very specific sets of facts. Thus, though the Supreme Court in *Boumediene* did not confine its holding to Guantanamo, it left open the issue of the extraterritorial application of the Constitution to US detention facilities in Iraq or Afghanistan, such as Bagram. Likewise, the Court said nothing about the substantive question of the authority to detain on preventive grounds and the law governing it, with both of these issues now percolating through the lower courts,

[203] *Boumediene*, at 2277.
[204] Ibid., at 2293 (per Roberts CJ, citations omitted).

which are forced to go about this business without guidance from above, often producing conflicting results.[205]

This tendency of political controversy to lead to narrow, even unprincipled rulings, incapable of providing guidance for the future and leading to conflicts in the jurisprudence, is manifest in the case law of the European Court. Consider, for example, how that case law was seen by the UK courts in *Al-Skeini*. In the Court of Appeal, Lord Justice Sedley somewhat euphemistically remarked that the 'decisions of the European Court of Human Rights do not speak with a single voice',[206] a remark echoed by Lord Rodger in the House of Lords.[207] All of the British judges exhibited a considerable degree of frustration with Strasbourg, and resented the fact that it produced an incoherent jurisprudence from which they had to pick and choose without any certain criteria. Hence, they were naturally predisposed to give as narrow an interpretation of the territorial scope of the ECHR as possible under Strasbourg jurisprudence. In the words of Lord Brown:

> There seems to me, indeed, a greater danger in the national court construing the Convention too generously in favour of an applicant than in construing it too narrowly. In the former event the mistake will necessarily stand: the member state cannot itself go to Strasbourg to have it corrected; in the latter event, however, where Convention rights have been denied by too narrow a construction, the aggrieved individual *can* have the decision corrected in Strasbourg.[208]

Baroness Hale likewise remarked that

> [w]hile it is our task to interpret the Human Rights Act 1998, it is Strasbourg's task to interpret the Convention. It has often been said that our role in interpreting the Convention is to keep in step with Strasbourg, neither lagging behind nor leaping ahead: no more, as Lord Bingham said in *R (Ullah) v Special Adjudicator* [2004] UKHL 26; [2004] 2 AC 323, para 20, but certainly no less: no less, as Lord Brown says at para 106, but certainly no more. If Parliament wishes to go further, or if the courts find it appropriate to develop the common law further, of course they may. But that is because they choose to do so, not because the Convention requires it of them.[209]

Nowhere is this 'let Strasbourg fix it' attitude more apparent than in the opinion of Lord Bingham. Recall that unlike the other law lords, he was of the view that the Human Rights Act did not apply extraterritorially, relying on a presumption against extraterritorial application of domestic legislation, and that thus the question of the ECHR's extraterritorial application did not arise domestically.[210] That was at least the (unconvincing) reason he formally gave. As the other law lords pointed out, why should Parliament have legislated a narrower scope of application for the HRA in domestic law than the one of the ECHR in international law, when this would in no way have violated the sovereignty of other states, and when the

[205] Thus, for example, post-*Boumediene* a US District Court found that habeas did extend to Bagram, but was overruled on that point by the DC Circuit Court of Appeals—see *Maqaleh v. Gates*, No. 09-5265, 21 May 2010. See also Keitner, above note 4, at 22 *et seq*; Raustiala, above note 22, at 214 *et seq*.

[206] *Al-Skeini CA*, para. 192. [207] *Al-Skeini HL*, para. 67.

[208] Ibid., para. 106. [209] Ibid., para. 90. [210] Ibid., para. 26.

whole purpose of the HRA was to provide domestic remedies for ECHR violations, so that applicants did not have to, and could not, immediately petition Strasbourg?[211]

The subtext of his opinion was simply that he did not wish to have to sort through Strasbourg's mess. He thought it 'not only unnecessary but unwise' to express an opinion on the extraterritorial application of the ECHR,[212] as there 'could scarcely be a more fundamental question, nor one more obviously suitable for resolution (in a doubtful case) by a supranational rather than a national court'.[213] He concluded that

> [i]f any of these claimants pursues an application against the UK at Strasbourg, as it is of course open to them to do, the court there will rule on the admissibility of the applications. I do not think that any useful purpose is served by seeking to predict what its decision will be or to suggest what it should be.[214]

Lord Bingham's sentiment—'passing the buck' in the American idiom—is perfectly understandable, in the context of political controversy coupled with a lack of legal certainty. Why should, after all, a British judge applying the ECHR have to decide as a matter of first impression on issues that Strasbourg itself wants to avoid?[215]

It is almost trite to say that courts tend to be more cautious and deferential in times of war or other public emergency, or that they are never truly insulated from the tides of public opinion. This tendency is frequently embodied in the fixation on preliminary questions, such as standing, admissibility, or jurisdiction, through the medium of which courts can avoid addressing the merits of hotly contested disputes. This, in my view, goes a long way in explaining the incoherence of the Strasbourg jurisprudence on Article 1 ECHR. It is understandable that, in the wake of 9/11, the European Court in *Bankovic* was reluctant to open the floodgates to litigation coming from places like Afghanistan or Iraq, particularly because examining such cases would frequently involve the application of international humanitarian law, in which the Court lacks both experience and expertise. The practical

211 Ibid., paras 50, 54–60 (per Lord Rodger); para. 88 (per Baroness Hale); para. 96 (per Lord Carswell); paras 140–50 (per Lord Brown).

212 Ibid., para. 27.

213 Ibid., para. 28.

214 Ibid., para. 32.

215 We can see a similar tendency in *R. (Smith) v. Secretary of State for Defence* [2010] UKSC 29, a case before the UK Supreme Court dealing with the ECHR rights of UK soldiers stationed in Iraq, that we will be examining in more detail in Chapter IV, Section 3.C.3. below. Thus, for example, Lord Phillips remarked that (para. 60):

> We are here dealing with the scope of the Convention and exploring principles that apply to all contracting States. The contention that a State's armed forces, by reason of their personal status, fall within the jurisdiction of the State for the purposes of article 1 is novel. I do not believe that the principles to be derived from the Strasbourg jurisprudence, conflicting as some of them are, clearly demonstrate that the contention is correct. The proper tribunal to resolve this issue is the Strasbourg Court itself.

Similarly, see also the judgments of Lord Hope (paras 90–93, esp. 92: 'A decision of that kind is best left to Strasbourg') and Lord Brown (para. 147).

difficulties in obtaining a reliable factual record could likewise be immense.[216] As a distinguished commentator put it, 'are these really questions which are fitting for a Court of Human Rights to address?'[217]

However, that the Court's reluctance is understandable does not mean that it is justifiable, and it has inevitably led to an Article 1 jurisprudence which is lacking in principle. I agree with Rick Lawson[218] that it would have been more intellectually honest for the Court in *Bankovic* to have invented on the spot a political question or justiciability doctrine that would have allowed it to keep the floodgates closed, if this is what it feared—in that regard, the American approach is to an extent more open and transparent. Yet, of course, because such doctrines do not fit very well with notions of human dignity and universal human rights, by doing so the Court would have openly gone against the basic values that it is meant to protect. The middle path it chose was understandably appealing.

My main suggestion is this: there is ample room for various forms of deference and flexibility on the merits. This is where these disputes are to be resolved, and where the hard questions lie. Denying the extraterritorial applicability of human rights treaties on the basis of wholly arbitrary criteria only serves to push these hard questions under the carpet, but they will not stay there forever. Though the temptation to delay deciding on such controversial questions as overseas security detention or the use of drones for targeted killings is understandable, giving into it only creates more problems in the future. Eventually, the buck must stop somewhere.

9. Effectiveness and Norm Conflict

Another factor which leads courts to limit the extraterritorial applicability of human rights instruments is their desire to avoid norm conflicts. Consider, for instance, the circumstances of the *Afghan Detainees* case. Assume that the Canadian Charter of Rights and Freedoms applied extraterritorially, and that it prohibited Canadian soldiers from transferring detainees in their custody to Afghan authorities if a

[216] Thus, in an interview given after his retirement, the former President of the European Court Luzius Wildhaber (who *inter alia* presided over the *Bankovic* Grand Chamber) remarked that

> ...from the point of view of evidence, extra-territoriality poses a problem. How can the Court function effectively as a court when there is no prospect of it acquiring reliable evidence concerning the situation beyond the frontiers of Member States? Does expecting the Court to act in such circumstances not risk turning it into a campaigning organisation making allegations without solid evidence? For judges like me seeking to avoid this risk this is a compelling reason to be very careful about extending the notion of extra-territoriality too far and to be wary about departing too much from the *Bankovic* judgment.

S. Greer, 'Reflections of a Former President of the European Court of Human Rights,' (2010) *EHRLR* 169, 174. On the other hand, one could observe that the Court has to grapple with such evidentiary difficulties even within Europe, and not without success, particularly with regard to disappearances and other human rights violations in Chechnya.

[217] See M. O'Boyle, 'The European Convention on Human Rights and Extraterritorial Jurisdiction: A Comment on "Life After Bankovic"', in Coomans and Kamminga 125, at 135.

[218] Lawson, above note 3, at 116.

serious risk was established that they would be subjected to inhuman treatment. At the same time, under its agreement with the Afghan government that allowed for the presence of Canadian troops on Afghan soil, Canada undoubtedly has an international obligation to comply with a request by the Afghan government to transfer any prisoners to its custody. If a serious risk of ill-treatment was in fact established, there would exist a conflict between Canada's obligations under a norm of its own constitutional law and a norm of international law. As far as international law is concerned, that conflict can only be resolved in favour of international law, since compliance with rules of domestic law—even domestic *constitutional* law—cannot be an excuse for failing to abide by rules of international law.[219] Domestic constitutional law can of course see things differently.[220] It is precisely because Canadian authorities would have been faced with the choice of complying with their treaty obligation and violating the Charter, or complying with their own constitution and violating the treaty, that it was an attractive option for courts to say that the Charter did not apply extraterritorially.

But that is not the end of the matter. If the Charter can apply extraterritorially to the Afghan detainees, so can the human rights treaties to which Canada is a party, notably the ICCPR and the CAT. These treaties can also create *non-refoulement* obligations, i.e. obligations not to surrender or return a person, when there is a serious risk of torture or inhuman treatment, as for example in the famous *Soering* case.[221] It would then be not just Canadian constitutional law and a norm of international law that would be conflicting—there would be a conflict between two norms of international law equally binding on Canada.

In international law, such situations of norm conflict are a part of the broader phenomenon of fragmentation.[222] I have examined the implications of such norm conflicts on human rights elsewhere in more detail,[223] while Chapter V of this study will deal more exhaustively with norm conflicts, human rights, and humanitarian law. Briefly, there are three basic ways of handling norm conflicts. First, they can be avoided through interpretation, in essence by interpreting one of the two conflicting norms downwards so that the conflict is only apparent, but not genuine. Secondly, if a norm conflict cannot be *avoided* through the harmonious interpretation of the two norms, it could be *resolved* by giving priority to one norm over another, on the basis of hierarchy or a hierarchy-like rule, such as *jus cogens*, but this is rarely a viable option.[224] Finally, in international law it is entirely possible that

[219] See Art. 27 VCLT; Art. 3 ILC ASR; *Reparation for Injuries Suffered in the Service of the United Nations*, Advisory Opinion, ICJ Reports 1949, 174, at 180.

[220] See, e.g., *Reid v. Covert*, 354 U.S. 1, 15–18 (1957) (holding that, from the standpoint of US law, the Constitution and its constraints on the actions of the government prevail over contrary treaties).

[221] *Soering v. United Kingdom*, App. No. 14038/88, Judgment, 7 July 1989.

[222] See generally ILC, 'Fragmentation of International Law: Difficulties Arising from the Diversification and Expansion of International Law', UN Doc. A/CN.4/L.682 (13 April 2006).

[223] See M. Milanovic, 'Norm Conflict in International Law: Whither Human Rights?', (2009) 20 *Duke J. Comp. & Int'l L.* 69.

[224] Hierarchical rules generally and *jus cogens* specifically are very few in number, and are of little practical relevance. For example, that the prohibition of torture and (possibly also) inhuman treatment

there are norm conflicts that are both unavoidable and unresolvable. States are, like human beings, entirely capable of entering into mutually contradictory commitments. In such cases, a court can do nothing but affirm the validity of both norms and apply the norm that it was asked to apply in the first place, while the conflict itself could only be resolved by political means, for instance by renegotiating one of the conflicting treaties. If that is impossible, the state that bound itself by two conflicting norms has a political choice to make—which norm to abide by, and which norm to breach and incur responsibility for doing so.

To see norm conflict avoidance at work, we need only turn back to our *Afghan Detainees* example. As for the Canadian Charter, we have already seen that the courts avoided the norm conflict by denying that the Charter applied extraterritorially. If we were to look at the case from a human rights treaty standpoint, it could be argued that though the treaty between Afghanistan and Canada on the surrender of detainees contained no provisions on *non-refoulement*, such a provision should be implicitly read into it, because of the wide acceptance of the *non-refoulement* principle in international law, including human rights treaties to which both Afghanistan and Canada are parties. This would be a human rights-friendly form of avoidance, but it would involve a risky assumption of authority on the part of courts, which would be reading human rights content into treaties when it is actually not there. Alternatively, it is the *non-refoulement* prohibition that could be interpreted downwards, either by denying that the human rights treaty applied extraterritorially, or by saying that the *non-refoulement* prohibition should be qualified in an extraterritorial setting.

Courts are plainly concerned with norm conflicts—it is natural for judges to abhor antinomies. Rather than face the possibility that the law as created by states is incoherent, it may be easier to simply deny that one of the conflicting norms, specifically a human rights treaty, applies altogether. This is really where the link between the conceptually distinct issues of extraterritorial application of human rights treaties and norm conflict is at its strongest, as the (generally laudable) goal of advancing cohesion in international law by avoiding norm conflict might work against extraterritorial application. One notable recent example is the *Al-Saadoon* case litigated before UK courts and the European Court, dealing with the UK's *non-refoulement* obligations under the ECHR with regard to detainees in Iraq, in which UK courts denied that the ECHR applied in Iraq precisely so that they could avoid a norm conflict. I will deal with this case in more detail below.[225] For the time being, however, my point is this: the possibility of norm conflict is not reason enough to deny the extraterritorial applicability of human rights treaties, as such conflicts can occur even when the treaties are being applied in the state's own

is *jus cogens* does not automatically entail that the *non-refoulement* obligation arising from this prohibition is also *jus cogens*.

[225] See *R. (Al-Saadoon and Another) v. Secretary of State for Defence* [2008] EWHC 3098 (Divisional Court); [2009] 3 WLR 957, [2009] EWCA Civ 7 (Court of Appeal); *Al-Saadoon and Mufdhi v. United Kingdom* (dec.), App. No. 61498/08, 30 June 2009; *Al-Saadoon and Mufdhi v. United Kingdom,* App. No. 61498/08, Judgment, 2 March 2010.

territory, as in *Soering*. The wisest method of avoiding such conflicts is to act pre-emptively, when the potentially conflicting rules are being created. For example, had the treaty between Afghanistan and Canada contained a provision stating that their obligations under the treaty were subject to their obligations under international human rights law, the conflict could not have arisen in the first place. When, however, such a conflict *does* arise, when states assume contradictory obligations either through lack of adequate caution or because of purposeful vagueness or ambiguity when negotiating a treaty text, then states should at least in some cases be expected to bear the consequences.

10. Between Universality and Effectiveness

A. What does and what should matter

What are the conclusions that can be drawn from this examination of the case law? Perhaps the most important and the most obvious such conclusion is that the extraterritorial applicability of individual rights depends foremost on the underlying ideological normative foundation of these rights. In international human rights law, that basis is universality, and universality sets a benchmark against which all other considerations are to be tested. In other systems, such as the US one, universality has not attained such status, and the system itself is normatively torn between it and other justifications for individual rights, such as the various theories of social compact.

From this universality baseline, which requires at least a rational justification for a wholesale denial of rights, we can say the following about the policy considerations examined above.

First, title over territory should be essentially irrelevant for assessing the territorial scope of application of human rights treaties. The state's capacity either to violate or to protect human rights in a given territory does not depend one whit on whether it possesses title or *de jure* sovereignty over it. Nor does it depend on whether the state is lawfully present in the territory, say pursuant to a lease or generally with the consent of the territorial state. It is only the state's actual *control* over the territory that matters. With the possible exception of the ICCPR, human rights treaties only require the state's lawful or unlawful exercise of jurisdiction over territory, but not title, to be applicable. This is neither a break with the traditional Westphalian order, nor with some general position in international law on the extraterritoriality of treaties (as there is none), nor is it a consequence of globalization. It is the logical conclusion of adopting universality as the foundation of human rights. Conversely, however, this also means that when a state *loses* control over its own territory, when it in fact no longer *has* jurisdiction even though it might have a *right* to exercise it, it should not be expected to secure or ensure the rights of the territory's inhabitants.[226] Hence, the human rights-friendly result in

[226] Thus, for example, in its 1994 Concluding Observations on Cyprus the Human Rights Committee noted that Cyprus 'is not in a position to exercise control over all of its territory and

Ilascu that Moldova has positive obligations toward persons present in Transnistria, over which it has title but no control, should in my view be rejected, as it undermines the broader principle that it is only the state's capacity to act, and not its territorial title, that should matter.[227]

Secondly, when a state in fact exercises effective overall control over the territory of another state, whether lawfully or unlawfully, the sovereignty of the latter is as such irrelevant for the territorial scope of human rights instruments. It is simply mistaken to say that imposing international human rights obligations on, say, UK troops in Iraq or Afghanistan, would somehow *ipso facto* violate the sovereignty of these states. The same would go for domestic individual rights protections. It is of course true that the international law doctrines of prescriptive and enforcement jurisdiction have something to say as to when domestic law can apply extraterritorially, but international law in no way prohibits the regulation by a state of the conduct of its own nationals or organs or agents abroad. Requiring Canadian troops in Afghanistan, or Canadian police in the Turks and Caicos, to abide by the Canadian Charter no more infringes on the sovereignty of Afghanistan or the Turks and Caicos than does requiring Canadian agents to abide by Canadian criminal law. Where the extraterritorial context *can* matter is on the merits, by adding some flexibility to rules that were originally developed for purely domestic application.

Likewise, it is perfectly possible that the extraterritorial application of a human rights treaty may lead to a norm conflict with the state's other international obligations. That might not be altogether *desirable*, but avoidance of norm conflict, or coherence in international law generally, is not some sort of paramount concern which all other considerations should be subject to. Rejecting extraterritorial application just for the sake of avoiding a norm conflict (as was done by the English Court of Appeal in *Al-Saadoon*),[228] is not in my view a position tenable in the long term.[229]

Thirdly, as for citizenship, as I have said above its relevance depends entirely on the ideological foundation of a given individual rights instrument. As we have seen from US case law, social compact theories are still dominant, and courts there almost naturally distinguish between the Constitution's extraterritorial applicability to citizens and to non-citizens. Whether this approach will hold in the future

consequently cannot ensure the application of the Covenant in areas not under its jurisdiction'. Concluding Observations of the Human Rights Committee: Cyprus, UN Doc. CCPR/C/79/Add.39 (1994), sec. 2.

[227] See also *Ilascu and others v. Moldova and Russia* [GC], App. No. 48787/99, Judgment, 8 July 2004; Partially Dissenting Opinion of Judge Bratza, joined by Judges Rozakis, Hedigan, Thomassen, and Pantiru, para. 8. See further *Treska v. Albania and Italy* (dec.), App. No. 26937/04, 29 June 2006, where a Chamber of the Court seemed to interpret *Ilascu* as holding that '[e]ven in the absence of effective control of a territory *outside its borders*, the State still has a positive obligation under Article 1 of the Convention to take the diplomatic, economic, judicial or other measures that it is in its power to take and are in accordance with international law to secure to applicants the rights guaranteed by the Convention' (emphasis added).

[228] See further Chapter V, Section 3 below.

[229] See further Milanovic, above note 223.

remains to be seen, considering that aliens are indeed protected within US territory, and in light of the fact that citizens can pose a danger to the polity as easily as aliens, as the UK example of home-grown terrorism shows. However, in international human rights law this is at least normatively no longer an issue, with the exception of a narrow category of political rights. Not only would there be no basis in treaty text, but it would be entirely antithetical to universality to say that, in whatever scenario, a human rights treaty would apply extraterritorially to citizens, but not to non-citizens.

Fourthly, with regard to regionalism the situation is somewhat more complicated. As we have seen, it is a major theme in ECHR jurisprudence, particularly in *Bankovic* where the European Court held that the 'Convention was not designed to be applied throughout the world, even in respect of the conduct of Contracting States' and that it operates 'in an essentially regional context and notably in the legal space (*espace juridique*) of the Contracting States'.[230] Somewhat more poetically, the then President of the Court, Luzius Wildhaber, remarked that the Convention was never 'intended to cure all the planet's ills'.[231] Put aside for a moment the fact that, with all due respect to the authority of those who made them, both of these statements about what the Convention was meant, designed, or intended to do are entirely unsupported—there is little evidence either in the *travaux* or elsewhere that the Convention's drafters gave any serious thought to the issue of its extraterritorial application—these statements are still only half-true, if that. Yes, of course, one could say that the Convention was not meant to apply throughout the world, nor to cure all of its ills—nor *could* it do so even if it was *meant* to do so. But why should it not apply to the conduct of contracting states, whether in 'Europe' (now presumably including all of Russia, Turkey, Georgia, Azerbaijan, and so forth) or outside it? A justification must be offered for this position, beyond just saying that the ECHR is a regional instrument, pulling out of the hat a concept such as the *espace juridique*, and wrapping it up in French for good measure just so it could sound a tad more sophisticated and respectable.

To my mind, only one such principled justification exists—relativism. It is only if one openly embraces the position that human rights are *not* universal, whether because some societies are 'not ready' for them or because they do not want them, that regionalism makes any sense as grounds for denying the extraterritorial applicability of human rights treaties. Whatever the validity of the position in philosophical terms, it is very hard, if not impossible, for courts to overtly adopt it. Though the gap between the factual and the normative is evident, human rights treaties *are* undoubtedly based in universality. It is nothing less than their driving force. And if that is so, then the regional nature of treaties such as the ECHR and the ACHR does not distinguish them in any way from universal human rights treaties with regard to their extraterritorial application. Nor could the ECHR, for example, not apply to the bombing of Belgrade, but apply to the bombing of Brussels—or Tbilisi for that matter—merely because the latter are in the Convention's '*espace juridique*'. If double standards for 'us' and 'them' are openly

[230] *Bankovic*, para. 80.
[231] Quoted according to O'Boyle, above note 217, at 125.

espoused—as is the case with the reliance on citizenship in the United States—then such a result could be defended. If, on the other hand, human rights really are *human* rights, then regionalism becomes essentially irrelevant. As I have said above, this is not to minimize the practical difficulties arising from applying human rights treaties extraterritorially in a foreign land, particularly with regard to positive obligations. However, such difficulties can be and can *only* be addressed on the merits, not artificially downplayed by denying the applicability of the treaty in the first place.

We have seen what does not, or should not matter, but what does? Surely it would be simply utopian to say that universality alone determines all outcomes, or that it should do so? Indeed it would be. What matters most for courts dealing with extraterritorial human rights cases are their own fears: fear of complexity, fear of imposing constraints on the government which are simply unrealistic, fear of norm conflict, fear of institutional incompetence, fear of treading on what are traditionally the provinces of the political branches, and, last but not least, fear of the potential political fallout and adverse consequences for their own position and status. Some of these fears are descriptively relevant, but should normatively be disregarded. Others are unwarranted or exaggerated, as is for instance the case with norm conflict. Yet others, however, are both real and legitimate, and need to be addressed if a persuasive argument in favour of extraterritorial application is to be made. It is these fears that I will generally consider under the rubric of effectiveness.

Effectiveness can mean many things. In a limited sense, it can go hand in hand with universality and favour the extraterritorial application of a domestic or international instrument protecting human rights. It cuts to the reality of things, by dispensing with formal categories such as sovereignty, which have precisely zero effect on the state's capacity to either secure or violate human rights, in favour of tangibles like actual control, as, for example, in *Loizidou* or in *Boumediene*, and forces the state to provide a real justification for a denial of rights. In most cases, however, considerations of effectiveness will not favour extraterritorial application unless they are given due regard. A tension thus builds between universality and effectiveness—a reflection, if you will, of the twin demands of normativity and concreteness in international law generally.[232]

In my view, four such considerations of effectiveness are absolutely vital: (1) flexibility—the extraterritorial application of human rights protection must not tie the hands of the state behind its back; the interpretation of applicable rules must be realistic, and take fully into account the extraordinary circumstances in which the treaty is being applied, potentially requiring significant adjustments from normal conditions; (2) impact—while the normal human rights regime needs to be adjusted for extraterritorial application, it must also contribute something useful so that the whole exercise is worth the bother; (3) regime integrity—likewise, though the normal regime may need to be watered down, it must not be watered down so much that the regime as a whole is potentially compromised; (4) clarity and predictability—arguments offered by the parties, and the rules ultimately

[232] See Koskenniemi, above note 2, at 16 *et seq*.

adopted, must be reasonably clear and predictable, both on preliminary and on substantive issues.

Universality notwithstanding, an argument in favour of extraterritorial application which does not take these considerations into account will not persuade either courts or governments. I will address each in turn.

B. Effectiveness: flexibility

Fears that applying human rights guarantees extraterritorially would prove to be wholly impracticable permeate all of the case law that we have examined. With regard to extraterritorial searches, for example, Justice LeBel in *Hape* put much emphasis on 'the theoretical and practical difficulties arising out of an attempt to apply *Charter* standards outside Canada',[233] arguing that extending Charter guarantees extraterritorially would hamper cooperative investigations, precisely at a time when criminal activity has become globalized.[234] Such was also the focus of Justices Rehnquist[235] and Kennedy[236] in *Verdugo-Urquidez*, with the latter stressing that applying Fourth Amendment requirements for a reasonable search and seizure in an extraterritorial context would be anomalous and impracticable.[237]

Similarly, when it comes to overseas detention, other than citizenship the main theme of Justice Jackson in *Eisentrager* was the practical difficulties that extending habeas corpus and other constitutional rights extraterritorially would produce, especially in wartime or during military occupation.[238] Likewise, in the Guantanamo cases the major theme of the judges arguing against the extraterritorial application of the Constitution was that judicial interference would severely hamper the Executive's ability to effectively wage war,[239] while the judges on the other side downplayed these difficulties.[240] And in *Al-Skeini* judges openly declared that

[233] *Hape*, para. 86 (per LeBel J).

[234] Ibid., paras 96 *et seq* (per LeBel J).

[235] *Verdugo-Urquidez*, at 273–5.

[236] Ibid., at 275–9. Justice Kennedy was in particular relying on Justice Harlan's view in *Reid v. Covert* that 'there is no rigid and abstract rule that Congress, as a condition precedent to exercising power over Americans overseas, must exercise it subject to all the guarantees of the Constitution, no matter what the conditions and considerations are that would make adherence to a specific guarantee altogether impracticable and anomalous'. *Reid v. Covert*, 354 U.S. 1, 74 (1957) (Harlan, J., concurring).

[237] See also G. Neuman, 'Understanding Global Due Process', (2009) 23 *Geo. Immigr. L.J.* 365, at 373–5.

[238] *Eisentrager*, at 779, 782–3.

[239] See, e.g., *Rasul*, at 498–9 (per Scalia, J); *Boumediene*, at 2294 (per Scalia, J.).

[240] Thus, for instance, the majority in *Boumediene* argued that the protections granted by habeas corpus to suspected terrorists held overseas need not resemble habeas corpus in regular criminal trials, and so unreasonably tie the hands of the executive:

> Although we hold that the DTA is not an adequate and effective substitute for habeas corpus, it does not follow that a habeas corpus court may disregard the dangers the detention in these cases was intended to prevent. *Felker, Swain,* and *Hayman* stand for the proposition that the Suspension Clause does not resist innovation in the field of habeas corpus. Certain accommodations can be made to reduce the burden habeas corpus proceedings will place on the military without impermissibly diluting the protections of the writ.

Boumediene, at 2276 (per Kennedy, J).

it would be 'utterly unreal'[241] and 'manifestly absurd'[242] to expect of the UK to secure all ECHR rights in occupied Iraq.

Forget red herrings such as comity or deference to the sovereignty of the territorial state, or unpersuasive invocations of the international law of prescriptive jurisdiction—*effectiveness* is paramount.[243] Judges are simply loathe to place burdens on governments that they could not realistically be expected to fulfil.[244] Whatever the dictates of universality, the case law on extraterritorial application will become sensible and rest on a principled basis only if courts are persuaded that a finding of such application as a preliminary matter would not *ipso facto* lead to a loss for the government on the merits (as has indeed mostly been the case so far), and it is only then that the governments themselves would be willing to accept that human rights instruments do in fact apply.[245]

Flexibility in the substantive application of the rules is therefore necessary if they are to be applied extraterritorially, and the reason why extraterritorial application does not happen more often in practice is precisely because courts and governments are concerned that flexibility will be lacking on substance. There are two deeper reasons why this is so. First, the text of the instrument in question may itself be rigid and inflexible. This is particularly the case, for instance, with the US Constitution, which is a much older document than other individual rights instruments and tends to use very absolutist language, thus posing serious difficulties of interpretation, in essence requiring the judges to significantly adapt the text if it were to apply extraterritorially in a realistic manner. Human rights treaties are much more flexible because most guaranteed rights are subject to express limitation clauses, and because they explicitly allow for derogation from some rights in situations of emergency.[246]

The second, more fundamental, problem is that the vast preponderance of the case law and doctrine that has built up on top of the texts of individual rights instruments was developed during times of normalcy. It is because the straightforward application of substantive rules derived from this judicial gloss of the text would frequently lead to results widely seen as unreasonable or impracticable in an extraterritorial context that extraterritorial application itself becomes vulnerable to what Martti Koskenniemi would term the critique of utopia.[247]

[241] *Al-Skeini CA*, para. 124 (per Brooke, LJ).

[242] *Al-Skeini HL*, para. 78 (per Lord Rodger).

[243] See also Neuman, above note 237, at 391: 'what a state owes should depend upon what it can deliver'.

[244] Similarly, see D. McGoldrick, 'Extraterritorial Application of the International Covenant on Civil and Political Rights', in Coomans and Kamminga 41, at 46.

[245] See also R. Lawson, 'Really out of Sight? Issues of Jurisdiction and Control in Situations of Armed Conflict under the ECHR', in A. Buyse (ed.), *Margins of Conflict: The ECHR and Transitions to and from Armed Conflict* (Intersentia, 2010), 57.

[246] See also Wilde, above note 195, at 93: 'Beyond non-derogable rights, human rights norms are conceived to mean different things in different contexts. Flexibility and contextualization are integral components of the very meaning of the obligations themselves, via the operation of limitation clauses, derogation provisions and, in the case of the ECHR, the "margin of appreciation" doctrine applied by the Strasbourg court.'

[247] Koskenniemi, above note 2, esp. at 58 *et seq*.

This is undoubtedly a real and difficult issue. One could of course say (and I will do so) that judge-made rules developed during normalcy can be further developed or modified to take into account the abnormal circumstances that usually arise in extraterritorial application, but this is easier said than done. It is naturally difficult for judges to overturn or radically modify precedent or a constant jurisprudence that has most often accumulated quite slowly through a process of accretion, and has now become entrenched.

That adapting human rights norms to extraordinary circumstances is easier said than done is undeniable. But, I submit, it can be worth it, and a principled approach to extraterritorial application will never be possible without it. Several tools can assist courts in that regard:

(1) Balancing is by far the most important. Thankfully, with a very few exceptions, human rights treaties and their judge-made gloss are replete with balancing or proportionality tests. There is little preventing courts from tweaking these tests a bit to accommodate extraterritorial application. At a more general level, one can always see the current content of human rights law as a larger balance between individual and community or state interest. In an extraterritorial context, the scales would weigh somewhat more heavily in favour of state interest than they would otherwise. Balancing is of particular importance when it comes to positive obligations under human rights law, specifically the overarching positive obligation to secure or ensure human rights even against third parties. As we have seen above, though this obligation can certainly at first glance appear onerous during armed conflict or occupation, it does not require states to do the impossible, but only that which they are reasonably able to.

(2) Taking into account other applicable legal regimes, such as international humanitarian law (IHL), while interpreting human rights law can be a valuable tool for assuring flexibility. Not only is this warranted by Article 31(3)(c) VCLT and the principle of systemic integration, it is only sensible to refer to rules that were designed to regulate the specific situation. This, however, is not a magic bullet, since it is entirely possible that norms from different regimes are actually conflicting.[248] I will be examining the relationship between human rights and IHL in particular in Chapter V of this study.

(3) Derogations are a crucial adaptive mechanism foreseen by the human rights treaties themselves. Their use in an extraterritorial context would provide much clarity. However, by derogating from some of their obligations states would explicitly say what they would be doing—for example, engaging in preventive detention—and would thereby not only concede that human rights treaties do apply extraterritorially, but would perhaps also incur a significant political cost domestically. This is precisely why states have so far not derogated when it comes to extraterritorial situations, choosing instead

[248] See further M. Milanovic, 'A Norm Conflict Perspective on the Relationship between International Humanitarian Law and Human Rights Law', (2009) 14 *JCSL* 459.

to deny that the treaties apply extraterritorially altogether. Yet this ultimately runs against long-term state interest, as the consequent loss in flexibility impairs their chances on the merits if extraterritorial application is in fact established, as has now been the case, for example, with regard to the UK's actions in Iraq. Derogations would in the longer run benefit both states and courts, and would be of great use in avoiding possible norm conflicts, but states still need the inducement to use them that further litigation will probably provide.[249]

(4) Finally, as a matter of litigation strategy, counsel arguing on behalf of plaintiffs relying on extraterritorial application should also prepared to be flexible. If the opportunity arises to address the merits of a given case, a maximalist argument could simply prove to be counterproductive. Rather than leading the court to find a reasonable substantive middle-ground, it could be an incentive for the court to declare that the human rights instrument does not apply extraterritorially, and thereby avoid the expected difficulties on the merits.

C. Effectiveness: impact

Flexibility is undoubtedly crucial. But if the substantive rules being applied extraterritorially must not be overly rigid, they also must not be thinned out so much that there is no useful purpose in the exercise. The only reason why we wish to apply human rights treaties to extraterritorial situations is to *change* things, to contribute something new to previously un- or under-regulated events. In other words, the extraterritorial application of human rights treaties must have a *real-world impact.* Otherwise it would not only be completely useless—it would be worse than useless. It would degenerate into nothing more than apology, a legitimization of extraterritorial state acts against individuals, such as the use of force, as not only lawful but lawful in a nice, human rights-friendly sort of way.

Though balancing, for example, is indispensable in providing flexibility when applying human rights norms in an abnormal context, whether when applying a specific rule to a particular set of facts or when searching for a broader balance between individual and state interest, it must not go too far. As well explained by Gerald Neuman:

> This emphasis on the countervailing necessities of overseas action may suggest that all of these models can be collapsed into a brand of harmless universalism: recognize constitutional rights as potentially applicable worldwide, and then balance them away. One might engage in ad hoc balancing in the individual case, or balance more categorically; the

[249] A preliminary question, of course, is whether extraterritorial derogations are even possible in situations where the 'life of the nation' doing the derogating is not itself in danger (as, for instance, with the UK in Iraq or Afghanistan). No definitive answer to this question has yet been given, though it has been raised by Lord Bingham in *R. (Al-Jedda) v. Secretary of State for Defence* [2007] UKHL 58, [2008] 1 AC 332. To my mind, that answer must be 'yes', but I will not explore that issue further in this study.

balancing process may be intrusive or highly deferential.... This approach suggests that, ultimately, extraterritorial constitutional rights boil down to a single right: the right to 'global due process'.[250]

Neuman points out a real danger. What would be the use of an ICCPR or an ECHR (or a constitution) that is widely applicable extraterritorially, if state action is to be judged under a standard so flexible and lenient that the state could almost never act unlawfully? It is better for universality to remain utopian, than for it to become harmless.

Finding the right balance in balancing is by any account a delicate task, which can only be performed on a case-by-case basis, if with an overarching need for consistency. This does not mean simply that one form of casuistry on the preliminary question of extraterritorial application that I find so unappealing is to be replaced with another, this time on the merits. As I see it, the real problem is not so much casuistry but the lack of principle behind it. Even though I can and will offer reasonably clear rules on *when* human rights treaties should apply extraterritorially I am obviously incapable of saying *how* they should substantively apply in every given factual scenario. In the abstract I can only say that although in large swathes of extraterritorial cases the balancing exercise would come out so strongly in favour of the state that it would amount to no more than what Neuman calls 'global due process'—say, for example, with extraterritorial searches or other interferences with privacy, which would probably not amount to violations in all but the most exceptional of circumstances—the extraterritorial application of the human rights regime would still be worth having with regard to torture, fair trial, or arbitrary deprivations of liberty or life.

D. Effectiveness: regime integrity

A human rights regime thus must not be watered down so much in an extraterritorial setting as to be rendered toothless. Another danger in making the substantive application of a human rights regime overly flexible in an extraterritorial context is that the regime as a whole may become compromised. Recall, for example, the European Court's rejection in *Bankovic* of the idea that Convention rights can be divided and tailored, and its ruling that the ECHR is an all or nothing package.[251] Similar concerns may underlie the Court's reluctance so far to explicitly take into account IHL norms when applying the ECHR to situations of armed conflict. As Michael O'Boyle put it, '[t]he rules of international humanitarian law sit uneasily with the Court's case law on the right to life.'[252] Allowing the state to kill combatants or insurgents *under human rights law* without showing the absolute necessity for doing so, or to detain preventively during armed conflict, might lead to allowing the state to do the same outside armed conflict, with one precedent

[250] Neuman, above note 109, at 920.
[251] *Bankovic*, para. 75.
[252] O'Boyle, above note 217, at 135.

leading to another, and then another, and yet another.[253] The potential of emergencies or states of exception to balloon out and become the norm rather than the exception has of course long been recognized. This is, *inter alia*, why there are derogation clauses in human rights treaties, and why they impose such strict requirements on any derogation.

It is natural for courts to resist the importation of values and rules foreign to the system in which they are normally operating. It is likewise understandable for them to fear that by nominally expanding the regime's coverage they would actually be diminishing it in substance, as well as overstepping the bounds of their own authority.[254] And it is because these concerns are valid and real that they must be addressed as openly as possible. With extraterritorial application as with many other problems, the best might turn out to be the enemy of the good. A 'perfect', oh-so-human-rights-friendly solution to the preliminary issue of extraterritorial application—much like the one I will be offering later in this study—might when applied in practice turn out to be positively *harmful* to the human rights regime as whole, so that all the participants in that regime are ultimately worse off than before.

E. Effectiveness: clarity and predictability

Finally, as Lord Justice Brooke aptly put it in *Al-Skeini*: '[i]t is essential... to set rules which are readily intelligible.'[255] Today, the rules regarding the extraterritorial applicability of human rights instruments are everything but. Although clarity does not necessarily require a jurisprudence of principle, it is greatly aided by it. This study is largely devoted precisely to establishing such principles. However, clarity regarding threshold rules will be difficult to achieve if the subsequent application of the substantive rules remains utterly uncertain. If extraterritorial application is to be effective, both the threshold and the substantive rules need to be reasonably clear and workable. This is not to say that these rules necessarily need to go along bright lines. But unless the parties offer the court

[253] For such regime integrity-inspired concerns with regard to drone targeted killings, see the Report of the Special Rapporteur on extrajudicial, summary, or arbitrary executions, Philip Alston: Study on targeted killings, Human Rights Council, UN Doc. A/HRC/14/24/Add.6, 28 May 2010, esp. paras 85 and 86. For commentary on this report, see M. Milanovic, 'More on Drones, Self-Defense, and the Alston Report on Targeted Killings', *EJIL: Talk!*, 5 June 2010, available at <http://www.ejiltalk.org/more-on-drones-self-defense-and-the-alston-report-on-targeted-killings/>.

[254] Thus, for example, in *Reid v. Covert*, at 14, while delivering the opinion of the Court, Justice Black opined that

> [t]he concept that the Bill of Rights and other constitutional protections against arbitrary government are inoperative when they become inconvenient or when expediency dictates otherwise is a very dangerous doctrine and, if allowed to flourish, would destroy the benefit of a written Constitution and undermine the basis of our Government. If our foreign commitments become of such nature that the Government can no longer satisfactorily operate within the bounds laid down by the Constitution, that instrument can be amended by the method which it prescribes. But we have no authority, or inclination, to read exceptions into it which are not there.

[255] *Al-Skeini CA*, para. 110.

viable standards, it is far easier for the court to deny extraterritorial applicability in the first place.

F. Conclusion

The question of *whether* a human rights instrument applies in principle to a given extraterritorial situation is conceptually distinct from *how* it should substantively apply. In reality, however, considerations of effectiveness which pertain to the merits influence how courts will approach the question of extraterritorial application. In a way, disputes about extraterritorial application serve as a substitute for an examination of the merits. It is largely because courts fear that if they so much as touch the substance of very controversial and complex cases these will explode in their faces that they choose to unduly restrict the extraterritorial application of human rights instruments. Thus, the result of *Bankovic* is in my view explained not by the European Court's exegesis on the general international law doctrine of jurisdiction, but by the Court's belief that the killing of the individuals concerned may well have been justified in terms of IHL, and that in any case it lacked both the ability and a reliable factual record to apply IHL rules on targeting, distinction, or precautions in attack. Did, for instance, the Serbian authorities use the RTS equipment for military communications, as claimed by NATO, and how would the Court establish that, or not? Likewise, the result in *Al-Skeini*, where persons killed by UK troops on patrol were not within its jurisdiction, while a person killed in detention was, is not explained merely by their Lordships application of inconsistent Strasbourg precedent, or by their apparent belief that prisons are like embassies for Article 1 ECHR purposes. Rather, they thought that while a killing by troops on patrol in a country embroiled in an insurgency may well be justified, the killing of a helpless prisoner could not, and required judicial intervention. In short, the treaty applied to Baha Mousa simply because it would have been morally intolerable for it not to.

It is easier for courts to dismiss hard cases *in limine*, on preliminary issues, rather than go into their merits. As time goes by, however, this jurisprudence becomes more and more exposed as unprincipled and unworkable. The only way of detaching the issue of extraterritorial applicability from an unstated assessment of the merits is if the courts are made to believe that the government is not unfairly and unreasonably constrained by rules that were not developed for these particular kinds of circumstances, and that the courts themselves will not have to micromanage controversies that go far beyond their competence. It is of course impossible to produce a magic formula on how to construct a persuasive legal argument, except in the broadest and accordingly least useful of generalities. Nonetheless, text and precedent aside, extraterritorial application of human rights treaties will not rest on any principle until courts are persuaded that: (1) such application could be sufficiently flexible and will not produce unreasonable results; (2) that, at the same time, it will not be so flexible as to no longer be meaningful; (3) that the added flexibility will not compromise the integrity of the human rights regime as a whole; and (4) that it is possible to have clear and workable rules on both threshold issues and those of substance. Universality aside, it is only then that the courts will think that the risks they assume by opening the regime further will be worth taking—and it is only then that governments themselves might follow suit.

IV

Models of Extraterritorial Application

1. Basic Models and Scenarios

A. Introduction

Having examined the policy considerations which courts take into account when deciding on the extraterritorial applicability of human rights instruments, whether in an international or in a domestic context, I will now turn back to the human rights treaties themselves. Recall that in Chapter II of this study, we examined the concept of state jurisdiction in the treaties' jurisdiction clauses, which ultimately governs their territorial scope of application. As I have endeavoured to explain, that concept of state jurisdiction refers to state control over territory, and perhaps also individuals, not to the prescription or enforcement of municipal law, as the European Court thought in *Bankovic*.

If the word 'jurisdiction' in Article 1 ECHR and similar jurisdiction clauses of other human rights treaties basically means 'effective overall control of an area', as the European Court held in *Loizidou*, then it of course still needs to be determined what degree of control suffices, and how exactly to geographically or spatially define an 'area'.[1] The rule about the territorial scope of application of human rights treaties would then be relatively simple—they apply whenever a state has effective overall control of an area, regardless of whether it has title over that area or not, and of whether its control over the area was obtained lawfully or unlawfully. Conversely, however, the treaties would not apply when such control is lacking. In my view, *contra Ilascu*, this would hold even for territory over which the state actually does have title. As I have argued above, there should be no difference in principle between intra- and extraterritorial application of a human rights treaty, in the absence of the requisite degree of state control.[2]

Yet, even if this territorial concept of jurisdiction was generally accepted, there would still be numerous cases where a state acts extraterritorially and thereby violates an individual's rights, while not actually controlling the territory in which it does so. The *Bankovic* scenario provides one example—aerial bombardment

[1] See below Sections 2.B and 2.C.

[2] See, however, K. Mujezinovic Larsen, '"Territorial Non-Application" of the European Convention on Human Rights,' (2009) 78 *Nord J Int'l L* 73, who argues in favour of a notion of 'residual jurisdiction' flowing from title to territory, out of which only a limited set of positive obligations would arise.

without occupation—but as we shall see there are many others. If all 'jurisdiction' meant was control over territory, then persons in such scenarios would simply not fall within it. Alternative strands of Strasbourg jurisprudence, the case law of other human rights bodies, and several authors thus advocate another conception of state jurisdiction, as authority and control over *individuals*. Such a personal, rather than territorial notion of jurisdiction would greatly expand the scope of human rights treaties. However, it remains to be seen whether it is tenable.

This chapter of the study will examine the possible models of extraterritorial application, the two prime candidates being jurisdiction as control over territory[3] and jurisdiction as control over individuals.[4] In my view, neither of these is entirely satisfactory. The former simply does not go far enough. If applied too strictly, it is apologetic of state power, and allows many human rights abuses to slip through the cracks. The latter runs against the language of some treaties, which explicitly conceive of state jurisdiction in territorial terms. More importantly, it cannot be limited on any principled basis and loses any meaning as a threshold, since it simply collapses into the position that a state has human rights obligations whenever it can actually violate the rights of the individuals concerned.

I therefore prefer a third model, where state jurisdiction is conceived of only territorially, but where that threshold criterion applies only to the positive obligation of states to secure or ensure human rights, because it is only when states possess a sufficient degree of control over territory that these obligations can be realistically kept.[5] When it comes to the negative obligation to respect human rights, no threshold criterion should apply because states can always control the actions of their organs or agents. In my view, this model fits best with the object and purpose of human rights treaties and is superior in reconciling the various policy considerations relevant to their extraterritorial application. It is, however, not free of all weaknesses, and is lacking in explicit textual support at least with regard to some treaties.

Besides these three basic models which concern treaties that possess dedicated jurisdiction clauses, I will also be examining treaties without such clauses,[6] as well as the ICCPR, with its perhaps more restrictive territoriality requirement.[7] Before moving on to do so, however, I wish to outline several factual scenarios in which the issue of extraterritorial application may present itself. Having criticized the jurisprudence of the European Court for casuistry, I have no wish to engage in it myself. Rather, my purpose in doing so is to be able to test the various models of extraterritorial applications on the facts of these scenarios. This will simply allow us to see more clearly what is at stake, as a matter of policy as much as a matter of law, in opting for one model over another.

Most of the scenarios that I will now present are not actually hypothetical but have already occurred in practice, whether they have been adjudicated on or not. I will group them with regard to the nature of the state act which would arguably

[3] See below, Section 2.
[4] See below, Section 3.
[5] See below, Section 4.
[6] See below, Section 6.
[7] See below, Section 5.

amount to a human rights violation, for example deprivation of life or liberty. Where the scenarios within each group differ from one another is in the various contextual elements, for example in whether the killing by a state occurred in a territory under the state's control, and it is precisely these contextual elements which matter the most for extraterritorial application. A caveat: there certainly is a degree of overlap between some of these scenarios, and I am not saying that they are all necessarily distinguishable. Nor am I saying that this list of scenarios is in any way exhaustive.

B. Extraterritorial deprivation of life

1. Killing in a territory within the state's control

There is no greater exercise of state power than a deprivation of life—and there is no issue which attracts greater controversy. Legally, perhaps the easiest scenario is when a state kills—or generally violates the rights of individuals—within a territory under its effective overall control, for instance during a belligerent occupation. The jurisprudence of human rights bodies, as well as the ICJ, is clear that in such situations the human rights treaty would apply. *How* it would apply is a different matter.[8]

As we have seen above, whether a particular treaty is regional or universal should have no bearing on the application of the effective overall control of an area test. Also irrelevant for the threshold issue of extraterritorial application is whether the killing takes place during an armed conflict, whether within or outside a territory under a state's control. This of course may have an impact on the *substantive* application of the treaty.

2. Killing in a territory outside the state's control

More interesting is the issue of a killing by state agents in a territory *not* under the state's control. This killing can take place during a large-scale use of armed force, for example in an aerial bombardment as in *Bankovic*, or on the battlefield, or for instance during patrol by troops in an area over which their control is arguably tenuous, as in *Al-Skeini*.

But again, the existence *vel non* of large-scale violence seems irrelevant for the threshold question of extraterritorial application. Killing can also take place in more limited, 'targeted' circumstances, as with the US use of drones in Pakistan to target Taliban or Al-Qaeda elements.[9] Then of course there is the classical assassination

[8] For example, in situations of armed conflict the lawfulness of the killing could be assessed by interpreting human rights law by taking into account the applicable rules of IHL—see Chapter V below.

[9] For general background, see, e.g., 'Drones Are Weapons of Choice in Fighting Qaeda', *The New York Times*, 16 March 2009, available at <http://www.nytimes.com/2009/03/17/business/17uav.html?_r=1>. For an insightful defence of the use of drones for targeted killings in US counter-terrorism policy, see K. Anderson, 'Predators Over Pakistan,' *The Weekly Standard*, 8 March 2010, available at <http://ssrn.com/abstract=1561229>.

scenario, as with the 2010 killing of a Hamas military leader in Dubai by suspected agents of the Mossad.[10] What all of these cases have in common is that they would fall outside a human rights treaty's scope of application under a strict effective overall control of an area model of state jurisdiction.

3. *Killing in a territory outside the state's control, but within the territory of a state party to the human rights treaty in question (*espace juridique*)*

But what if a *Bankovic*-type scenario were to occur in a territory of a state party to the treaty, for example, when Russia bombed parts of Georgia in 2008? Similarly, consider the assassination of Alexander Litvinenko in London, and assume that it was done by Russian agents. Was Litvinenko within Russia's jurisdiction? Did Russia have an ECHR obligation not to kill him, or the obligation to investigate his death?

Litvinenko's family has lodged an application with the European Court, so we will eventually see what the Court will say. To my mind, the only thing distinguishing the Litvinenko-type scenario from *Bankovic* is that the killing took place within the ECHR's *espace juridique*. It is irrelevant that the killing was done by radioactive sushi, rather than by a missile from a drone or a bomb from an airplane. However, while I am sure that Litvinenko's family's counsel will precisely try to distinguish *Bankovic* on *espace juridique* grounds, as we have seen that notion should also be irrelevant for the question of extraterritorial application. In other words, if *Bankovic* governs state action outside Europe, then it should also govern it inside Europe.

4. *Killing by third parties*

Finally, so far we have only examined killing by state organs or agents, which engages the state's *negative* obligation to respect human rights. However, killings by third parties can engage the state's positive obligation to do all it reasonably can to prevent such killings, and the obligation to investigate them. A good example would be insurgent violence in occupied Iraq. The issue, of course, is when such positive obligations arise. One could also envisage various complicity scenarios, dealing with the state's responsibility connected to acts by third parties. Would, for example, states selling weapons to Sudan that they know or have reason to suspect Sudan will use against civilians in Darfur have duties under human rights treaties towards the inhabitants of Darfur?[11]

[10] See, e.g., 'Dubai Points Finger at Mossad over Hamas Assassination,' *The Sunday Times*, 18 February 2010, available at <http://www.timesonline.co.uk/tol/news/world/middle_east/article7031749.ece>; 'Dubai Hamas Assassination: How It Was Planned', *The Telegraph*, 17 February 2010, available at <http://www.telegraph.co.uk/news/worldnews/middleeast/dubai/7251960/Dubai-Hamas-assassination-how-it-was-planned.html>.

[11] See, e.g., 'China Defends Arms Sales to Sudan,' *BBC News*, 22 February 2008, available at <http://news.bbc.co.uk/2/hi/asia-pacific/7258059.stm>.

C. Extraterritorial deprivation of liberty

1. Detention in a territory within the state's control

When it comes to the deprivation of liberty, as with deprivation of life, the easiest scenario would be detention by state agents in a territory under the state's effective overall control. Perhaps the most common example would be the detention of civilians or combatants during belligerent occupation. Of course, there are complexities in defining what actually amounts to effective overall control over territory, and even more interestingly what counts as a territory. This brings me to my next example.

2. Detention in a place within the state's control

One can easily imagine detention by a state which does not take place in a *territory* under its control, but in something more approaching a *place*. Take as one example a full-fledged military base, such as Guantanamo or Bagram. For my present purposes, perhaps the most intriguing example is offered by the so-called 'black sites' that the CIA operated in the war on terror, in order to render, detain, and interrogate high-value detainees, until they were shut down by the Obama administration. Though many specifics are still unknown, credible sources have asserted that the CIA operated such sites in Lithuania, Poland, and Romania among others.[12]

This example brings the territorial conception of state jurisdiction as the *Loizidou* 'effective overall control of an area' to its breaking point. Does a former riding school in the suburbs of Vilnius—one of such black sites—qualify as an 'area' under US jurisdiction, even though it is located in a *territory* under the control of Lithuania? Forget enormous military bases such as Guantanamo—what about lone buildings, or apartments or basements within such buildings? Do they fall within the 'jurisdiction' of the detaining state?

Or consider various abduction scenarios, as for instance with the capture of Adolf Eichmann by Israeli agents in Argentina in 1960. To briefly recount the facts, on 11 May 1960, as he was returning home from work, the disguised Eichmann was intercepted by Israeli agents who knocked him unconscious, put him in a car, and took him to a safe house in Buenos Aires. He was kept there until 21 May 1960 when he was smuggled out of Argentina on an El Al flight to Israel.[13]

[12] See, e.g., 'Lithuania Hosted Secret CIA Prisons,' *BBC News*, 22 December 2009, available at <http://news.bbc.co.uk/1/hi/8426028.stm>; 'Secret Prisons: Obama's Order to Close "Black Sites"', *The Guardian*, 23 January 2009, available at <http://www.guardian.co.uk/world/2009/jan/23/secret-prisons-closure-obama-cia>; 'A Window Into C.I.A.'s Embrace of Secret Jails', *The New York Times*, 12 August 2009, available at <http://www.nytimes.com/2009/08/13/world/13foggo.html?pagewanted=1&_r=2&hp>. See also M. Danner, 'US Torture: Voices from the Black Sites', *The New York Review of Books*, Vol. 56(6), 9 April 2009, available at <http://www.nybooks.com/articles/22530> (giving details on a confidential ICRC report on prisoner abuse at such sites).

[13] See further N. Bascomb, *Hunting Eichmann: How a Band of Survivors and a Young Spy Agency Chased Down the World's Most Notorious Nazi* (HMH, 2009).

Now, because the Israeli agents operated in Argentina without its knowledge or consent, they violated Argentinean sovereignty.[14] But that aspect of the legality, *vel non*, of Eichmann's capture is for our present purposes beside the point. The question I am interested in, on the facts of this scenario, is at what point Eichmann would have had enforceable human rights vis-à-vis Israel. Would it be at the moment of capture, when he was as an individual within the Israeli agents' control? Or when he was brought to the safe house, arguably a place under Israel's control, and thus within its jurisdiction? Or was it when he was put on the El Al plane, or rather when that plane entered Israeli airspace or touched down on the tarmac? This brings me to my next example.

3. *Detention on a ship or aircraft*

What about persons detained aboard a ship or aircraft? Does it matter, for example, that Eichmann was drugged, disguised as flight crew, and brought to Israel on an El Al plane, which flies Israel's flag, since general international law recognizes the flag jurisdiction of states? Or would the result be any different if Eichmann was taken on a British Airways plane? Is it, in other words, the flag that matters, or is it the actual control over the individual by the state?

In more contemporary terms, the United States has held a number of prisoners aboard ships in its 'war on terror'.[15] But where this particular issue is of the greatest practical relevance is probably with respect to piracy off the coast of Somalia. When do, for example, captured pirates enjoy the *non-refoulement* protections of the ECHR or the ICCPR? And when do refugees or economic migrants aboard ships trying to reach a better life on European or American shores?

4. *Killing, torture, or ill-treatment during extraterritorial detention*

I have so far talked about detention pure and simple. Yet, as we well know, deprivation of liberty is frequently followed by various forms of violence to the person. This, after all, was what the US programme of interrogation of high-value detainees, such as Khalid Sheikh Mohammed, the mastermind of 9/11, was all about. The political importance of such scenarios is not in doubt. However, the threshold question of extraterritorial application probably does not and should not depend on the particular act of violence concerned. What does is the fact that such violence takes place in a situation where the individual is under state control, or is in an area or place under state control. In that regard, the violence as such is inseparable from the deprivation of liberty—but what if the torturer is an agent of a different state than the one holding the detainee? And what of the situations where one state is merely complicit in a human rights violation by another?

[14] See UNSCR 138 (1960).

[15] See, e.g., 'US Accused of Holding Terror Suspects on Prison Ships', *The Guardian*, 2 June 2008, available at <http://www.guardian.co.uk/world/2008/jun/02/usa.humanrights>.

D. Complicity scenarios

1. Territorial complicity

There are two basic complicity scenarios, the first being when the territorial state acquiesces or is complicit in an extraterritorial human rights violation by another state. Thus, for example, if Lithuania allowed the United States to operate the CIA black site near Vilnius, knowing that it would be used for the incommunicado detention and coercive interrogation of suspected terrorists, it may be responsible under the relevant human rights treaties, such as the ECHR, because the individuals concerned were within Lithuanian jurisdiction. The same might go, for example, for states knowingly allowing rendition flights to overfly their territories, or to use their airports for refuelling.

2. Extraterritorial complicity

But complicity itself can also be extraterritorial. Consider, for example, the actions of various intelligence services in the 'war on terror', such as the involvement of Canadian agents in the rendition to Syria of Maher Arar,[16] or in the case of Omar Khadr, or that of UK agents in the case of Binyam Mohamed, both of whom were detained and interrogated by the US in Guantanamo, with the interrogations being conducted with the connivance, presence, or participation of Canadian and UK intelligence services.[17] This was of course not an isolated practice—there have been numerous allegations of UK complicity in torture in Pakistan, Egypt, and Uzbekistan, in addition to Guantanamo.[18]

What distinguishes these cases from territorial complicity is that the individual concerned is not located within a territory or even a place under the complicit state's control, but is rather under the control of the principal. The actions of the principal state may themselves be committed extraterritorially, as with the US in Guantanamo. The issue, of course, is whether complicity in such circumstances is actually prohibited. Did, for example, Binyam Mohamed have ECHR rights vis-à-vis the UK while he was being held and mistreated by the US in Guantanamo, with UK collusion?

3. Distinguishing between primary and secondary rule complicity

In that regard, a distinction must be made between complicity based on primary rules of international law, such as those in human rights treaties, and complicity

[16] See, e.g., the 2006 Report of the Commission of Inquiry into the Actions of Canadian Officials in Relation to Maher Arar, available at <http://www2.pch.gc.ca/cs-kc/arar/index_e.htm>.

[17] See, e.g., 'MI5 Telegrams "Fed Interrogation" ', *BBC News*, 7 March 2009, available at <http://news.bbc.co.uk/2/hi/uk_news/7930708.stm>.

[18] See, e.g., Joint Committee on Human Rights, 'Allegations of UK Complicity in Torture', 21 July 2009, available at <http://www.publications.parliament.uk/pa/jt200809/jtselect/jtrights/152/152.pdf>; 'Torture—new claim of secret UK complicity', *The Guardian*, 26 July 2009, available at <http://www.guardian.co.uk/world/2009/jul/26/alam-ghafoor-torture-uk-intelligence>.

based on the general, secondary rules of state responsibility. The content of the latter has been codified by the ILC in Article 16 ASR:

A State which aids or assists another State in the commission of an internationally wrongful act by the latter is internationally responsible for doing so if:

(a) that State does so with knowledge of the circumstances of the internationally wrongful act; and
(b) the act would be internationally wrongful if committed by that State.

Note that Article 16 requires that both the principal state and the complicit state be bound by the same legal obligation. As the ILC puts it:

The third condition limits article 16 to aid or assistance in the breach of obligations by which the aiding or assisting State is itself bound. An aiding or assisting State may not deliberately procure the breach by another State of an obligation by which both States are bound; a State cannot do by another what it cannot do by itself. On the other hand, a State is not bound by obligations of another State vis-à-vis third States. This basic principle is also embodied in articles 34 and 35 of the Vienna Convention on the Law of Treaties. Correspondingly, a State is free to act for itself in a way which is inconsistent with obligations of another State vis-à-vis third States. Any question of responsibility in such cases will be a matter for the State to whom assistance is provided vis-à-vis the injured State. Thus it is a necessary requirement for the responsibility of an assisting State that the conduct in question, if attributable to the assisting State, would have constituted a breach of its own international obligations.[19]

This creates a two fold problem. Let us take the ill-treatment of Binyam Mohamed in Guantanamo, in which the UK was allegedly complicit, as an example. Assume that Binyam Mohamed lodges an application against the UK with the European Court, alleging that the UK's complicity was a violation of Article 3 ECHR. The first problem is that, unlike say with the CAT, the US and the UK were not bound by the same legal obligation, since the US is not a party to the ECHR.[20] Secondly, even if the first problem were to be resolved, it would be far from clear that the UK was bound to respect or secure Mohamed's rights as he was not located in a territory under its jurisdiction. Therefore, even if the case was looked at under the CAT, the two states might not have had the same obligations towards the same individual.

The requirement for parallel state obligations thus poses a great difficulty in relying on Article 16 ASR to deal with complicity scenarios.[21] Rather, recourse must be had to complicity derived from primary rules, i.e. from the positive and negative state obligations under the human rights treaties themselves. In other words, it is only if Article 3 ECHR is interpreted as prohibiting state party involvement in torture or ill-treatment by non-parties that complicity would in

[19] ILC ASR Commentary, at 157, para. 6.

[20] One could argue, however, that they were bound by the same obligation in *substance*, even though it did not come from the same *source*.

[21] For the relevance of this distinction between primary and secondary complicity in the *Bosnian Genocide* case before the ICJ, see M. Milanovic, 'State Responsibility for Genocide: A Follow-Up', (2007) 18 *EJIL* 669, at 680–4.

fact be unlawful. This, of course, still leaves the question of extraterritorial application. If, per *Bankovic*, killing a person in a territory outside the state's control would not engage the ECHR, why would helping a third state torture a prisoner do so? And even if the notion of state jurisdiction was not solely spatial, but was also personal in nature, i.e. if the jurisdiction threshold would be satisfied if a state had authority and control over an individual, would the mere provision of information or questions to third-party torturers qualify as such control?

E. Extraterritorial law enforcement

This brings me to my next category of scenarios, which I will put under the general rubric of extraterritorial law enforcement. We have already seen one example in *Hape* and *Verdugo-Urquidez*—extraterritorial searches and seizures for the purposes of criminal investigations.[22] Such acts can be done under colour of law, for instance under a warrant and/or with the consent of the authorities of the territorial state, but they can also be done without such legal authority. Likewise, abductions, which we have already examined, could also have a place in a criminal law context as the means of obtaining custody over a fugitive, as was indeed the case with Eichmann.[23] The question is whether this law enforcement context is of any relevance.

Consider also two further examples which are far from infrequent in practice. First, in absentia trials of persons who are not located in the territory of the prosecuting state, either because extradition has been refused or because the fugitive is simply very good at running. Would a person subjected to such a trial, as is possible in most civil law systems, actually enjoy the right to a fair trial under human rights instruments? After all, he is not within the jurisdiction, if jurisdiction is conceived of territorially. Or, rather, is the fact that an in absentia trial is a legal process dispositive for some sort of personal notion of jurisdiction?

Secondly, what about the seizure of domestic assets, such as bank accounts or immovable property, of a person located outside the state's territory, for instance because of tax evasion, or because he is suspected of involvement in terrorism? Would that person, even though he is outside the jurisdiction, be entitled to due process, or the protection of his property rights, say under Article 1 of Protocol No. 1 to the ECHR? Should the fact that the seizure is a legal process matter? What if, say, the individual's house was simply burnt down by state authorities as a punitive measure, without any legal process? The intuition of most people would probably be that human rights guarantees should apply in such circumstances. But why? What (if anything) distinguishes these acts from, say, an extraterritorial search or seizure, or even a killing à la *Bankovic*?

F. Transboundary environmental harm

Finally, let us consider a scenario of transboundary environmental harm. This is of course something that international law has increasingly dealt with, but it has

[22] See above, Chapter III, Sections 4 and 5.
[23] Similarly, see *United States v. Alvarez-Machain*, 504 U.S. 655 (1992).

mostly done so either by considering such harm to violate the sovereignty of the state on which it encroaches, a general principle such as *sic utere tuo ut alienum non laedas*, or specific instruments of international environmental law.[24] But what about the individuals affected by such harm?

Currently pending before the ICJ is the *Aerial Herbicide Spraying* case between Ecuador and Colombia. Ecuador alleges that Colombia has sprayed toxic herbicides within its borders, but near Ecuador, in its campaign to suppress coca cultivation, and that these toxins have caused significant transboundary harm in Ecuador itself. According to Ecuador's application, this harm has not been confined to the environment in the abstract:

> During and after each of Colombia's spraying campaigns, for instance, Ecuador's population in the northern boundary areas has reported serious adverse health reactions including burning, itching eyes, skin sores, intestinal bleeding and even death. Because of the non-discriminating nature of the herbicide used by Colombian authorities, there has also been serious and widespread damage to non-target plant species, including key local crops such as yucca, plantains, rice, coffee, hay and others. The consequences of the crop damage have been serious in the context of the subsistence farming needs of the local population.[25]

Assuming that the facts as alleged are true, would the Ecuadorians affected by the Colombian campaign have rights such as the right to life, bodily integrity, health, or food vis-à-vis Colombia under, say, the ICCPR, the ICESCR, or the ACHR?[26]

2. The Spatial Model: Jurisdiction as Control of an Area

A. Introduction

Having now examined some of many possible scenarios of extraterritorial application, let us move to the first possible model of such application, that would conceive of state jurisdiction in *spatial* terms, as effective overall control of an area. This is undoubtedly the model with the most textual support, and as we have seen in Chapter II, such a model also fits with the general treaty practice of states, which use the term 'jurisdiction' to denote control over territory. Likewise, several human rights treaties, such as the CAT, explicitly refer to jurisdiction in spatial terms—e.g. '[e]ach State Party shall take effective legislative, administrative, judicial or other measures to prevent acts of torture in any territory under its jurisdiction', in Article 2(1) thereof.

This model also fits best with the current state of jurisprudence. The golden standard was of course set in *Loizidou:*

> Bearing in mind the object and purpose of the Convention, the responsibility of a Contracting Party may also arise when as a consequence of military action—*whether lawful*

[24] See, e.g., *Trail Smelter (United States v. Canada)* (1941) 3 RIAA 1905.

[25] *Aerial Herbicide Spraying (Ecuador v. Colombia)*, Application Instituting Proceedings, 31 March 2008, para. 4.

[26] Ecuador at least says yes—see ibid., para. 38(B)(ii).

or unlawful—it exercises *effective control of an area* outside its national territory. The obligation to secure, in such an area, the rights and freedoms set out in the Convention, derives from the fact of such control whether it be exercised directly, through its armed forces, or through a subordinate local administration.[27]

The spatial model of jurisdiction has likewise been adopted by other human rights bodies,[28] as well as by the ICJ in the *Wall* and *Congo v. Uganda* cases.[29] Its primary benefit, from a judicial perspective, is that it requires very little overruling of prior case law. Even *Bankovic* can stand on that count, if we forget its misguided reliance on the *other* ordinary meaning of the word 'jurisdiction' in international law, that which delimits the municipal legal orders of states. Likewise, the spatial model seems to reconcile the normative demands of universality and the factual demands of effectiveness, as extraterritorial application would happen when it is realistically possible, in the circumstances of state control over territory.[30]

The principal problem with the spatial model, however, is that on a deeper look it does not reconcile universality and effectiveness all that well. Adhering to it strictly would lead to numerous morally intolerable situations—intolerable from the standpoint of universality—in which a state acts extraterritorially but the relevant human rights treaty would not apply, as with most of the scenarios that

[27] *Loizidou v. Turkey*, App. No. 15318/89, Judgment (preliminary objections) of 23 February 1995, para. 62 (citations omitted, emphasis added).

[28] For example, the Human Rights Committee has found that the ICCPR was applicable to Occupied Palestinian Territories—see Concluding Observations of the Human Rights Committee: Israel, UN Doc. CCPR/C/79/Add.93, 18 August 1998, para. 10: 'the Covenant must be held applicable to the occupied territories and those areas . . . where Israel exercises effective control', as did the Committee on Economic, Social, and Cultural Rights with regard to the ICESCR—see Concluding Observations of the Committee on Economic, Social and Cultural Rights: Israel, UN Doc. E/C.12/1/Add.27, 4 December 1998, para. 8: 'The Committee is of the view that the State's obligations under the Covenant, apply to all territories and populations under its effective control.' The Committee Against Torture did the same with regard to the CAT in Iraq and Afghanistan—see Conclusions and Recommendations: United Kingdom of Great Britain and Northern Ireland, UN Doc. CAT/C/CR/33/3, 10 December 2004, para 4(b): 'The Committee expresses its concern at: (b) the State Party's limited acceptance of the applicability of the Convention to the actions of its forces abroad, in particular its explanation "that those parts of the Convention which are applicable only in respect of territory under the jurisdiction of a State party cannot be applicable in relation to actions of the United Kingdom in Afghanistan and Iraq"; the Committee observes that the Convention protections extend to all territories under the jurisdiction of a State party and considers that this principle includes all areas under the *de facto* effective control of the State party's authorities.' Similarly, the Committee on the Rights of the Child considered that the CRC applied to the Occupied Palestinian Territories, and also seemed to have though that it applied to Israeli army activities with regard to de-mining in Southern Lebanon—see Concluding Observations of the Committee on the Rights of the Child: Israel, UN Doc. CRC/C/15/Add.195, 4 October 2002, paras 2, 5, 57–8.

[29] *Legal Consequences of the Construction of a Wall in the Occupied Palestinian Territory*, Advisory Opinion, 9 July 2004, ICJ Reports 2004, at 136, paras 109–13; *Armed Activities on the Territory of the Congo* (Congo v. Uganda), Judgment, 19 December 2005, ICJ Reports 2005, at 168, paras 179, 216–17. The primary basis for extraterritorial application in both cases seems to have been the occupied status of the territories in question—see especially *Wall*, para. 112: 'the Court cannot accept Israel's view. It would also observe that the territories occupied by Israel have for over 37 years been subject to its territorial jurisdiction as the occupying Power.' However, the Court did not in any way limit the possibility of extraterritorial application to the spatial model alone.

[30] For the best general overview of the spatial model, see R. Wilde, 'Triggering State Obligations Extraterritorially: The Spatial Test in Certain Human Rights Treaties', (2007) 40 *Israel L Rev* 503.

I have just outlined above. Even if it is better than just saying that human rights treaties cannot apply extraterritorially at all, it is still simply far too rigid. As we will see, this has invariably led either to the rejection of the spatial model in favour of other approaches by the more adventurous human rights bodies, or to its attenuation and the carving out of relatively unprincipled exceptions by the more faint of heart.

I will now first examine the object of the spatial test—what actually counts as an 'area' susceptible to being subjected to a state's jurisdiction. Secondly, I will address the requirement of state control over an area. Thirdly, I will move on to control over places and the special problems posed by embassies and consulates, as well as by ships and aircraft. Finally, I will assess the viability of the spatial model more generally, and by reference to the scenarios discussed above.

B. What is an 'area'?

The *Loizidou* test of effective overall control of an area makes perfect sense on the facts of *Loizidou* itself. After all, Turkey had occupied a third of the island of Cyprus, and it is such large chunks of land that we generally consider to be 'areas' or 'territories'. But problems arise as the area in question spatially diminishes. Is a city an 'area' that can be subjected to state jurisdiction? How about a mere village? Or a military base? Or a prison? Or just any building, or an apartment within that building?

The more an area diminishes in size, the more artificial it becomes to our ears to call it an area. *Obviously*, a mere square foot cannot qualify as an area for the purpose of the *Loizidou* test. But why not? What *exactly* distinguishes a building, or an apartment within that building, or a room within that apartment, from your northern Cypruses or southern Iraqs? Why should size alone matter, and to what extent should it matter?

The limits of the spatial conception of jurisdiction certainly do matter, as we have seen above from the examples of CIA black sites and extraterritorial abductions. Indeed, the United States under the Bush administration denied the applicability of human rights treaties such as the CAT to the black sites *precisely* on the grounds that they were *places* located in a *territory* controlled by a third state. Recall that it was the Bush administration's general position that human rights treaties, including the CAT, did not apply extraterritorially *at all*, thus rejecting even the effective overall control of an area notion of jurisdiction.[31] However, though it had some support in the text and the *travaux* of the ICCPR, the language of the jurisdiction clauses of the CAT was different. In 2005 the CIA asked the Office of Legal Counsel (OLC), the division of the US Department of Justice which provides authoritative legal advice for the executive branch of government, to advise

[31] See, e.g., Memorandum for William J. Haynes, II, from Jay S. Bybee, re The President's power as Commander in Chief to transfer captured terrorists to the control and custody of foreign nationals, US Department of Justice, Office of Legal Counsel, 13 March 2002, available at <http://www.justice.gov/opa/documents/memorandumpresidentpower03132002.pdf>, at 22–5.

on whether the Article 16 CAT prohibition on cruel, inhuman, or degrading treatment protected persons detained by the CIA in the so-called black sites.[32] Under Article 16(1):

> Each State Party shall undertake *to prevent in any territory under its jurisdiction* other acts of cruel, inhuman or degrading treatment or punishment which do not amount to torture as defined in article 1, when such acts are committed by or at the instigation of or with the consent or acquiescence of a public official or other person acting in an official capacity. In particular, the obligations contained in articles 10, 11, 12 and 13 shall apply with the substitution for references to torture or references to other forms of cruel, inhuman or degrading treatment or punishment. (emphasis added)

The 'any territory under its jurisdiction' clause clearly precludes an interpretation that would confine its application only to a territory over which the state has title, as the OLC realized:

> By its terms, Article 16 is limited to conduct within "territory under [United States] jurisdiction." We conclude that territory under United States jurisdiction includes, at most, areas over which the United States exercises at least de facto authority as the government. Based on CIA assurances, we understand that the interrogations do not take place in any such areas. We therefore conclude that Article 16 is inapplicable to the CIA's interrogation practices and that those practices thus cannot violate Article 16.[33]

So, the OLC accepts that Article 16 is not confined merely to those territories over which the state has title, but also to those over which the state has *de facto* authority. However, says the OLC, the CIA has informed it that it is not conducting the interrogations of high-value detainees in any such territory. Therefore, Article 16 CAT does not apply.[34]

Let me translate that for you: because we are holding these people in some undisclosed secret prison—be it someplace in Afghanistan, Poland, Lithuania, or wherever—in a *place*, but not in a *territory* over which we have effective overall control, the treaty does not apply on its own terms. A prison is not a territory. Like it or not, that's not a bad textual argument, morally repugnant though it might be, and it brings the spatial conception of state jurisdiction to a breaking point.

There are only two ways around this argument, neither of which is immediately apparent as correct, at least on the text of the CAT. Either the spatial conception of jurisdiction can extend to something as small as a *place* under state control, or the negative obligation of state agents not to engage in ill-treatment does not depend on

[32] Memorandum for John A. Rizo from Steven G. Bradbury, re Application of United States Obligations Under Article 16 of the Convention Against Torture to Certain Techniques that May Be Used in the Interrogation of High Value al Qaeda Detainees, US Department of Justice, Office of Legal Counsel, 30 May 2005, available at <http://www.fas.org/irp/agency/doj/olc/article16.pdf>.

[33] Ibid., at 1–2, developed in detail at 16 *et seq.*

[34] Alternatively, if this argument fails, the OLC turns to the US reservation to Art. 16 CAT, providing that cruel, inhuman, and degrading treatment within the meaning of the CAT will be understood as cruel and unusual punishment as defined by the applicable provisions of the Bill of Rights of the US Constitution. Since, in the OLC's view, the US Constitution does not apply extraterritorially to aliens, neither does the CAT.

a territorial jurisdiction threshold at all, even if the positive obligation to prevent ill-treatment does. I personally do not find it easy to accept that a riding school in Vilnius can be a territory under US jurisdiction, as Article 16 CAT requires, but this is not impossible, considering in particular the nature of the acts that the CAT prohibits, and that they invariably take place in a custodial setting.

Indeed, we should take note of Article 2 of the European Convention for the Prevention of Torture, which seems to have been drafted precisely to take such situations into account:

> Each Party shall permit visits, in accordance with this Convention, *to any place within its jurisdiction* where persons are deprived of their liberty by a public authority.

And also Article 4(1) of the Optional Protocol to the CAT, providing that

> [e]ach State Party shall allow visits, in accordance with the present Protocol, by the mechanisms referred to in articles 2 and 3 *to any place under its jurisdiction and control* where persons are or may be deprived of their liberty, either by virtue of an order given by a public authority or at its instigation or with its consent or acquiescence (hereinafter referred to as places of detention). These visits shall be undertaken with a view to strengthening, if necessary, the protection of these persons against torture and other cruel, inhuman or degrading treatment or punishment. (emphasis added)

This now brings me to *Al-Saadoon*, a case that ran its course through English courts[35] and the European Court at a most hectic pace, that we have already mentioned above in the context of norm conflict.[36] The applicants in that case were individuals detained by UK troops in Iraq. They were initially held as security detainees on the basis of Resolution 1546. While they were detained on this basis, per *Al-Jedda*, the Article 5 ECHR prohibition on preventive detention could be deemed to have been overridden by virtue of Article 103 of the Charter.[37] But then the legal basis for the applicants' detention changed. They were no longer held as security detainees, but were charged with specific crimes by the Iraqi judiciary.[38] Their detention by British forces continued, but now with the consent and at the behest of the Iraqi government, in effect in a sort of pre-trial detention that is in principle permitted by Article 5(1)(c) ECHR.

As the expiry of their UN mandate and the withdrawal of UK forces from Iraq drew near, the Iraqi authorities requested the transfer of the applicants to Iraqi custody. They challenged their impending transfer, first before the UK domestic courts and then before the European Court, arguing, *inter alia*, that there was a serious risk that they would be subjected to the death penalty by hanging if they

[35] *R. (Al-Saadoon and Mufdhi) v. Secretary of State for Defence* [2009] EWCA Civ 7 (hereinafter *Al-Saadoon CA*); *R. (Al-Saadoon and Mufdhi) v. Secretary of State for Defence* [2008] EWHC 3098 (hereinafter *Al-Saadoon HC*).

[36] See Chapter III, Section 9. See also Chapter V below for more discussion.

[37] *R. (Al-Jedda) v. Secretary of State for Defence* [2007] UKHL 58, [2008] 1 AC 332; see further M. Milanovic, 'Norm Conflict in International Law: Whither Human Rights?', (2009) 20 *Duke J. Comp. & Int'l L.* 69.

[38] *Al-Saadoon HC*, paras 24–31.

were transferred to Iraqi custody, and that this triggered the UK's *non-refoulement* obligations under the ECHR.

To make this argument, the applicants of course first had to establish that the ECHR applied to them extraterritorially. In *Al-Skeini*[39] the UK government had conceded that the ECHR applies to persons detained by UK forces in Iraq.[40] It thus had to come up with something new to deny the extraterritorial applicability of the ECHR in *Al-Saadoon*, and that it did. It argued that the applicants were held by UK forces at the order of an Iraqi court, and that the UK had a legal obligation to abide by the ruling of that court and transfer them to Iraqi authorities since UK forces were present in Iraq with Iraq's consent. Because it had no independent legal authority to detain the *Al-Saadoon* applicants, so argued the government, they were not within the UK's jurisdiction within the meaning of Article 1 ECHR.

In its decision, the Divisional Court did not accept this argument, ruling that the applicants were indeed within the UK's jurisdiction.[41] It held, however, that the ECHR *non-refoulement* principle had to be qualified because the UK had a legal obligation to transfer the applicants to Iraq, pursuant to a Court of Appeal ruling[42] that the Divisional Court thought was wrongly decided, but had to follow anyway.[43] On appeal, the Court of Appeal agreed entirely with the government's argument, finding that the ECHR did not apply because the UK had no independent legal authority to detain the applicants.[44] The applicants promptly moved to the European Court, and obtained a provisional measures order prohibiting their transfer to Iraqi authorities. And, for the first time in many years, the UK government decided to disobey such an order by the European Court, and transferred the applicants to Iraqi custody.[45]

Al-Saadoon presents one of the best examples of unresolvable norm conflict, specifically between the UK's obligations towards Iraq on one side, and its obligations under the ECHR on the other. I will now however deal only with the issue of extraterritorial applicability, which the Court of Appeal denied so that it could avoid the norm conflict. Note the UK government's argument, which the Court of Appeal accepted, that Article 1 ECHR jurisdiction requires the exercise of a *legal* authority, something that the UK in Iraq supposedly lacked. As we have seen in Chapter II above, this argument results from the confusion, particularly in *Bankovic*, between the various concepts of state jurisdiction in international law: the one

[39] *R. (Al-Skeini and Others) v. Secretary of State for Defence* [2007] UKHL 26, [2007] 3 WLR 33, [2007] 3 All ER 685 (hereinafter *Al-Skeini HL*); *R. (Al-Skeini and Others) v. Secretary of State for Defence* [2005] EWCA Civ 1609, [2007] QB 140 (hereinafter *Al-Skeini CA*).

[40] For commentary on the UK courts' reasoning regarding the extraterritorial applicability issues in *Al-Skeini*, see R. Wilde's note in (2008) 102 *AJIL* 628, as well as T. Thienel, 'The ECHR in Iraq: The Judgment of the House of Lords in *R. (Al-Skeini) v Secretary of State for Defence*' (2008) 6 *JICJ* 115.

[41] *Al-Saadoon HC*, para. 82.

[42] *R. (B) v. Secretary of State for Foreign and Commonwealth Affairs* [2004] EWCA Civ 1344, [2005] QB 643; see further Chapter V below.

[43] *Al-Saadoon HC*, paras 89–96.

[44] *Al-Saadoon CA*, paras 32–40.

[45] See T. Thienel, 'Cooperation in Iraq and the ECHR: An Awful Epilogue', *Invisible College*, 21 January 2009, available at <http://invisiblecollege.weblog.leidenuniv.nl/2009/01/21/cooperation-in-iraq-and-the-echr-an-awfu>.

that we are dealing with here most certainly does not require the exercise of a legal authority over a territory or an area. This *Bankovic*-induced confusion was to an extent also reflected in the admissibility decision in *Al-Saadoon* by a Chamber of the European Court,[46] which ultimately quite rightly rejected the UK's argument against extraterritorial application:

During the first months of the applicants' detention, the United Kingdom was an occupying power in Iraq. The two British-run detention facilities in which the applicants were held were established on Iraqi territory through the exercise of military force. The United Kingdom exercised control and authority over the individuals detained in them initially solely as a result of the use or threat of military force. Subsequently, the United Kingdom's *de facto* control over these premises was reflected in law. In particular, on 24 June 2004, CPA Order No. 17 (Revised) (see paragraph 13 above) provided that all premises currently used by the MNF should be inviolable and subject to the exclusive control and authority of the MNF. This provision remained in force until midnight on 31 December 2008 (see paragraphs 20–21 above).

The Court considers that, given the total and exclusive *de facto*, and subsequently also *de jure*, control exercised by the United Kingdom authorities over the premises in question, the individuals detained there, including the applicants, were within the United Kingdom's jurisdiction (see *Hess v. the United Kingdom*, no. 6231/73, Commission decision of 28 May 1975, Decisions & Reports vol. 2, p. 72). This conclusion is, moreover, consistent with the dicta of the House of Lords in *Al-Skeini* and the position adopted by the Government in that case before the Court of Appeal and House of Lords (see paragraph 62 above).

In the Court's view, the applicants remained within the United Kingdom's jurisdiction until their physical transfer to the custody of the Iraqi authorities on 31 December 2008. The questions whether the United Kingdom was under a legal obligation to transfer the applicants to Iraqi custody and whether, if there was such an obligation, it modified or displaced any obligation owed to the applicants under the Convention, are not material to the preliminary issue of jurisdiction (see, *mutatis mutandis*, *Bosphorus*, cited above, § 138) and must instead be considered in relation to the merits of the applicants' complaints.[47]

Note the several moves that the Court makes. First, though it refers, if not in so many words, to the general theory of Article 1 jurisdiction offered by the applicants as *de facto* authority and control over individuals, it does not explicitly accept it. Secondly, what it does in the second paragraph quoted above is to posit jurisdiction as control not over a territory or a wider geographical area, but control over *places or premises*, here the detention facility in which the applicants were kept. This is of course exactly the issue raised before with regard to the CIA black sites. Thirdly, like the High Court, the Chamber expressly disagrees with the Court of Appeal's position that Article 1 jurisdiction requires an exercise of legal authority, and finds that the question of what impact the UK's legal obligation to surrender the applicants to Iraqi authorities had on its ECHR obligations, if any, belongs to the merits.[48]

46 *Al-Saadoon and Mufdhi v. United Kingdom* (dec.), App. No. 61498/08, 3 July 2009, paras 84–5.
47 Ibid., paras 87–9.
48 Note also that the Court in *Al-Saadoon* relies on the former European Commission's decision in *Hess v. the United Kingdom* (dec.), App. No. 6231/73, 2 DR 72, 28 May 1975, which at first glance

So, what are we to make of this? If we conceive of state jurisdiction in human rights treaties in spatial terms, we can observe that space or area to which it refers on a continuum from something that we would broadly call a 'territory,' such as Northern Cyprus, to what we would generally call a 'place,' such as a UK-run prison in Iraq or that riding school in Vilnius. The question is whether that continuum extends even further, to even smaller areas or places. I, for one, cannot discern a clear cut-off one way or the other. What is certainly true is that there is a degree of artificiality to this approach, and that the artificiality increases as the size of the area decreases. For example, I would personally find it artificial in the extreme to argue that Israel had human rights obligations towards Eichmann because it controlled the *apartment* in which Eichmann was held before he was transferred to Israel, rather than Eichmann himself. In other words, the spatial concept of state jurisdiction as control over an area tends to collapse into the personal model of jurisdiction as control over individuals, or indeed into the absence of any threshold at all.

How then can we define an 'area' in a principled way? The only possible definition is in my view a *functional* one: only something over which state can exercise a sufficient degree of control can count as an area. The obvious drawback of

concerned the lawfulness of the continued detention of the former Nazi leader Rudolf Hess in the Spandau prison in UK-occupied Berlin, who was held after the Nuremberg trials under the joint authority of the four occupying powers, of which only the Soviets continued to oppose Hess's release. However, it was Hess's wife who actually lodged the application, complaining that *her* rights were violated. The threshold question was therefore whether Mrs Hess was within UK jurisdiction, but the Commission examined instead whether Hess himself fell within the scope of Article 1:

> The Commission first observes that in the present case the exercise of authority by the respondent Government takes place not in the territory of the United Kingdom but outside its territory, in Berlin.... The Commission is of the opinion that there is in principle, from a legal point of view, no reason why acts of the British authorities in Berlin should not entail the liability of the United Kingdom under the Convention. The question therefore arises whether the Commission, in the particular circumstances of this case, is competent to receive an application against the United Kingdom in respect of the continued imprisonment of Rudolf Hess...
>
> The Commission concludes that the responsibility for the prison at Spandau, and for the continued imprisonment of Rudolf Hess, is exercised on a Four Power basis and that the United Kingdom acts only as a partner in the joint responsibility which it shares with the three other Powers. The Commission is of the opinion that the joint authority cannot be divided into four separate jurisdictions and that therefore the United Kingdom's participation in the exercise of the joint authority and consequently in the administration and supervision of Spandau Prison is not a matter 'within the jurisdiction' of the United Kingdom, within the meaning of Art. 1 of the Convention.

Ibid., at 73–4.

This was undoubtedly a complex case. Even if we accept that the threshold issue was whether Mr Hess, rather than Mrs Hess, was within UK jurisdiction, I can only note that several different concepts are mixed up in the Commission's approach, namely attribution and jurisdiction as the threshold question; the content of the substantive obligation if the threshold is reached; and the responsibility *vel non* of the UK in connection with the acts of other states. Then there is also the question of possible norm conflict between the UK's obligations to the other three occupying powers, and its obligations under the ECHR. At any rate, in my view the Commission's decision is as such not particularly instructive.

this definition—if it can even be called a definition—is that we must establish what this sufficient degree of control is. This is what I will proceed to do next.

C. What amounts to 'control'?

1. *Lawful or unlawful*

We may remind ourselves once more of the European Court's holding in *Loizidou* that it is the *fact* of state control over territory which is the basis of state jurisdiction, whether obtained lawfully or unlawfully. As I have explained in Chapter II, it might at first glance seem a bit odd to say that a word as familiar as 'jurisdiction' actually refers to a factual concept that does not require the exercise of legal competence. Yet this is precisely how the word is used in the treaty practice of states, as a somewhat overdone way of saying that a state has *de facto* authority or control over a given piece of territory. It may or may not have the *right* to exercise such control, but it is the *fact* of control that matters.

This result is of course not simply warranted by rules of treaty interpretation, but by sound considerations of policy, since any other conception of jurisdiction would allow ample room for abuse, as we have already seen in a number of cases. A state can obtain control over the territory of another state in only two basic scenarios—with or without the territorial state's consent. In the former scenario, so long as the consent of the territorial state remains valid, and so long as the acts of the foreign state remain within the limits of that consent, the foreign state will have acted lawfully. The consent may be expressed in a treaty, e.g. a lease, or a foreign intervention may be invited somewhat less formally. All that matters, however, is that the consent is given. In the latter scenario, the control over another state's territory may or may not be lawful. As it involves the use of force, the lawfulness of obtaining such control is assessed on the basis of the relevant branch of international law, the *jus ad bellum*. Yet, just as with the law of occupation and the *jus in bello*, whose applicability is quite deliberately *not* dependent on the lawfulness *vel non* of the use of force as a matter of the *jus ad bellum*,[49] so should the applicability of human rights treaties be independent of such lawfulness.[50] Not only does this avoid frequently quite controversial and under-determinate preliminary questions regarding the use of force, but whether a state obtained control over a territory lawfully or not its ability to affect and protect the human rights of its inhabitants is the same.

However, we have also seen that post-*Bankovic* the Strasbourg jurisprudence has become somewhat more ambiguous. Note that even in *Al-Saadoon*, the Chamber considered that 'given the total and exclusive *de facto*, and subsequently also *de jure*, control exercised by the United Kingdom authorities over the premises in question,

[49] For an excellent recent defence of the wall of separation between the *jus in bello* and the *jus ad bellum*, see R. Sloane, 'The Cost of Conflation: Preserving the Dualism of Jus Ad Bellum and Jus in Bello in the Contemporary Law of War', (2009) 34 *Yale J. Int'l L.* 47.

[50] Similarly, see N. Lubell, *Extraterritorial Use of Force Against Non-State Actors* (Oxford University Press, 2010), at 214–15.

the individuals detained there, including the applicants, were within the United Kingdom's jurisdiction'.[51] It is the 'subsequently also *de jure*' part which is troubling, because it is not apparent why it should matter at all. The Chamber probably mentioned it to pre-empt any *Bankovic*-inspired assault on its decision. This would not make such a limitation any less arbitrary, particularly when we keep in mind that the *jus* referred to is nothing more than the law promulgated by the occupying powers, which they could switch on or off at will.

In short, any attempt to demand that the state's control over territory be lawful or anything other than purely factual should be resisted.[52] This, of course, still leaves the question of what the quality of the control should be as a matter of fact, to which I turn next.

2. *Effective overall control over territory*

In the preliminary objections stage of *Loizidou*, the European Court described Turkey's control over northern Cyprus only as 'effective'. At the merits stage, the Court added the word 'overall' to the test:

> It is not necessary to determine whether, as the applicant and the Government of Cyprus have suggested, Turkey actually exercises detailed control over the policies and actions of the authorities of the 'TRNC'. It is obvious from the large number of troops engaged in active duties in northern Cyprus that her army exercises effective overall control over that part of the island. Such control, according to the relevant test and in the circumstances of the case, entails her responsibility for the policies and actions of the 'TRNC'. Those affected by such policies or actions therefore come within the 'jurisdiction'—of Turkey for the purposes of Article 1 of the Convention. Her obligation to secure to the applicant the rights and freedoms set out in the Convention therefore extends to the northern part of Cyprus.[53]

The Court reiterated the 'effective overall control' formula in *Cyprus v. Turkey*[54] and subsequent cases. Readers will recall from my discussion of state responsibility issues in Chapter II that the Court's rulings in *Loizidou* and *Cyprus v. Turkey* can reasonably be interpreted in two ways: either the Court thought that the acts of the TRNC were *attributable* to Turkey, or it thought it unnecessary to deal with this question, finding instead that Turkey had a *positive obligation* to secure the human rights of the inhabits of northern Cyprus by virtue of its control over the territory. As I explained above, I prefer the latter option, because it reconciles the European Court with the ILC's and the ICJ's approach to state responsibility. Hence, while this Article 1 ECHR 'effective overall control' test bears resemblance to the ICJ's effective control test in *Nicaragua*, the two are conceptually distinct—the former

[51] *Al-Saadoon* (dec.), para. 88.

[52] See also A. Zimmermann, 'Extraterritorial Application of Human Rights Treaties—The Case of Israel and the Palestinian Territories Revisited', in I. Buffard, J. Crawford, A. Pellet, and S. Wittich (eds), *International Law Between Universalism and Fragmentation: Festschrift in Honour of Gerhard Hafner* (Brill, 2008), 747, at 758.

[53] *Loizidou* (merits), para. 56, internal citations omitted.

[54] *Cyprus v. Turkey*, para. 77.

refers to state control over *territory* for the purpose of establishing whether the state has jurisdiction over the territory, the latter to state control over *actors* and their specific acts for the purpose of attributing these acts to the state. Of course, the control by a state over territory must be exercised by its agents, i.e. persons whose acts are attributable to it.[55]

In these terms, the 'overall' prong of the *Loizidou* test serves an important purpose—applicants do not need to show that the state controlling a territory 'exercise[d] detailed control over the policies and actions' of the (possibly non-state) actor whose conduct directly violated their rights. As the Court put it in *Cyprus v. Turkey*, '[h]aving effective overall control over northern Cyprus, [Turkey's] responsibility cannot be confined to the acts of its own soldiers or officials in northern Cyprus but must also be engaged by virtue of the acts of the local administration which survives by virtue of Turkish military and other support.'[56]

But what exactly does it mean that the state's control needs to be 'effective'? In the most general terms, the state needs to have enough power over the territory and its inhabitants to broadly do as it pleases. That said, control over territory is a fluid thing, and is not limitless even under the best of conditions. To move from the abstract to the concrete we would need to examine specific cases, and we would then see that the threshold of control required by courts has generally been high. In the northern Cyprus cases, Turkey had thousands of troops on the ground, and the TRNC administration was at least initially little more than its puppet. In the ICJ *Wall* and *Congo v. Uganda* cases, the situation was again one of belligerent occupation. And in *Bankovic*, of course, the Court affirmed that the threshold of control should be strict, holding that air strikes alone were unable of satisfying it, and that

> its recognition of the exercise of extra-territorial jurisdiction by a Contracting State is exceptional: it has done so when the respondent State, through the effective control of the relevant territory and its inhabitants abroad as a consequence of military occupation or through the consent, invitation or acquiescence of the Government of that territory, exercises all or some of the public powers normally to be exercised by that Government.[57]

The 'public powers' requirement in particular appears to set the bar very high, as it seems to require the establishment of a structured (military) administration in the territory.[58] On the other hand, the Court backtracked from this stringent requirement of territorial control in *Issa*, a Chamber case, and in *Ilascu*, decided by a Grand Chamber. *Issa* dealt with an incursion by Turkish troops into northern Iraq, and the possible killing by these troops of certain Iraqi nationals. In one paragraph of the judgment, the Chamber seemed to have allowed for conceiving jurisdiction in *personal* terms, as authority and control over individuals,[59] while it said the

[55] See above, Chapter II, Section 3.
[56] *Cyprus v. Turkey*, para. 77.
[57] *Bankovic and Others v. Belgium and Others* [GC] (dec.), App. No. 52207/99, 12 December 2001, para. 71.
[58] See Wilde, above note 30, at 516.
[59] *Issa v. Turkey*, App. No. 31821/96, Judgment, 16 November 2004, para. 71.

following with regard to Turkey's *territorial* jurisdiction over the area of northern Iraq:

In this connection, the Court notes that it is undisputed between the parties that the Turkish armed forces carried out military operations in northern Iraq over a six-week period between 19 March and 16 April 1995 (see paragraphs 58 and 63 above). It transpires from the parties' submissions and the documentary evidence contained in the case-file that the cross-border operation conducted at that time was extensive and was aimed at pursuing and eliminating terrorists who were seeking shelter in northern Iraq (see paragraphs 36, 43, 45, 58 and 63).

The Court does not exclude the possibility that, as a consequence of this military action, the respondent State could be considered to have exercised, temporarily, effective overall control of a particular portion of the territory of northern Iraq. Accordingly, if there is a sufficient factual basis for holding that, at the relevant time, the victims were within that specific area, it would follow logically that they were within the jurisdiction of Turkey (and not that of Iraq, which is not a Contracting State and clearly does not fall within the legal space (*espace juridique*) of the Contracting States (see the above-cited *Banković* decision, § 80).

However, notwithstanding the large number of troops involved in the aforementioned military operations, it does not appear that Turkey exercised effective overall control of the entire area of northern Iraq. This situation is therefore in contrast to the one which obtained in northern Cyprus in the *Loizidou v. Turkey* and *Cyprus v. Turkey* cases (both cited above). In the latter cases, the Court found that the respondent Government's armed forces totalled more than 30,000 personnel (which is, admittedly, no less than the number alleged by the applicants in the instant case—see § 63 above—but with the difference that the troops in northern Cyprus were present over a very much longer period of time) and were stationed throughout the whole of the territory of northern Cyprus. Moreover, that area was constantly patrolled and had check points on all main lines of communication between the northern and southern parts of the island. The essential question to be examined in the instant case is whether at the relevant time Turkish troops conducted operations in the area where the killings took place.[60]

The Court then found that the evidence for such operations was lacking. However, the Court allowed for the possibility that Turkey exercised its jurisdiction in Iraq: (1) over a much more limited area than the whole of northern Iraq—indeed, the immediate area of a village; (2) over a much shorter period of time—some six weeks in all—than in the northern Cyprus cases; and (3) consequently with control with significantly less permanence, stability or exercise of any 'public powers'.

While *Issa* relaxes the *Loizidou* threshold somewhat, *Ilascu* perhaps goes so far as to redefine it. In addition to the question of Russia's control or jurisdiction over the territory of Transdniestria in Moldova, that case was plagued with severe problems regarding the Convention's temporal scope of application, and the Court's methodological approach is generally hard to decipher. Thus, with regard to the period before the entry of the Convention into force with respect to Russia, the Court considered that

[60] Ibid., paras 73–6.

...the Russian Federation's responsibility is engaged in respect of the unlawful acts committed by the Transdniestrian separatists, regard being had to the military and political support it gave them to help them set up the separatist regime and the participation of its military personnel in the fighting. In acting thus the authorities of the Russian Federation contributed both militarily and politically to the creation of a separatist regime in the region of Transdniestria, which is part of the territory of the Republic of Moldova.

The Court next notes that even after the ceasefire agreement of 21 July 1992 the Russian Federation continued to provide military, political and economic support to the separatist regime (see paragraphs 111 to 161 above), thus enabling it to survive by strengthening itself and by acquiring a certain amount of autonomy vis-à-vis Moldova.[61]

It was at this time that the applicants in the case were arrested by Russian troops in Transdniestria, who then transferred them to the Transdniestrian separatist authorities.[62] According to the Court,

...on account of the above events the applicants came within the jurisdiction of the Russian Federation within the meaning of Article 1 of the Convention, although at the time when they occurred the Convention was not in force with regard to the Russian Federation.

This is because the events which gave rise to the responsibility of the Russian Federation must be considered to include not only the acts in which the agents of that State participated, like the applicants' arrest and detention, but also their transfer into the hands of the Transdniestrian police and regime, and the subsequent ill-treatment inflicted on them by those police, since in acting in that way the agents of the Russian Federation were fully aware that they were handing them over to an illegal and unconstitutional regime.

In addition, regard being had to the acts the applicants were accused of, the agents of the Russian Government knew, or at least should have known, the fate which awaited them.

In the Court's opinion, all of the acts committed by Russian soldiers with regard to the applicants, including their transfer into the charge of the separatist regime, in the context of the Russian authorities' collaboration with that illegal regime, are capable of engaging responsibility for the acts of that regime.

It remains to be determined whether that responsibility remained engaged and whether it was still engaged at the time of the ratification of the Convention by the Russian Federation.[63]

The Court here says that Russia had jurisdiction over the territory of Transdniestria even though the Convention was, for Russia, not yet in force. There is in principle nothing wrong with that conclusion. But note the lack of clarity in the Court's formulation that the responsibility of Russia was 'engaged'—does the Court here think that the conduct of the Transdniestrian authorities was *attributable* to Russia, and if so on what grounds? Note also that whatever Russia did at the time was not *wrongful* under the ECHR, as the ECHR was not yet in force for Russia. The Court then moved on to the period after Russia had ratified the ECHR on 5 May 1998:

The Russian army is still stationed in Moldovan territory in breach of the undertakings to withdraw them completely given by the Russian Federation at the OSCE summits in

[61] *Ilascu and others v. Moldova and Russia* [GC], App. No. 48787/99, Judgment, 8 July 2004, para. 382.

[62] Ibid., para. 383. [63] Ibid., paras 384–5.

Istanbul (1999) and Porto (2001). Although the number of Russian troops stationed in Transdniestria has in fact fallen significantly since 1992 (see paragraph 131 above), the Court notes that the ROG's weapons stocks are still there.

Consequently, in view of the weight of this arsenal (see paragraph 131 above), the ROG's military importance in the region and its dissuasive influence persist.

All of the above proves that the 'MRT', set up in 1991–1992 with the support of the Russian Federation, vested with organs of power and its own administration, *remains under the effective authority, or at the very least under the decisive influence, of the Russian Federation, and in any event that it survives by virtue of the military, economic, financial and political support given to it by the Russian Federation.*

That being so, the Court considers that there is a continuous and uninterrupted link of responsibility on the part of the Russian Federation for the applicants' fate, as the Russian Federation's policy of support for the regime and collaboration with it continued beyond 5 May 1998, and after that date the Russian Federation made no attempt to put an end to the applicants' situation brought about by its agents, and did not act to prevent the violations allegedly committed after 5 May 1998.

Regard being had to the foregoing, it is of little consequence that since 5 May 1998 the agents of the Russian Federation have not participated directly in the events complained of in the present application.

In conclusion, the applicants therefore come within the "jurisdiction" of the Russian Federation for the purposes of Article 1 of the Convention and its responsibility is engaged with regard to the acts complained of.[64]

Put aside for one moment the lack of rigour in the Court's approach to issues of state responsibility, both with regard to attribution and with regard to the question of continuous wrongful acts and the ECHR's temporal application. If the Court's judgment is interpreted as holding that the actions of the MRT have at all times been attributable to Russia, then it is clear that Russia still has jurisdiction and control over the territory of Transdniestria. If, on the other hand, the acts of the MRT are *not* generally attributable to Russia, but Russia is being held responsible for its own failure to comply with its positive obligations, then the Court uses a pretty lax standard to establish Russia's jurisdiction over Transdniestria—all that mattered was that Russia had 'decisive influence' over it. This is a far cry from requiring the exercise of 'public powers' by Russia or a *de facto* government or administration of the territory.

The Court justified its approach by referring to Russia's involvement in the creation of the MRT, the far more extensive military control that it had in the past, and perhaps most importantly, the *potential* for such control that has remained. This is of course a remarkably similar situation to northern Cyprus, where Turkey has likewise gradually downgraded its military presence, while the local administration has become more independent. In other words, Russia could still be said to exercise effective overall control over Transdniestria, because *if it wanted to* it could easily make its power felt more overtly.[65]

[64] Ibid., paras 387, 392–4, emphasis added.

[65] Similarly, on the importance of potential for control, specifically in the Israeli/Palestinian context, see O. Ben-Naftali and Y. Shany, 'Living in Denial: The Application of Human Rights Treaties in the Occupied Territories', (2003) 37 *Israel L Rev* 17, at 63–4.

So what are we to make of the case law? First, the threshold of effective overall control of a territory is set relatively high. As a general matter, it requires boots on the ground. Secondly, though the threshold is set high, that level of control still need not be as high as the one that the state has over its own territory in peacetime or during normalcy. Thirdly, effective overall control is itself a spectrum, ranging from the more entrenched and visible exercise of *de facto* government, administration, or public powers, to the more borderline cases of less permanent or overt state control as in *Issa* and *Ilascu*.

More fundamentally, the case law aside, I think it is fair to say that the test of effective overall control of an area is also a functional one. Its stringency in the degree of effectiveness required depends foremost on the consequences that attach to the fact of such effective control. Recall the European Court's regime integrity concern in *Bankovic* that the ECHR cannot be cut up into little pieces, some of which would apply extraterritorially and some not, and that accordingly the ECHR is an all or nothing package. If the obligation incumbent on states is to secure *all* Convention rights in an area where they exercise effective overall over that area, then the degree of control required must be such as to allow states to realistically comply with this obligation.[66]

On the other hand, two considerations militate against having a threshold which is overly strict. First, the state's ability to comply with its negative obligation to respect human rights does not depend on its control over territory. Rather, the state by definition has control over its own agents. Secondly, even with respect to the state's positive obligation to secure human rights, that obligation is not as onerous as is it sometimes made out to be. It is an obligation of due diligence, of the state doing all that it could reasonably be expected to do to protect a territory's inhabitants even from third parties. Accepting, say, that the UK had jurisdiction over the parts of southern Iraq that it had occupied post-2003 does not entail that the UK had the obligation to turn Iraq into a simulacrum of idyllic English countryside.[67]

This brings me to the next issue I will address, the relationship between the effective overall control of an area threshold for establishing state jurisdiction and the threshold of control required for belligerent occupation.

3. *Relationship with the threshold of belligerent occupation*

In a scenario where a state obtains control over the territory of another state with that state's consent, the resulting occupation is pacific in nature. When, however, such control is obtained through the use of force (even if that use of force need not necessarily be opposed), IHL and the law of belligerent occupation apply concurrently with human rights treaties.[68] One issue that needs to be considered is whether the state jurisdiction threshold for the application of human rights treaties

[66] Similarly, see *Ilascu*, Dissenting Opinion of Judge Bratza.
[67] See above, Chapter III, Section 10.B.
[68] See further Chapter V below.

is the same as or different than the threshold required for the establishment of a belligerent occupation.[69]

Like the concept of jurisdiction in human rights treaties, belligerent occupation also depends on purely factual criteria. As provided in Article 42 of the Hague Regulations: 'Territory is considered occupied when it is actually placed under the authority of the hostile army. The occupation extends only to the territory where such authority has been established and can be exercised.' This standard is generally interpreted in the literature as being one of 'effective control'.[70]

But is this the same degree of control that we have examined above with regard to human rights treaties? Could there, in other words, be occupation without jurisdiction, or jurisdiction without occupation?

For the latter possibility, we might take another look at *Issa*. Recall that the Turkish military operation in northern Iraq lasted for a mere six weeks, and that the control of portions of Iraq by Turkish troops was fluid and intermittent. Assuming that Turkey had jurisdiction over parts of Iraq at the relevant time, was it also their belligerent occupier? On that point, IHL has always made a distinction between an invasion and an occupation.[71] Occupation may follow an invasion, but it requires that a territory is solidly or firmly seized and that its sovereign has actually been displaced—a mere vacuum in authority does not suffice. Though the precise line between the invasion and the occupation phase can be hard to draw, occupation requires a degree of stability.[72] As to whether the facts of *Issa* disclose the existence of a belligerent occupation, it is at best a borderline case.

Or take another example, that of Gaza after Israel's 2005 unilateral disengagement. Even though Israel no longer has troops on the ground in Gaza (at least not on a regular basis), nor exercises a military administration there, it has been argued, particularly by the human rights community broadly conceived, that Gaza still remains occupied by Israel.[73] Most of the arguments in favour of this position are to my mind unpersuasive. For example, control of the border crossings into a territory is not the occupation of that territory, but a siege. Nor is control over the sea and the airspace tantamount to the actual control of the territory itself. There has not been a single instance of belligerent occupation that I am aware of where a state was considered to have been the occupant of a territory without having troops on the ground. Likewise, that Israel remains the occupying power in the West Bank does not mean that it could not have terminated its occupation over Gaza, or that it could not have done so unilaterally.[74]

To my mind, the only good argument that Gaza remains occupied is that Israel still possesses an active *potential* for regaining effective control—cf. our discussion

[69] See also Wilde, above note 30.

[70] See, e.g., Y. Dinstein, *The International Law of Belligerent Occupation* (Cambridge University Press, 2009), at 40, 42 *et seq.*

[71] Ibid., at 38 *et seq.*

[72] Ibid.

[73] See, e.g., ibid., at 177–8, 277–9.

[74] See further M. Milanovic, 'Is Gaza Still Occupied by Israel?,' *EJIL: Talk!*, 1 March 2009, available at <http://www.ejiltalk.org/is-gaza-still-occupied-by-israel/>.

of *Ilascu* above. Occupation does not necessarily mean the control of all of the territory, all of the time, at the same level of intensity. Many occupations are met with significant and active resistance. Consider, for example, the US and UK-led occupation of Iraq. The strength of the insurgency there was such that at any given time these states lacked control over, say, some parts of the Sunni Triangle or some parts of Basra. That does not mean that their occupation of those parts of Iraq ceased, because the occupants were able to re-establish their control at will, as they did. The lapse in control was merely temporary, due to the fact that the level of resistance was such that the occupants could not control all of Iraq, all of the time. Another well-known example is the German occupation of Yugoslavia during the Second World War. The partisans fighting against the Nazis were from time to time able to temporarily liberate a bit of territory here or there, but that did not mean that Germany's occupation ceased. This was so because, as the military tribunal at Nuremberg ruled in *United States v. List*, 'the Germans could at any time they desired assume physical control of any part' of Yugoslavia.[75] The ICTY in *Naletilic* likewise ruled that an occupation exists so long as the occupying army has the 'capacity to send troops within a reasonable time to make the authority of the occupying power felt'.[76]

In any event, it may be *arguable* that Gaza remains occupied by Israel, but it is far from clear that this is the case, and it is certainly not so clear for this claim to be made in such forceful terms as it is by some in the human rights community.[77] Why is this argument then made so emphatically? The answer to that question is simple. Under IHL, a belligerent occupant has *positive obligations* to ensure the well-being of the civilian population, including the provision of food and other supplies. In a state of siege without occupation, however, the party to the conflict only has *negative obligations* not to interfere with relief consignments and so forth, and even these can be subject to military necessity. It does *not* have to provide food and supplies to the civilian population of its adversary.[78] Indeed, to impose such a requirement would be manifestly absurd. The problem is of course precisely that the civilian population of Gaza is heavily dependent on Israel and needs Israel not just to let humanitarian aid through, but also to provide electricity and other supplies of its own. This is why the gentler souls among us need to argue that Israel is the belligerent occupier of Gaza. It is the only legally certain way of assigning some positive obligations on Israel to provide for the civilians of Gaza—something that I agree with entirely as a matter of policy. But the certainty is unfortunately only deceptive.

[75] *United States of America v. Wilhelm List* et al., 8 Law Reports of Trials of Major War Criminals 38, 55–6.

[76] *Prosecutor v. Naletilic*, IT-98-34-T, Trial Chamber Judgment, 31 March 2003, para. 217.

[77] See, e.g., Report of the United Nations Fact Finding Mission on the Gaza Conflict (Goldstone Report), A/HRC/12/48, 15 September 2009, paras 274–80, esp. para. 276: 'Israel has *without doubt* at all times relevant to the mandate of the Mission exercised effective control over the Gaza Strip. The Mission is of the view that the circumstances of this control establish that the Gaza Strip remains occupied by Israel' (emphasis added).

[78] Cf. Arts 69 and 70 of Additional Protocol I.

So what are we to do then, if Israel is no longer the occupying power in Gaza, as has in fact been held by the Israeli Supreme Court?[79] One way of dealing with the issue would be to say that the state of dependency of Gaza on Israel created by Israel's occupation is a consequence of Israel's wrongful failure to abide by its duties as an occupying power, and that it has the obligation to provide full reparation for this failure. Likewise, Yuval Shany has made the intriguing argument that Israel might not have effective control over Gaza for the purposes of the law of occupation and thus has no positive obligations under IHL, but that it might have *jurisdiction* over Gaza under the relevant human rights treaties, such as the ICCPR, and would hence have positive obligations towards Gaza's inhabitants.[80] In other words, there could be jurisdiction without occupation, just as in *Ilascu* Russia had jurisdiction simply because it had 'decisive influence' over Transdniestria.

Consider, on the other hand, the possibility of a converse scenario of occupation without jurisdiction. Let us take a look, for example, at the insurgency in Iraq that followed the 2003 invasion, whose strength and magnitude was such that the coalition forces could not maintain total control over all of Iraq, all of the time. As we have seen above, this does not necessarily mean that the US and the UK ceased to be the occupying powers in Iraq, even if only temporarily, since they could always make their power felt. But does this mean that they had a sufficient degree of control to satisfy the jurisdiction threshold of human rights treaties?

In *Al-Skeini* both the Court of Appeal and the House of Lords thought that the UK lacked effective overall control over Basra for the purpose of establishing jurisdiction even though it was the occupying power in that part of Iraq.[81] In the Court of Appeal Lord Justice Brooke (joined by Lord Justice Richards) thought that it was

> ... quite impossible to hold that the UK, although an occupying power for the purposes of the Hague Regulations and Geneva IV, was in effective control of Basrah City for the purposes of ECHR jurisprudence at the material time. If it had been, it would have been obliged, pursuant to the *Bankovic* judgment, to secure to everyone in Basrah City the rights and freedoms guaranteed by the ECHR. One only has to state that proposition to see how utterly unreal it is. The UK possessed no executive, legislative or judicial authority in Basrah City, other than the limited authority given to its military forces, and as an occupying power it was bound to respect the laws in force in Iraq unless absolutely prevented. It could not be equated with a civil power: it was simply there to maintain security, and to support the civil administration in Iraq in a number of different ways.
>
> It would indeed have been contrary to the Coalition's policy to maintain a much more substantial military force in Basrah City when its over-arching policy was to encourage the Iraqis to govern themselves. To build up an alternative power base capable of delivering all the rights and performing all the obligations required of a contracting state under the ECHR at the very time when the IGC had been formed, with CPA encouragement, as a step

[79] *Gaber Al-Bassiouni v. Prime Minister*, H.C.J. 9132/07, Judgment, 30 January 2008.

[80] See Y. Shany, 'The Law Applicable to Non-Occupied Gaza: A Comment on *Bassiouni v. Prime Minister of Israel*', *Hebrew University International Law Research Paper No. 13-09*, 27 February 2009, available at <http://ssrn.com/abstract=1350307>, at 10 *et seq*.

[81] See also Wilde, above note 30, at 518 *et seq*.

towards the formation by the people of Iraq of an internationally recognized representative Government, would have run right against the grain of the Coalition's policies.[82]

This provoked the following response from Lord Justice Sedley:

In any event, the Court's explanation in *Bankovic* of the legal distinction between the bombing of Belgrade and the occupation of northern Cyprus places the British occupation of the Basra region of Iraq unequivocally in the latter class—subject to the single question of how much control is, for this purpose, effective control. International human rights law in its present phase does not answer this question. On the one hand, it sits ill in the mouth of a state which has helped to displace and dismantle by force another nation's civil authority to plead that, as an occupying power, it has so little control that it cannot be responsible for securing the population's basic rights. On the other, the fact is that it cannot: the invasion brought in its wake a vacuum of civil authority which British forces were and still are unable to fill. On the evidence before the Court they were, at least between mid-2003 and mid-2004, holding a fragile line against anarchy.

No doubt it is absurd to expect occupying forces in the near-chaos of Iraq to enforce the right to marry vouchsafed by Art. 12 or the equality guarantees vouchsafed by Art. 14. But I do not think effective control involves this. If effective control in the jurisprudence of the ECtHR marches with international humanitarian law and the law of armed conflict, as it clearly seeks to do, it involves two key things: the de facto assumption of civil power by an occupying state and a concomitant obligation to do all that is possible to keep order and protect essential civil rights. It does not make the occupying power the guarantor of rights; nor therefore does it demand sufficient control for all such purposes. What it does is place an obligation on the occupier to do all it can.

If this is right, it is not an answer to say that the UK, because it is unable to guarantee everything, is required to guarantee nothing. The question is whether our armed forces' effectiveness on the streets in 2003–4 was so exiguous that despite their assumption of power as an occupying force they lacked any real control of what happened from hour to hour in the Basra region. My own answer would be that the one thing British troops did have control over, even in the labile situation described in the evidence, was their own use of lethal force. Whether they were justified in using it in the situations they encountered, of which at least four of the cases before us are examples, is precisely the subject of the inquiry which the appellants seek. It is in such an inquiry that the low ratio of troops to civilians, the widespread availability of weapons and the prevalence of insurgency would fall to be evaluated. But, for reasons I now come to, I am not confident that this route is open in the present state of ECHR jurisprudence.[83]

Just like Lord Justice Brooke, the majority of the House of Lords thought that the UK did not have effective overall control and hence jurisdiction over Basra even though it had occupied it, although their primary holding was that the effective overall control of an area test could not even apply in Iraq, as it was outside the ECHR's *espace juridique*.[84]

[82] *Al-Skeini CA*, paras 124 and 125, citations omitted.

[83] Ibid., paras 194, 196, and 197.

[84] See *Al-Skeini HL*, para. 83 (per Lord Rodger):

I would not consider that the United Kingdom was in effective control of Basra and the surrounding area for purposes of jurisdiction under article 1 of the Convention at the

Al-Skeini is instructive on several points. First, most of the British judges thought that the UK's occupation of southern Iraq need not necessarily result in its jurisdiction over it. Secondly, their reasoning was that in their view the Article 1 ECHR positive obligation to secure human rights would be far too onerous under the circumstances. Among these circumstances were the strength of the insurgency against the occupation and the resulting 'near-chaos' in the area, the complicated institutional arrangements among the occupying powers, and the existence of a local administration that has been growing steadily more independent. Thirdly, the judges were worried in particular about the procedural positive obligations under Article 2 ECHR to investigate deprivations of life, which they thought the UK could not realistically fulfil, primarily because these requirements were developed by the European Court in times of normalcy, and would be difficult to depart from.[85]

These concerns are real, and we have already examined them above under the rubric of effectiveness. If these fears are not addressed, the extraterritorial application of human rights treaties beckons only as utopia. So let me try to address them. Most importantly, as Lord Justice Sedley well pointed out, the obligation to secure human rights is not, and should not be as onerous as the other judges made it seem.[86] It is not an obligation outside the realm of the possible. In its application we may certainly take into account the law of occupation, and have regard to the extraordinary circumstances involved, if not those that are solely of the UK's own making.[87]

That, for example, the UK chose to enter into complicated joint arrangements with the United States is quite frankly its own problem. So long as its military presence is such that it remains the occupant, why should it also not be expected to do all it reasonably can to secure the human rights of the population of the occupied territory? It cannot just say that it is there only to provide security, and that it is not even particularly good at that. It should also be borne in mind that IHL equally abhors 'paper occupations'.[88] A belligerent occupation without effective control, or at the very least the potential that such control can quickly be re-established, is simply no longer an occupation.

> relevant time. Leaving the other rights and freedoms on one side, with all its troops doing their best, the United Kingdom did not even have the kind of control of Basra and the surrounding area which would have allowed it to discharge the obligations, including the positive obligations, of a contracting state under article 2, as described, for instance in *Osman v. United Kingdom* (1998) 29 EHRR 245, 305, paras 115–116.

See also *Al-Skeini HL*, para. 90 (per Baroness Hale); para. 97 (per Lord Carswell); para. 129 (per Lord Brown).

[85] See *Al-Skeini CA*, paras 129–40 (per Brooke LJ); *Al-Skeini HL*, para. 128 (per Lord Brown): 'Bear in mind too the rigour with which the Court applies the Convention, well exemplified by the series of cases from the conflict zone of south eastern Turkey in which, the state's difficulties notwithstanding, no dilution has been permitted of the investigative obligations arising under articles 2 and 3.'

[86] Note, however, that Lord Justice Sedley's preferred solution was simply that the UK had control over the actions of its own troops, and that these troops brought individuals within their authority and control into the UK's jurisdiction or that no jurisdiction threshold applied, but that he thought this option was precluded under *Bankovic* (which it was).

[87] See also Wilde, above note 30, at 518–19.

[88] See Dinstein, above note 70, at 42–5.

To my mind at least, so long as the British occupation of southern Iraq was not terminated, the UK had effective overall control for Article 1 ECHR purposes over the area in question. In all but the most exceptional of cases, the two thresholds should be the same, so long as it is borne in mind that effective control implies the power of the state not only to affect the human rights of the population, but also to secure them, yet that the consequent obligation to secure or ensure human rights is a flexible one.

Conversely, turning back to the above example of Gaza, we should also resist accepting that jurisdiction may arise without occupation in situations where states obtain control over foreign territory by force. While it may seem attractive in some cases to relax the threshold of jurisdiction below that of occupation, doing so would actually expose extraterritorial application to the charge of utopia as an ideal completely divorced from practical considerations. Jurisdiction does require effectiveness, and a relatively stable presence of troops on the ground appears to be the only way of securing it, at least for the foreseeable future. Borderline cases will of course present themselves, either somewhere on the fuzzy line between invasion and occupation, as in *Issa*, or in situations where the state's grip on the territory is being challenged by another force. Likewise, the regime of belligerent occupation may be transformed into something else without the immediate loss of effective control and jurisdiction, as has arguably happened in Iraq. Though it is obviously impossible to provide a one-size-fits-all formula for dealing with such cases, one thing is certain—that they must be addressed with great caution.

4. Does control need to be exclusive?

A further question that arises is whether control over territory needs to be exclusive to be considered effective. The short answer is that effective control does normally, but not necessarily, exclude the exercise of control by the territorial state as the displaced sovereign.

When control over territory is obtained through the use of force, whether lawfully or unlawfully, effective control both for the purposes of state jurisdiction and the establishment of belligerent occupation presumes that the hostile army has established its own authority over the territory and substituted it for that of the displaced sovereign, and has successfully managed to suppress the resistance of the enemy, except perhaps in isolated pockets.

This simple scenario is perhaps the most common one, as for instance with regard to northern Cyprus. However, the legal situation can get more complicated, when (1) the occupant creates a subordinate local administration, and allows it a degree of autonomy; (2) allows the vestiges of the local administration of the displaced sovereign to operate normally; (3) an insurgency develops in the occupied territory; (4) the use of force and the occupation are conducted jointly by several states; (5) the use of force and the occupation are conducted under the auspices of an international organization; or (6) the belligerent occupation regime is terminated and evolves into some type of pacific occupation.

Scenarios (1) to (3) are perhaps the easiest to deal with. As the European Court held in *Loizidou*, the obligation to secure human rights arises whenever a state has effective control of an area, 'whether it be exercised directly, through its armed forces, or through a subordinate local administration'.[89] Indeed, occupants have frequently either set up their own proxy administrations or quisling governments, or have allowed some parts of the former administration to function so long as any military resistance is suppressed. This does not imply that everything these local actors do is attributable to the occupying power. However, while effective control of the occupant over the territory persists, both occupation and jurisdiction will remain even in the face of a mounting insurgency. It is of course entirely possible that the occupation forces are defeated and that they lose effective control, but it is only then that jurisdiction over the territory will also cease. With regard to the existence of a semi-autonomous local administration, particularly a *legitimate* one, its competences can certainly be taken into account in a due diligence analysis on the merits with respect to the state's positive obligation to secure or ensure human rights.

Scenario (4) is more complex, if not hopelessly so. States frequently take joint military action. They can do so in a variety of ways, from operating in a loose alliance, to putting their forces under unified military command. Depending on the exact institutional arrangements that they create, they can either exercise control over an area jointly, or they can divide a larger area into smaller units, each of which would be controlled by a specific national contingent, or perhaps even both. Post-2003 Iraq provides a good example. Iraq was under a regime of belligerent occupation from the completion of the invasion, with the United States and the United Kingdom being regarded by the UN Security Council as the occupying powers in Iraq, who had set up a unified command, the Coalition Provisional Authority.[90] The belligerent occupation lasted up until the end of June 2004 at the earliest, when the CPA transferred authority to the new Interim Government of Iraq—though the issue of the end of the occupation is controversial.[91] Let us consider at least the uncontested period of occupation, when the UK and the US were joint occupying powers over the whole of Iraq, but when in fact the UK mostly held the southern parts of Iraq, while the US occupied the rest. During this period, did the UK have jurisdiction over the whole of Iraq, since it was one of the joint occupying powers operating under unified command, or did it merely have jurisdiction over Basra and southern Iraq? This question is certainly a difficult one,

[89] *Loizidou*, para. 62.

[90] See UNSC Res. 1483 (2003), with the Council in a preambular paragraph 'recognizing the specific authorities, responsibilities, and obligations under applicable international law of these states as occupying powers under unified command (the "Authority")'.

[91] See UNSC Res. 1546 (2004), op. para. 2, in which the Council welcomed that 'by 30 June 2004, the occupation will end and the Coalition Provisional Authority will cease to exist, and that Iraq will reassert its full Sovereignty'. It is unclear whether the Council was merely recognizing the *fact* of the occupation ending, even though whether this factual appraisal may have been inaccurate, or whether the Council actually *terminated* the occupation on its own authority. See further R. Kolb, 'Occupation in Iraq since 2003 and the Powers of the UN Security Council', (2008) 90 *IRRC* 29; M. Sassoli, 'Legislation and Maintenance of Public Order and Civil Life by Occupying Powers', (2005) 16 *EJIL* 661.

and to answer it one would need to examine in detail the realities of the supposedly unified US/UK command. In principle, however, it is certainly possible that both the US and the UK had jurisdiction, i.e. effective overall control, over the same areas of Iraq.[92]

Similar to (4), but even more complex, is scenario (5), which involves joint state action though an international organization. Take as an example Kosovo after the 1999 NATO intervention against the then Federal Republic of Yugoslavia. The Security Council passed Resolution 1244 (1999), creating an international civil administration under UNMIK, a subsidiary organ of the UN, and KFOR, a NATO-led international military presence. Kosovo was not under a regime of belligerent occupation, since Yugoslavia had ultimately consented (if under some duress) to the deployment of the international forces in Kosovo, and since the Security Council had created a comprehensive regime of international administration over the territory. Though KFOR operated under unified command, Kosovo was divided up into several military districts, each of which was controlled by a particular national contingent.

The fundamental question posed by the Kosovo example is whether, say, the inhabitants of the area controlled by the French national contingent were within the jurisdiction of France within the meaning of Article 1 ECHR. Antecedent to that question is the issue whether the acts of French troops in Kosovo were, in fact, still attributable to France. In *Behrami*, the Grand Chamber of the European Court held that the conduct of French troops was not attributable to France, but to the United Nations, since French troops in Kosovo operated there pursuant to a UN Security Council mandate.[93] The customary rules regarding attribution of conduct to international organizations are admittedly less clear than those on state responsibility. Their codification is currently in the ILC's pipeline, which has so far produced and adopted on first reading a set of draft articles.[94] Having said so, in *Behrami* the European Court manifestly misapplied even the few clear rules that we actually have, by equating Security Council authorization with (exclusive) attribution of conduct to the UN, even when the UN itself lacks effective control over national contingents, and by ignoring the possibility of dual or multiple attribution. The *Behrami* decision has therefore attracted overwhelming criticism from numerous scholars,[95] including from the present author.[96] Most importantly, it has been rejected by the ILC in no uncertain terms.[97]

[92] The occupation of Germany by the Four Powers presented a similar factual scenario. See also the discussion of the *Hess* case in note 48 above.

[93] *Behrami and Behrami v. France, Saramati v. France, Germany and Norway* [GC] (dec.), App. Nos 71412/01 and 78166/01, 2 May 2007.

[94] ILC, Draft Articles on the Responsibility of International Organizations, as adopted on first readings, UN Doc. A/64/10 (2009) (hereinafter ILC DARIO).

[95] See, e.g. A. Sari, 'Jurisdiction and International Responsibility in Peace Support Operations: The Behrami and Saramati Cases', (2008) 8 *HRLR* 151; K. Mujezinovic Larsen, 'Attribution of Conduct in Peace Operations: The "Ultimate Authority and Control" Test', (2008) 19 *EJIL* 509.

[96] See M. Milanovic and T. Papic, 'As Bad As It Gets: The European Court of Human Rights' *Behrami and Saramati* Decision and General International Law', (2009) 58 *ICLQ* 267.

[97] See G. Gaja, Seventh Report on Responsibility of International Organizations, UN Doc. A/CN.4/610, 27 March 2009, at 10; ILC DARIO, at 62 *et seq.*

But though *Behrami* may have been wrongly decided, it is true that attribution matters for resolving the question of state jurisdiction. It is only if the conduct of French troops in Kosovo was attributable to France, although it may also have been attributable to NATO or even to the UN as separate legal persons, that France could have exercised jurisdiction over a part of Kosovo. Conversely, it is only if the conduct of the troops is *not* attributable to the troop contributing state, but *only* to the organization, that the possibility that the state is exercising jurisdiction can be excluded.[98]

In any case, as with attribution, the exercise of jurisdiction need not be exclusive. That UNMIK ran the civil administration of Kosovo, and that other parts of Kosovo were controlled by other national contingents, are both circumstances that can be taken into account when interpreting the substantive content of the positive obligation to secure human rights.

As for scenario (6), it poses unique challenges in that it involves the transformation of a regime which is relatively clear, that of belligerent occupation, into one of pacific occupation (if that),[99] the contours of which are far less defined. Consider the examples of Iraq and Afghanistan. In both cases a coalition of states used force against the two countries, whether lawfully or unlawfully. In both cases a regime of belligerent occupation ensued, whether more or less formalized. As time went on, that regime was supplemented and in parts overridden by UN Security Council action, while preparations were underway for the creation of new, internationally recognized governments. Both of these states eventually gained such new governments, which then gave their consent to the presence of foreign troops. At that point at the latest, the regime of belligerent occupation must have terminated, as such occupation can by definition be only non-consensual.[100]

From that moment onwards, the conduct of foreign troops was governed by the limits of the territorial state's consent, and by applicable Security Council resolutions. The situation on the ground, however, need not have changed dramatically, if it had changed at all. Although, for example, the nascent armed forces of the new legitimate government now started participating in joint military operations with foreign troops, the interveners could still be said to have had effective control over large swathes of territory. In my view, while such control persists, the intervening states can be said to exercise jurisdiction over the areas concerned within the meaning of the relevant human rights treaties. Likewise, the same could be said of the now independently governed territorial state, so long as it also had the requisite degree of actual control. There is no need, in principle, for the control of either the intervening states or the territorial state to be exclusive.

What is certainly true is that, after the termination of belligerent occupation, the new limits on what the intervening states may lawfully do can pose substantial

[98] See above, Chapter II, Section 3.D.

[99] See generally E. Benvenisti, 'Occupation, Pacific', in *Max Planck Encyclopedia of Public International Law*, available at <http://www.mpepil.com>. See also G. Fox, *Humanitarian Occupation* (Cambridge University Press, 2008).

[100] See, e.g., Dinstein, above note 70, at 35–7.

difficulties. However, the interpretation of the obligation to secure human rights can be sufficiently flexible to take into account both the dispositions of authority in Security Council mandates, and the need to respect the sovereignty of the territorial state. But, as I have discussed above, though flexible it cannot be watered down so much that it becomes completely ineffective. If, for example, the US and UK had knowledge that Iraqi government forces were engaging in human rights violations in areas that were still under US and UK effective overall control,[101] the US and the UK would (or should) have been faced with a choice: either put an end to Iraqi human rights abuses, by force if necessary and in likely violation of Iraqi sovereignty, or withdraw from Iraq. The situation, in other words, would be one of norm conflict.[102]

In sum, the scenarios that we have just examined show us that while jurisdiction over territory is normally exclusive, it need not necessarily be so. And while such cases can pose significant difficulties, these are not insurmountable. Further examples of joint or parallel jurisdiction can arise in cases, to which I will turn next, involving control over places or objects, rather than large areas of territory.

D. Control over places and objects

1. A general theory?

Readers will recall my discussion above of what constitutes an 'area' susceptible to state control for the purposes of establishing its jurisdiction. That discussion has ultimately proven to be incapable of providing any firm answers, except that the more an 'area' diminishes in size, the more it becomes a 'place', the more artificial it seems to apply the spatial concept of state jurisdiction to it. And of course the line between a 'place' proper, such as a building, or parts of that building or even a square foot or two within it, is also difficult to draw.

But assume that we can actually draw some reasonable lines that would enable us to prevent the collapse of the control over a place criterion into something resembling control over a mere speck or an atom of land. Does such control over a place present a *general* theory or grounds for establishing state jurisdiction for the purposes of human rights treaties? Or is it rather the *special nature* of some places, such as embassies, that brings them and the individuals located in them within the extraterritorial state's jurisdiction?

The first proposition is certainly the more intriguing one, as it has the potential of providing a principled approach rather than relying on some unavoidably arbitrary criteria of speciality. Explicit textual support for this proposition can be

[101] See, e.g., 'Iraq's Death Squads,' *Washington Post*, 4 December 2005, available at <http://www.washingtonpost.com/wp-dyn/content/article/2005/12/03/AR2005120300881.html>; 'New Order, Same Abuses: Unlawful Detentions and Torture in Iraq', *Amnesty International*, 13 September 2010, available at <http://www.amnestyusa.org/pdf/Iraqrpt_newordersameabuses.pdf>; 'Iraq: Detainees Describe Torture in Secret Jail', *Human Rights Watch*, 27 April 2010, at <http://www.hrw.org/en/news/2010/04/27/iraq-detainees-describe-torture-secret-jail>.

[102] See above, Chapter III, Section 9.

found in two jurisdiction clauses, Article 2 CPT[103] and Article 4(1) of the Optional Protocol to the CAT.[104] As for the CAT itself, the Committee Against Torture has in its General Comment No. 2 interpreted its jurisdiction clauses which refer to 'any territory under [a state's] jurisdiction' as also encompassing *places* under the state's control:

> The Committee also understands that the concept of 'any territory under its jurisdiction,' linked as it is with the principle of non-derogability, includes any territory *or facilities* and must be applied to protect any person, citizen or non-citizen without discrimination subject to the de jure or de facto control of a State party. The Committee emphasizes that the State's obligation to prevent torture also applies to all persons who act, de jure or de facto, in the name of, in conjunction with, or at the behest of the State party.[105]

Thus, in the Committee's view, the concept of 'territory' includes 'facilities'. The Committee further added that

> [a]rticle 2, paragraph 1, requires that each State party shall take effective measures to prevent acts of torture not only in its sovereign territory but also 'in any territory under its jurisdiction.' The Committee has recognized that 'any territory' includes all areas where the State party exercises, directly or indirectly, in whole or in part, de jure or de facto effective control, in accordance with international law. The reference to 'any territory' in article 2, like that in articles 5, 11, 12, 13 and 16, refers to prohibited acts committed not only on board a ship or aircraft registered by a State party, but also during military occupation or peacekeeping operations and in such places as embassies, military bases, detention facilities, or other areas over which a State exercises factual or effective control. The Committee notes that this interpretation reinforces article 5, paragraph 1 (b), which requires that a State party must take measures to exercise jurisdiction 'when the alleged offender is a national of the State.' The Committee considers that the scope of 'territory' under article 2 must also include situations where a State party exercises, directly or indirectly, de facto or de jure control over persons in detention.[106]

Note, however, how the Committee's approach tends to collapse to the notion of jurisdiction as control over individuals—in a textually problematic way, it must be said, with regard to the provisions of the CAT. As for the European Court, the case which came the closest to endorsing a general theory is of course *Al-Saadoon*:

> The Court considers that, given the total and exclusive *de facto*, and subsequently also *de jure*, control exercised by the United Kingdom authorities over the premises in question, the individuals detained there, including the applicants, were within the United Kingdom's jurisdiction. This conclusion is, moreover, consistent with the dicta of the House of Lords in

[103] 'Each Party shall permit visits, in accordance with this Convention, *to any place within its jurisdiction* where persons are deprived of their liberty by a public authority' (emphasis added).

[104] 'Each State Party shall allow visits, in accordance with the present Protocol, by the mechanisms referred to in articles 2 and 3 *to any place under its jurisdiction and control* where persons are or may be deprived of their liberty' (emphasis added).

[105] Committee Against Torture, General Comment No. 2, UN Doc. CAT/C/GC/2, 24 January 2008, para. 7 (emphasis added).

[106] Ibid., para. 16.

Al-Skeini and the position adopted by the Government in that case before the Court of Appeal and House of Lords.[107]

The Chamber in *Al-Saadoon* exhibited a high level of caution, aware as it was of contradictory case law all around it. Note, for example, how it treats the 'dicta' of the House of Lords in *Al-Skeini*, agreeing with the conclusion that Baha Mousa, who was detained by UK troops, was within the UK's jurisdiction, yet without endorsing the *reasoning* of the House of Lords. That reasoning, as we have seen, was precisely that Baha Mousa was within the UK's jurisdiction for a 'special' reason: that a UK military prison abroad was essentially analogous to a UK embassy.[108]

I, for one, find these special or exceptional explanations endorsed by the House of Lords to be entirely unsatisfactory. Not only does the analogy between a military prison and an embassy defy common sense, as there is simply no such thing as a 'special status' of a military prison in international law,[109] it introduces distinctions which are entirely arbitrary. For example, a person detained in an embassy or in a military prison might be entitled to protection under human rights treaties, but what about a person detained in a CIA 'black site', such as that riding school in Lithuania? Is that also sufficiently like an embassy or not, and if not, why not?

The *only* thing common both to a prison and to an embassy are that they operate on the basis of the territorial state's *consent*.[110] But if this was the basis for the extraterritorial state's jurisdiction, no justification is given for this position, aside from the fallacy that the application of human rights treaties would as such infringe on the territorial state's sovereignty.[111] Indeed, one would imagine that non-consensual interventions would generally be more likely to adversely affect the human rights of the population, while consent itself can be given to many things, such as to the presence of foreign forces in general, or to rendition overflights, or the operation of a CIA 'black site'.

If, on the other hand, we were to adopt the general theory of control over places as a basis for state jurisdiction, then many of the same considerations that we have already examined with regard to control over territory would apply accordingly, with one major difference. A *place* is by definition located within a *territory* under another state's jurisdiction, and control over such places can by its nature be much more fleeting than effective overall control over large areas of territory would normally be. In all but the most exceptional of circumstances, such control can be maintained with any stability only with the consent of the territorial state. In other words, without Lithuania's consent, the US could probably never have operated the black site on the outskirts of Vilnius. Its control over that place was only viable because the state controlling the territory acquiesced in it. Accordingly,

107 *Al-Saadoon* (dec.), para. 88, citations omitted.

108 See *Al-Skeini HL*, para. 97 (per Lord Carswell), para. 132 (per Lord Brown). See also *Al-Skeini DC*, para. 287: 'a British military prison, operating in Iraq with the consent of the Iraqi sovereign authorities, and containing arrested suspects, falls within even a narrowly limited exception exemplified by embassies, consulates, vessels and aircraft'.

109 See also T. Thienel, 'The ECHR in Iraq', (2008) 6 *JICJ* 115, at 127.

110 See also above note 50 and accompanying text.

111 See above, Chapter III, Section 4.

this is also a scenario of concurrent, rather than exclusive jurisdiction. If we accept that the US had jurisdiction over the black site near Vilnius, this does not mean that Lithuania did not have jurisdiction—its obligation to secure the human rights of all persons detained at the black sites was undiminished. The black site most certainly did not drill a black hole in Lithuania's own ECHR obligations.

On the other hand, if we turn back to our example of the abduction of Adolf Eichmann from Argentina, I would personally find it difficult to accept that the safe house in which the Israeli agents kept Eichmann until he was smuggled to Israel aboard an aircraft amounted to a place under Israel's jurisdiction. Again, it is simply far too artificial to say that it is the Israeli agents' control of a minuscule space between four walls, rather than their control of Eichmann's *person*, that would entitle Eichmann to the protection of his human rights vis-à-vis Israel. In any event, though the viability of any long-term control over a place would generally require the consent of the territorial state, this is simply a question of *fact*, rather than a legal requirement—and new facts can always surprise us.

That is all that can be said, I think, about control over a place as a general basis for state jurisdiction. I will now turn to the supposedly special cases of embassies, vessels, and aircraft on which the House of Lords relied so prominently in *Al-Skeini*. The root cause of this methodological confusion can, predictably, be traced back to *Bankovic*.

2. *Embassies and consulates*

In order to limit as much as possible the scope of the ECHR's extraterritorial application, the European Court in *Bankovic* held that such application can be justified only exceptionally, supposedly on grounds of general international law. One of the many problems with this holding is that it contradicted the prior case law of the Convention institutions, which the Court was now forced to 'clarify', rather than explicitly depart from. It thus said the following:

> Additionally, the Court notes that other recognised instances of the extra-territorial exercise of jurisdiction by a State include cases involving the activities of its diplomatic or consular agents abroad and on board craft and vessels registered in, or flying the flag of, that State. In these specific situations, customary international law and treaty provisions have recognised the extra-territorial exercise of jurisdiction by the relevant State.[112]

Note, first, that the Court here speaks about acts of diplomatic and consular *agents*, not diplomatic and consular *premises*, as the supposed 'embassy exception' was interpreted by UK courts. Even so, this paragraph is doubly wrong. First, to the extent that it claims to restate the prior case law of the Strasbourg institutions, it actually provides an anachronistic reinterpretation of these cases, as we will now see. Secondly, to the extent that it argues that public international law recognizes some sort of special or exceptional exercise of state jurisdiction over embassies, ships, or aircraft, it is incorrect, and again mixes up the various meanings of the word

[112] *Bankovic*, para. 73.

'jurisdiction' and accordingly the various *concepts* of jurisdiction in international law.

Let us first take a look at the case law—the case law being that of the now defunct European Commission, since the Court itself has to my knowledge never taken up an 'embassy' case. Perhaps the first such case, decided in 1965, was *X v. Federal Republic of Germany*,[113] which dealt with several complaints made by a German national residing in Morocco against the conduct of German diplomatic and consular officials there. In that case the Commission gave the briefest consideration to the issue of extraterritorial application, in the following terms:

> ... in certain respects, the nationals of a Contracting State are within its 'jurisdiction' even when domiciled or resident abroad; [...] in particular, the diplomatic and consular representatives of their country of origin perform certain duties with regard to them which may, in certain circumstances, make that country liable in respect of the Convention.[114]

As I have argued above, the nationality of the victim of a human rights violation, on which the Commission may have relied, should have no bearing on the question of extraterritorial application.[115] On the other hand, the Commission's second sentence might be taken as regarding as relevant the supposedly special nature of diplomatic and consular agents. Ten years later, in its first decision on northern Cyprus,[116] the Commission developed on its holding in *X v. Germany*:

> The Commission further observes that nationals of a State, including registered ships and aircraft, are partly within its jurisdiction wherever they may be, and that authorised agents of a State, including diplomatic and consular agents and armed forces, not only remain under its jurisdiction when abroad but bring any other persons or property 'within the jurisdiction' of that State, to the extent that they exercise authority over such persons or property. Insofar as, by their acts or omissions, they affect such persons or property, the responsibility of the State is engaged.[117]

Again the Commission refers to the relevance of nationality, now subsuming registered ships and aircraft under that heading. More importantly, however, it says that the authorized agents of a state, *including*, but not limited to its diplomatic and consular agents, remain under the state's jurisdiction, and can bring under its jurisdiction those persons over which they exercise authority. In other words, the Commission adopted a *personal*, rather than spatial conception of Article 1 ECHR jurisdiction, which it thought applicable to *all* state agents. This became the constant jurisprudence of the Commission.[118]

[113] *X v. Federal Republic of Germany* (dec.), App. No. 1611/62, 25 September 1965, 8 *Yearbook of the European Convention on Human Rights* 158. See also Chapter III, note 134 and accompanying text.

[114] Ibid., at 168.

[115] See above, Chapter III, Section 5. See also below, Section 3.D.3.

[116] *Cyprus v. Turkey* (dec.), App. Nos 6780/74 and 6950/75, 26 May 1975.

[117] Ibid, at 136, para. 8.

[118] See *X v. United Kingdom* (dec.), App. No. 7547/76, 15 December 1977, concerning the acts of British consular officials in Jordan. Though the application was held to be manifestly ill-founded, the

Viewed as a whole, the Commission's jurisprudence gave little weight to the supposedly special nature of diplomatic or consular officials, or of diplomatic and consular premises, for the purpose of establishing state jurisdiction within the meaning of Article 1. Nor does general international law, for that matter. It is true that diplomatic and consular officials are permitted to exercise certain public functions over individuals while in the territory of the receiving state—from issuing visas and entry permits, to protecting their nationals abroad. But the only reason why they can do so is because the receiving state has *consented* to their presence. In times past, diplomatic and consular officials in certain non-Western states had far greater powers under the rubric of consular jurisdiction, which entailed, for example, their exclusive capacity to try their own nationals for offences committed in the host state. Even then, however, this capacity was based on the host state's consent through so-called capitulations or unequal treaties.[119] In modern international law, which is premised on the sovereign equality of all states, the role of diplomatic and consular officials is far more limited.

Either way, it is entirely misleading to say that customary law recognizes the 'special' nature of the extraterritorial exercise of state jurisdiction by these officials. There is nothing special about it, except that it is regularized, in the sense that a defined set of customary and treaty rules, agreed upon in advance, applies to these officials once the territorial state gives its consent. But the basis for their acts is still exclusively in the consent itself. The territorial state can withdraw its consent at any time, or can provide consent which is much wider in scope, for example by inviting a foreign military intervention, or by allowing a foreign power to run a military prison. As with the case of effective control over territory generally, this consent is only relevant for assessing whether the extraterritorial act of a foreign state is *lawful* or not, i.e. whether it violates the sovereignty of the territorial state, but it has no bearing on the *fact* of the extraterritorial exercise of state power, which may affect individuals in the territorial state. Again, there is simply no valid explanation as to why consent by the territorial state should matter for the purposes of establishing Article 1 ECHR state jurisdiction.[120]

On the other hand, it is true that both customary and conventional international law provide for a limited *immunity* of diplomatic and consular officials, and

Commission thought that the applicant was within UK jurisdiction even though he was outside UK territory:

> It is clear, in this respect, from the constant jurisprudence of the Commission that authorised agents of a State, including diplomatic or consular agents, bring other persons or property within the jurisdiction of that State to the extent that they exercise authority over such persons or property. Insofar as they affect such persons or property by their acts or omissions, the responsibility of the State is engaged.

See also *WM v. Denmark* (dec.), App. No. 17392/90, 14 October 1993; *Gill and Malone v. The Netherlands and the United Kingdom* (dec.), App. No. 24001/94, 11 April 1996. See further below, Section 3.

[119] See generally C. Bell, 'Capitulations', as well as P. Czubik and P. Szwedo, 'Consular Jurisdiction', both in *Max Planck Encyclopedia of Public International Law*, available at <http://www.mpepil.com>, and the sources cited therein.

[120] See above, Section 2.C.1.

diplomatic and consular premises, from the host state's jurisdiction.[121] And, of course, the host state may well decide to grant similar immunities to other foreign agents on its territory, as is often the case, for instance, with foreign troop contingents and military bases, the deployment of which frequently requires the conclusion of a status of forces agreement that may provide for various privileges and immunities.[122] Does this mean that even though they are present in the host state's territory, or more precisely a territory under its control, these officials and premises are not actually within the host state's jurisdiction, within the meaning of the relevant human rights treaties?

The short answer is no. To develop that, let us take a closer look at the immunity rules of diplomatic and consular law. For example, Article 22 of the Vienna Convention on Diplomatic Relations provides that the premises of a diplomatic mission shall be inviolable; that they may not be entered by the agents of the receiving state except with the consent of the head of mission; that they will be immune from search, requisition, or execution; and that the receiving state has a special duty to protect them from third parties. With respect to diplomatic agents, Article 29 of the Convention provides that their persons shall be inviolable, while Article 31 stipulates that they shall be immune from the criminal jurisdiction of the receiving state, as well as from civil jurisdiction except in certain cases, such as disputes regarding private immovable property. In any event a diplomatic agent enjoys immunity from execution.

However, pursuant to Article 41(1) of the Convention, '[w]ithout prejudice to their privileges and immunities, it is the duty of all persons enjoying such privileges and immunities to respect the laws and regulations of the receiving State. They also have a duty not to interfere in the internal affairs of that State.' Article 41(3) further provides that '[t]he premises of the mission must not be used in any manner incompatible with the functions of the mission as laid down in the present Convention or by other rules of general international law or by any special agreements in force between the sending and the receiving State'.

So, for example, are the Serbian ambassador to the United Kingdom and the Serbian embassy in London within the UK jurisdiction for the purposes of Article 1 ECHR or the relevant jurisdiction clauses of other human rights treaties? Or are they rather within Serbian jurisdiction? Let us consider the following (purely hypothetical!) examples:

(1) The Serbian ambassador, while acting in his official capacity and under authorization from the Serbian government, kills somebody in London; was the victim of that deprivation of life within UK or Serbian jurisdiction, and could his family file an application against either the UK or Serbia with Strasbourg?

[121] See generally H. Fox, *The Law of State Immunity* (Oxford University Press, 2nd edn, 2008), at 665 *et seq.*

[122] See generally P. Conderman, 'Status of Armed Forces on Foreign Territory Agreements (SOFA)', in *Max Planck Encyclopedia of Public International Law*, available at <http://www.mpepil.com>.

(2) The Serbian ambassador, again while acting in his official capacity, kills someone on the premises of the Serbian embassy in London; did, for instance, the UK have the positive obligation to secure the right to life of the victim, even though he was present within the Serbian embassy?

(3) Conversely, the Serbian ambassador, now a complete innocent, is killed by some crazy person while on a stroll in Hyde Park; can his family file an application against the UK for failing to secure his right to life?

To address these scenarios, it is again important to recall the various meanings that the word 'jurisdiction' can have in international law. The 'jurisdiction' from which diplomatic agents have immunity within certain limits is the *jurisdiction to enforce*, the state's authority to enforce the laws that it has previously prescribed. However, diplomatic agents do *not* have immunity from the receiving state's *jurisdiction to prescribe*—they must obey and comply with applicable domestic law. On the other hand, *our* jurisdiction—the state jurisdiction referred to in human rights treaties—is *neither* of these. Rather, it refers to state control through its agents over territory, or perhaps a place, or maybe even an individual.[123]

So, to look at example (1), if the Serbian ambassador were to kill a Londoner, the victim would certainly be within the UK's jurisdiction, as the killing took place in London, an area under which the UK has effective overall control. The victim would only be within Serbia's jurisdiction if we consider that jurisdiction can also arise from the authority and control exercised by state agents over an individual, even if that control is exercised in an area outside the state's control. This is of course the *personal* rather than spatial conception of jurisdiction, the model which I will be examining shortly in some detail. However, whatever the general validity of that model, it seems to me to be perfectly irrelevant whether the killing was committed by a diplomatic agent, such as the Serbian ambassador, or by any other authorized agent of Serbia, i.e. a person whose acts are attributable to it.

Now, if the victim of the killing was within the UK's jurisdiction, this means that the UK had a positive obligation to secure the right to life of the victim. Normally, this obligation has two aspects: the UK would have to do all it could to prevent the killing from taking place, and if it was nonetheless unable to do so, it would have to investigate the crime and prosecute the person responsible.[124] On the other hand, Article 29 of the Vienna Convention provides that '[t]he person of a diplomatic agent shall be inviolable. He shall not be liable to any form of arrest or detention', while Article 31 exempts diplomatic agents from the territorial state's criminal jurisdiction. Imagine, therefore, that under the observant gaze of UK police officer, the Serbian ambassador accosts his victim on a London street, holding a gun in his right hand, while carrying in his left hand a sign saying 'I am the Serbian ambassador, I am acting in my official capacity, and my person is inviolable.' Would the

[123] See above, Chapter II, Section 2.

[124] See generally A. Mowbray, *The Development of Positive Obligations under the European Convention on Human Rights by the European Court of Human Rights* (Hart Publishing, 2004), at 7 *et seq.*

police officer have the positive obligation to tackle or maybe even shoot the ambassador to prevent the killing from taking place?

This situation is one of apparent norm conflict, with the ECHR commanding the UK to prevent the killing, and the Vienna Convention prohibiting it from taking any practical measures in doing so.[125] This norm conflict can perhaps be avoided. Thus, we could say either that the ECHR positive obligation to prevent the killing should be read down so that the UK does not have to act in such a way as to violate its other international obligations by infringing on the ambassador's person, or that it is the Vienna Convention that should be read down so that a limited exception is carved out of its prohibitions, allowing the authorities of the receiving state to prevent diplomatic agents from committing imminent unlawful acts (as a matter of policy, if not the treaty text, the most reasonable solution).[126] Alternatively, if such a reading down of either treaty is impossible, then we would have an unresolvable norm conflict on our hands—whatever the police officer did, the UK would violate one of its international obligations.

Turning now to example (2), if the Serbian ambassador killed someone in the Serbian embassy, could the victim now be said to be within Serbian jurisdiction? If we consider that Article 1 ECHR jurisdiction extends to control over places, and not just large areas of territory, *à la Al-Saadoon*, then one could say yes to that question. But again, the fact that the premises in question are an *embassy* to me seems to be perfectly irrelevant. If we recall for a moment the example of the CIA black site in Lithuania, it would seem entirely arbitrary to say that a person detained in the US embassy in Vilnius would be within US jurisdiction, but that a person detained at the Vilnius riding school black site a few miles away would not, just because the former is a diplomatic mission. Not only would such a position be devoid of any broader principle, but it would manifestly open the door to abuse.

But what about the positive obligations of the host country? If, say, UK police officers stationed near the Serbian embassy in London heard loud screaming coming from the embassy, with the poor victim shouting at the top of his lungs that he was being tortured inside, would the UK have a positive obligation under the ECHR to put an end to the crime, by entering the embassy by force if necessary, as it would in all other circumstances? One possible answer is that since the Serbian embassy in London is within Serbia's jurisdiction, it would be outside the UK's jurisdiction, and therefore that the UK would owe no positive obligations to persons in it. However, recall that jurisdiction need not necessarily be exclusive. There is to my mind nothing contradictory in saying that the Serbian embassy in London is a place under Serbia's control, and therefore within its

[125] See above, Chapter III, Section 9.

[126] In fact, this is what the ICJ did in the *Tehran Hostages* case, where it remarked that

> [n]aturally, the observance of this principle [of inviolability] does not mean—and this the Applicant Government expressly acknowledges—that a diplomatic agent caught in the act of committing an assault or other offence may not, on occasion, be briefly arrested by the police of the receiving State in order to prevent the commission of the particular crime.

United States Diplomatic and Consular Staff in Tehran (United States v. Iran), Judgment, ICJ Reports 1980, 3, at 40, para. 86.

jurisdiction, but that this place is located in an area under UK control, and therefore also within the UK's jurisdiction. And if this so, then just as with example (1) we are faced with an apparent norm conflict between the UK's obligations under the ECHR and its obligations under diplomatic law.[127]

As for example (3), that the Serbian ambassador to the UK is immune from UK jurisdiction in the sense that the UK may not subject him to criminal proceedings, detention, or arrest, does not mean that he is actually not within the UK's jurisdiction for the purposes of Article 1 ECHR. If he were actually to be killed while taking a walk in London, the UK would not only have an obligation to prevent and investigate his death under the Vienna Convention—it would also have such an obligation under Articles 1 and 2 ECHR.

These scenarios could of course be considered to be quite fanciful. After all, aside from the occasional traffic or hunting accident, ambassadors do not normally go about killing or torturing people. These examples do, however, convincingly expose the sheer folly of considering either diplomatic and consular agents, or the premises from which they operate, as somehow being 'special' for the purposes of establishing state jurisdiction under human rights treaties.[128] These agents are not distinguishable by their function from all other state agents, nor are these premises different from other places under state control. We should also not forget that states may grant similar immunities to foreign military forces or other agents. Nor, consequently, does an analogy between an embassy and any other place, such as a military prison in *Al-Skeini*, make any sense whatsoever. It is manifestly nothing more than a device used to narrow extraterritorial application to those situations which we substantively consider to be morally intolerable, such as the abuse and killing of a defenceless prisoner, while rejecting extraterritorial application in cases which are not so black and white.

3. Ships and aircraft

The *Bankovic* and *Al-Skeini* dicta on the supposedly special nature of ships and aircraft are similarly misguided. Again, while it is true that international law recognizes that states may exercise prescriptive jurisdiction over registered ships and aircraft, this is simply *not* the 'jurisdiction' referred to in Article 1 ECHR. Let me once more bring up the example of Article 9(1) of the UN Enforced Disappearances Convention, which provides as follows:

> Each State Party shall take the necessary measures to establish its jurisdiction over the offence of enforced disappearance:

[127] See also M. Shaw, *International Law* (Cambridge University Press, 6th edn, 2009), at 754; E. Denza, *Diplomatic Law* (Oxford University Press, 3rd edn, 2008), at 150: 'In the last resort, however, it cannot be excluded that entry without the consent of the sending State may be justified in international law by the need to protect human life.'

[128] Similarly, see Zimmermann, above note 52, at 759–60.

(a) When the offence is committed in any territory under its jurisdiction or on board a ship or aircraft registered in that State;
(b) When the alleged offender is one of its nationals;
(c) When the disappeared person is one of its nationals and the State Party considers it appropriate.

The first use of the word 'jurisdiction' in the *chapeau* of this article refers to the concept of the *jurisdiction to prescribe*. The article as a whole requires states to make enforced disappearances crimes under their domestic law, and that the scope of application of this domestic law should be based on the territorial and active personality and if the state wishes also the passive personality heads of prescriptive jurisdiction. It is this 'jurisdiction' of states that general international law specifically *recognizes* with regard to registered ships and aircraft, by assimilating them to state territory for the purpose of the state's authority to legislate. In other words, just as international law accepts that it is a fundamental prerogative of a state's sovereignty to prescribe legal rules for persons and their conduct within its territory, so does it accept the state's authority to do the same with regard to registered ships and aircraft or those flying its flag.[129]

But this 'jurisdiction' is not the jurisdiction referred to in Article 1 ECHR or in the similar clauses of other human rights jurisdiction. *That* 'jurisdiction' is the one that makes its appearance in subparagraph (a) quoted above: 'territory under its jurisdiction'. When Article 9(1)(a) of the Disappearances Convention stipulates that a state party shall have the duty to 'establish its jurisdiction... when the offence is committed in any territory under its jurisdiction', it is not setting out a tautology. Rather, the meaning of the two 'jurisdictions' is not the same. The former refers to the authority to prescribe legal rules, the latter to effective overall control by a state over a territory. Therefore, in plain English, the article as a whole requires states to criminalize the Convention offences in all territories under their control.

The question thus is not whether the former concept of jurisdiction can apply to registered ships and aircraft—it clearly does. Rather, the issue is whether the spatial concept of jurisdiction as control over territory and places can also extend to cover ships and aircraft, as distinct man-made objects. And, if it can, what should be its parameters? Should it really matter whether the ship or aircraft is *registered* in a specific state, or whether it flies its flag, or should jurisdiction over such objects rather depend on state *control* over them?

To answer this question, we now need to take a closer look at some of the case law. With regard to ships, perhaps the European Court's earliest case is *Rigopoulos v. Spain*.[130] The applicants were the crew of a Panamanian-flagged ship that was on the high seas when it was boarded by a Spanish vessel on suspicion of drug trafficking, pursuant to a warrant by a Spanish court and with the consent of Panamanian authorities. They complained that they were unlawfully detained.

[129] See above, Chapter II, Section 2.
[130] *Rigopoulos v. Spain* (dec.), App. No. 37388/97, 12 January 1999.

Though the Court declared the application inadmissible as manifestly ill-founded, it seems to have accepted that the ECHR could apply to a Panamanian ship on the high seas, though it did not explicitly consider Article 1 ECHR.

In *Xhavara*,[131] the applicants were Albanian nationals who were attempting to get to Italy on an Albanian-registered ship, when that ship was rammed by an Italian warship on the high seas. The applicants alleged that the Italian warship's action was intentional, and that it therefore engaged Articles 2 and 3 of the Convention. The Court declared the application inadmissible for failure to exhaust domestic remedies, but it made no mention of the fact that the applicants were located on an Albanian ship on the high seas at the time the violation occurred. In other words, it again did not seem to have considered that there was an Article 1 ECHR issue at all.

Article 1 ECHR did make an appearance in *Medvedyev and Others v. France*, the facts of which remarkably resembled *Rigopoulos*: the applicants were crew members on a merchant ship infelicitously named the *Winner*, flying the Cambodian flag. As part of the international effort to combat drug trafficking, the French authorities were informed that the ship might be carrying large quantities of illegal drugs. France requested, and obtained, permission from Cambodian authorities to search and seize the ship, and detain those aboard, which the French navy swiftly did. The ship was then brought to the French port of Brest, a voyage of some thirteen days, where the applicants were brought before a magistrate. In Strasbourg, they complained *inter alia* that their detention on board the ship was unlawful, because it had no legal basis, and because they were not promptly brought before a judge.

The case was first decided by a Chamber.[132] France accepted that it had Article 1 ECHR jurisdiction, while the Court itself said the following:

> . . . the Court notes, on the one hand, that it is not disputed that between 13 June 2002 (the date on which the *Winner* was intercepted) and 26 June 2002 (when it arrived in Brest harbour) the *Winner* and its crew were under the control of French military forces, so that even though they were outside French territory, they were within the jurisdiction of France for the purposes of Article 1 of the Convention. It further notes that the parties agree that throughout that period on board the *Winner*—and the subsequent police custody—the applicants were deprived of their liberty within the meaning of Article 5 of the Convention, 'for the purpose of bringing them before the competent legal authority' (Article 5 § 1 (c)).
>
> That is also the opinion of the Court, which also refers to the *Rigopoulos* decision cited above.[133]

Note that the Chamber did not attach any importance to the supposedly 'special' nature of ships. Nor could it have done so, as the *Winner* was a *Cambodian, and not a French-flagged ship*. It therefore did not base France's jurisdiction over the applicants on some alleged general principle of public international law, but on

[131] *Xhavara c. l'Italie et l'Albanie* (dec.), App. No. 39473/98, 11 January 2001.
[132] *Medvedyev and Others v. France*, App. No. 3394/03, Judgment, 10 July 2008.
[133] Ibid., paras 50–1.

the fact that the ship and the crew 'were under the control of French military forces'.[134]

The case was then referred to the Grand Chamber, which delivered its judgment in 2010.[135] It addressed the Article 1 issue in much more detail than the Chamber before it. First, unlike the Chamber, which did not even mention *Bankovic*, the Grand Chamber took great pains in restating its *Bankovic* position on the exceptional nature of extraterritorial application:

In keeping with the essentially territorial notion of jurisdiction, the Court has accepted only in exceptional cases that acts of the Contracting States performed, or producing effects, outside their territories can constitute an exercise of jurisdiction by them for the purposes of Article 1 of the Convention (see *Banković,* cited above, § 67, and *Ilaşcu and Others v. Moldova and Russia* [GC], no. 48787/99, § 314, ECHR 2004-VII). In its first Loizidou judgment (*preliminary objections*), for example, the Court found that bearing in mind the object and purpose of the Convention, the responsibility of a Contracting Party might also arise when as a consequence of military action—whether lawful or unlawful—it exercised effective control of an area outside its national territory (see *Loizidou v. Turkey* (*preliminary objections*) [GC], 23 March 1995, § 62, Series A no. 310). This excluded situations, however, where—as in the *Banković* case—what was at issue was an instantaneous extraterritorial act, as the provisions of Article 1 did not admit of a 'cause-and-effect' notion of 'jurisdiction' (*Banković*, § 75).

Additionally, the Court notes that other recognised instances of the extraterritorial exercise of jurisdiction by a State include cases involving the activities of its diplomatic or consular agents abroad and on board aircraft and ships registered in, or flying the flag of, that State. In these specific situations, customary international law and treaty provisions have clearly recognised and defined the extraterritorial exercise of jurisdiction by the relevant State (see *Banković*, cited above, § 73).[136]

Note, first, the new gloss that the Grand Chamber puts on *Bankovic*—that it only concerned an 'instantaneous extraterritorial act'. With an eye on the future, this could have been meant to distinguish a killing from detention, though not on any principled ground that I can see. The Grand Chamber then reiterates the supposedly special nature of the jurisdiction exercised by states over ships registered in them or flying their flag. But, of course, even if true, this is entirely irrelevant on the facts of *Medvedyev*, as the ship on which the applicants were detained was not French, but Cambodian. The Grand Chamber then moved to applying (or not) these principles to the actual facts of *Medvedyev*:

In the instant case, the Court notes that a French warship, the frigate *Lieutenant de vaisseau Le Hénaff*, was specially instructed by the French naval authorities to intercept the *Winner*, and that the frigate sailed out of Brest harbour on that mission carrying on board the French Navy commando unit *Jaubert*, a special forces team specialised in boarding vessels at sea. When the *Winner* was spotted off Cape Verde on 13 June 2002, the frigate issued several

[134] See also T. Thienel, 'Oral Argument in *Medvedyev v. France*', *Invisible College Blog*, 13 May 2008, available at <http://invisiblecollege.weblog.leidenuniv.nl/2008/05/13/oral-argument-in-medvedyev-v-france>.

[135] *Medvedyev and Others v. France* [GC], App. No. 3394/03, Judgment, 29 March 2010.

[136] Ibid., paras 64–5.

warnings and fired warning shots, before firing directly at the merchant ship, under orders from France's Maritime Prefect for the Atlantic. When they boarded the *Winner*, the French commando team were obliged to use their weapons to defend themselves, and subsequently kept the crew members under their exclusive guard and confined them to their cabins during the journey to France, where they arrived on 26 June 2002. The rerouting of the *Winner* to France, by decision of the French authorities, was made possible by sending a tug out of Brest harbour to tow the ship back to the French port, escorted by another warship, the frigate *Commandant Bouan*, all under orders from the Maritime Prefect and at the request of the Brest Public Prosecutor.

That being so, the Court considers that, as this was a case of France having exercised full and exclusive control over the *Winner* and its crew, at least *de facto*, from the time of its interception, in a continuous and uninterrupted manner until they were tried in France, the applicants were effectively within France's jurisdiction for the purposes of Article 1 of the Convention (contrast *Banković*, cited above).[137]

So, as with the Chamber, what really matters is that France had 'full and exclusive control over the *Winner* and its crew, at least *de facto*', not some considerations of general international law. Note again that it is unclear whether the Court applies a spatial model (control over the ship) or a personal one (control over the crew). And note how, without the slightest hint of self-irony, the Court instructs us to contrast this approach with *Bankovic*—as if the two were somehow related or compatible![138] Nor does it seem to have been relevant that Cambodia had consented to the seizure of the ship—should, after all, the outcome have been any different otherwise?

With regard to aircraft, the first Strasbourg case is the Commission inadmissibility decision in *Freda v. Italy*,[139] which concerned an applicant arrested in Costa Rica and surrendered by Costa Rican authorities to Italian agents at an airport, who put him on an Italian plane back to Italy. The Commission considered that the applicant was within Italy's jurisdiction:

Or, il est établi que le requérant a été pris en charge par des agents de la force publique italienne et privé de liberté dans un avion militaire italien. Dès lors, le requérant, à partir du moment de la remise, relevait effectivement de l'autorité de l'Italie, et donc de la 'juridiction' de ce pays, même si cette autorité s'est exercée en l'occurence à l'étranger.

The Commission relied, *inter alia*, on its prior decision in *Cyprus v. Turkey* (1975), where it adopted the personal conception of Article 1 jurisdiction as control over individuals by state agents, and that is exactly what it did here: '. . . le requérant, à partir du moment de la remise, relevait effectivement de l'autorité de l'Italie, et donc de la "juridiction" de ce pays'. The Commission repeated this same reasoning in *Illich Sanchez Ramirez v. France*,[140] one of the two Strasbourg cases dealing with

[137] *Medvedyev and Others v. France* [GC], App. No. 3394/03, Judgment, 29 March 2010, paras 66–7.

[138] Similarly, see D. Guilfoyle, 'ECHR Rights at Sea: Medvedyev and others v. France', *EJIL: Talk!*, 19 April 2010, available at <http://www.ejiltalk.org/echr-rights-at-sea-medvedyev-and-others-v-france/>.

[139] *Freda v. Italy* (dec.), App. No. 8916/80, 7 October 1980.

[140] *Illich Sanchez Ramirez v. France* (dec.), App. No. 28780/95, 24 June 1996. Note that the later decisions concerning Carlos are correctly labelled as *Ramirez Sanchez*, rather than *Sanchez Ramirez*, but I will use the Court's own nomenclature.

the notorious terrorist Carlos the Jackal.[141] The (very colourful) facts of the case were these: in the early 1990s, Carlos sought refuge in Sudan as he was being actively hunted by the intelligence agencies of several Western states. In 1994, the Sudanese authorities made a deal with French and US intelligence services to hand Carlos over. Unfortunately for Carlos, he was at the time scheduled to have minor surgery in a Sudanese hospital on a varicose vein on his scrotum. After the surgery, he was moved to a villa and assigned bodyguards. During the night of 14–15 August 1994, his own bodyguards turned against him. He was promptly tied up and tranquilized, surrendered to French agents waiting at the airport, and was immediately flown on a French military plane back to Paris, where he was later tried and convicted for various offences. He complained, *inter alia*, that his arrest was unlawful under Article 5 ECHR.

At what point, therefore, did Carlos come within French jurisdiction? The Commission held as follows:

> According to the applicant, he was taken into the custody of French police officers and deprived of his liberty in a French military aeroplane. If this was indeed the case, from the time of being handed over to those officers, the applicant was effectively under the authority, and therefore the jurisdiction, of France, even if this authority was, in the circumstances, being exercised abroad.

Again, what mattered was that Carlos was handed over to French officers, not that he was put on a French-flagged plane.

Very similar is *Öcalan v. Turkey*, the case filed by Abdullah Öcalan, the leader of the Workers' Party of Kurdistan (PKK), considered by Turkey to be the arch-terrorist. The facts of the case could be briefly summarized as follows: after having fled Turkey to several countries, in 1999 Öcalan found himself in Nairobi, Kenya. On 15 February 1999 he was arrested, and was flown to Turkey aboard a Turkish military plane. It is unclear whether he was arrested with the permission and help of (some) Kenyan authorities, or whether the Turkish forces simply engaged in an extraterritorial abduction. Before the European Court, he complained *inter alia* about the conditions and the lawfulness of his detention, and the lack of fairness of his subsequent trial, after which he was sentenced to the death penalty. What concerns us here, of course, is the detention—at what point was Öcalan to be considered to have been within Turkey's jurisdiction? Was it already in Kenya upon his arrest, or was it aboard the Turkish aircraft, or was it when that aircraft entered Turkish airspace?

The Court's admissibility decision made no mention of the issue.[142] During the merits phase before the Chamber, however, Turkey raised the Court's freshly minted *Bankovic* decision to disclaim that it had jurisdiction at the time of Öcalan's arrest in Kenya. The Chamber rejected this argument as follows:

> In the instant case, the applicant was arrested by members of the Turkish security forces inside an aircraft in the international zone of Nairobi Airport. Directly after he had been

[141] The other being *Ramirez Sanchez v. France* [GC], App. No. 59450/00, Judgment, 4 July 2006.
[142] *Öcalan v. Turkey* (dec.), App. No. 46221/99, 14 December 2000.

handed over by the Kenyan officials to the Turkish officials the applicant was under effective Turkish authority and was therefore brought within the 'jurisdiction' of that State for the purposes of Article 1 of the Convention, even though in this instance Turkey exercised its authority outside its territory. The Court considers that the circumstances of the present case are distinguishable from those in the aforementioned *Banković and Others* case, notably in that the applicant was physically forced to return to Turkey by Turkish officials and was subject to their authority and control following his arrest and return to Turkey.[143]

As in the Carlos case, the Chamber clearly adopted the personal conception of jurisdiction as the exercise of 'effective Turkish authority' or the subjection of the applicant to its 'authority and control'. Jurisdiction was most emphatically *not* based on the flag of the aircraft that transported Öcalan.

As the Chamber generally decided the case in Öcalan's favour, it was then referred to the Grand Chamber at Turkey's request. This time, however, Turkey did not dispute that it had jurisdiction over Öcalan at the time of his arrest, and the Grand Chamber merely reproduced some of the Chamber's reasoning:

It is common ground that, directly after being handed over to the Turkish officials by the Kenyan officials, the applicant was effectively under Turkish authority and therefore within the 'jurisdiction' of that State for the purposes of Article 1 of the Convention, even though in this instance Turkey exercised its authority outside its territory. It is true that the applicant was physically forced to return to Turkey by Turkish officials and was under their authority and control following his arrest and return to Turkey (see, in this respect, the aforementioned decisions in *Sánchez Ramirez* and *Freda*, and, by converse implication, *Banković and Others v. Belgium and Others* (dec.) [GC], no. 52207/99, ECHR 2001-XII).[144]

Note how both the Chamber and the Grand Chamber in *Öcalan* try to make it seem that there is no contradiction between their holding that jurisdiction attached because of authority and control over an individual and the *Bankovic* rejection of such a conception of jurisdiction. It is unclear what exactly in the Court's view makes *Bankovic* distinguishable 'by converse implication'. Is it because 'authority and control' over an individual means *physical custody* of that individual, rather than just the ability to *kill* that individual, and if so, why? Or is it rather that Turkey exercised some legal power, i.e. that it was trying to *arrest* Öcalan on a criminal charge that rendered him within its jurisdiction, and if so, why?

In my view, at play was simply the Court's policy consideration that extraterritorial abductions, even if somewhat extraordinary, could be effectively dealt with without much controversy or political fallout, while the use of force in an armed conflict is a wholly different game. Be that as it may, we can see that in many of the cases that we have now examined neither the state concerned, nor the Court or the Commission, which should after all have minded *ex officio* whether the ECHR applied and accordingly whether they had subject-matter jurisdiction, thought that there was an Article 1 ECHR issue.[145] Even in cases where Article 1 ECHR was

[143] *Öcalan v. Turkey*, App. No. 46221/99, Judgment, 12 March 2003, para. 93.

[144] *Öcalan v. Turkey* [GC], App. No. 46221/99, Judgment, 12 May 2005, para. 91.

[145] The best recent example is *Women on Waves and Others v. Portugal*, App. No. 31276/05, Judgment, 13 January 2009. The applicants were reproductive rights NGOs which hired a ship to take

considered, they simply did not turn on any supposedly 'special' nature of ship and aircraft in general international law. Rather, it was the authority and control exercised by state agents over an individual that was the basis of the state's jurisdiction. In other words, it was not the spatial but the personal model of jurisdiction that was applied.

In sum, the European cases cited above most certainly do *not* support the proposition that because international law recognizes that the flag state or the state in which ships and aircraft are registered may prescribe legal rules regulating conduct occurring on them, this means that ships and aircraft somehow present a special justification for extending to them the spatial concept of jurisdiction in human rights treaties. Indeed, except perhaps for *Medvedyev*, the cases do not even stand in favour of a purely factual notion of jurisdiction as *control* over ships and aircraft, regardless of the flag or registration. Rather, it is the control over individuals than forms the basis of state jurisdiction. No amount of post-*Bankovic* reinterpretation or 'clarification' of these cases can change this—the cases say what they say.

As for the case law of other institutions, recall that the Committee Against Torture in its General Comment No. 2 stated that the 'reference to "any territory" in article 2, like that in articles 5, 11, 12, 13 and 16, refers to prohibited acts committed not only on board a ship or aircraft registered by a State party, but also during military occupation or peacekeeping operations and in such places as embassies, military bases, detention facilities, or other areas over which a State exercises factual or effective control'.[146] The Committee here endorsed a general theory of jurisdiction as control over places or objects, and explicitly referred to ships and aircraft. It had the opportunity to apply this theory in the *Marine I* case,[147] which concerned a ship, the *Marine I*, which capsized in international waters with 369 immigrants from various Asian and African countries on board. In response to a distress call, a Spanish ship reached the *Marine I* and towed it to Mauritanian waters, where it was anchored for eight days, while negotiations between Spain and several other states were underway with the purpose of repatriating the migrants. They were then transported to a former fish plant in Mauritania, and most of them were subsequently repatriated voluntarily. The complainants refused repatriation, and complained *inter alia* of the inhumane conditions while on board the ship and in the repatriation centre.

Spain argued that it did not have jurisdiction over the complainants, and the Committee held as follows:

them to Portugal, for the purpose of staging activities that would promote the decriminalization of abortion in Portugal. The ship was intercepted by a Portuguese warship, which prohibited it from entering Portuguese territorial waters. The Court unanimously found that this amounted to a violation of the Article 10 guarantee of the freedom of expression. Thus, even though the event took place on the high seas and the applicants never entered Portugal, the Court did not think that an Article 1 ECHR issue arose.

[146] CAT General Comment No. 2, above note 105, para. 16.

[147] *P.K. et al. v. Spain*, Communication No. 323/2007, UN Doc. CAT/C/41/D/323/2007, 21 November 2008 (hereinafter the *Marine I* case).

The Committee takes note of the State party's argument that the complainant lacks competence to represent the alleged victims because the incidents forming the substance of the complaint occurred outside Spanish territory. Nevertheless, the Committee recalls its general comment No. 2, in which it states that the jurisdiction of a State party refers to any territory in which it exercises, directly or indirectly, in whole or in part, de jure or de facto effective control, in accordance with international law. In particular, it considers that such jurisdiction must also include situations where a State party exercises, directly or indirectly, de facto or de jure control over persons in detention. This interpretation of the concept of jurisdiction is applicable in respect not only of article 2, but of all provisions of the Convention, including article 22. In the present case, the Committee observes that the State party maintained control over the persons on board the Marine I from the time the vessel was rescued and throughout the identification and repatriation process that took place at Nouadhibou. In particular, the State party exercised, by virtue of a diplomatic agreement concluded with Mauritania, constant de facto control over the alleged victims during their detention in Nouadhibou. Consequently, the Committee considers that the alleged victims are subject to Spanish jurisdiction insofar as the complaint that forms the subject of the present communication is concerned.[148]

Clearly, in *Marine I* the Committee generally conceived of jurisdiction as 'de facto or de jure control over persons' or 'constant de facto control over the alleged victims'.[149] It did not rely on jurisdiction as control over the ship itself. And this raises a fundamental question. If we accept control over individuals as a basis for jurisdiction, do we even need any concept of jurisdiction as control over vessels? What good would it actually do?

Perhaps the two most practically relevant types of cases which involve ships today are the concerted actions by a number of states against Somali pirates in the Gulf of Aden, and state attempts to curb illegal immigration. The former raise several potential human rights issues, from the lawfulness of the detention of any suspected pirates, to the transfer of the captured pirates to a nearby state for trial, which could implicate the *non-refoulement* obligations of the capturing state, for instance with regard to the risk of inhumane treatment in the receiving state or the lack of adequate fair trial guarantees.[150]

When it comes to illegal immigration, we have already seen one example in the *Marine I* case examined above. Attempts by migrants from, say, North Africa to reach Europe, or those from Cuba to reach the United States, frequently involve the use of ships and the interdiction of these ships by the state that the migrants are attempting to reach. It has been widely reported, for example, that Italy chooses to interdict migrant ships before they reach Italian territorial waters and return them to their harbour of origin, rather than assess the claims of the migrants to asylum on an individual basis.[151] This again, of course, raises questions about the lawfulness of

[148] Ibid., para. 8.2.

[149] See also K. Wouters and M. Den Heijer, 'The *Marine I* Case: A Comment', (2009) 22 *International Journal of Refugee Law* 1, 8–11.

[150] For an excellent overview of the various issues, as well as for factual background, see D. Guilfoyle, 'Counter-Piracy Law Enforcement and Human Rights', (2010) 59 *ICLQ* 141.

[151] See, e.g., Wouters and Den Heijer, above note 149, at 9.

detention, as well as *non-refoulement*.[152] As of the time of writing, one such case, *Hirsi and others v. Italy*,[153] is pending before the European Court.[154]

Turning back to the question whether the spatial conception of jurisdiction as control over ships and aircraft would have any use if the personal conception of jurisdiction as control over individuals is accepted, we could only say that such usefulness would be very limited, if it existed at all. In most cases, it would be impossible to distinguish as a matter of fact between the control over a ship and the control over individuals on it. To the extent that a personal notion of jurisdiction would govern, the spatial notion would be redundant.

If, however, the personal notion is rejected, either because it is limited artificially, or because it is at odds with the text of the relevant treaty (as is arguably the case with the CAT), or indeed, as is my view, because it is impossible to maintain a personal notion of jurisdiction which would not collapse into the absence of any jurisdictional threshold at all, then there could be some use to a concept of jurisdiction as control over objects, particularly with regard to delineating the applicability of positive obligations. What if, say, a murder occurs on a British Airways flight from London to France, either while the UK-registered aircraft was in international airspace, or indeed just after the plane had touched the tarmac at Charles de Gaulle. Which of the two states has the Article 2 ECHR obligation to investigate the crime? Similarly, what of a scenario where one shipmate kills another on a private ship registered in the UK, and the murder takes place on the high seas? Would the UK have an Article 2 obligation to investigate?

In both cases there is no direct control of a state over the ship or aircraft. All we have is the fact that the vessel is *registered* in a state, which certainly grants the state concerned the *right* in customary law to prescribe offences aboard these vessels and enforce these laws, as well as the *obligation* to do so under certain treaties,[155] but it is not immediately apparent that this would constitute jurisdiction in the spatial sense within the meaning of human rights treaties. In the cases we have examined, that notion of jurisdiction has been predicated on factual control, rather than purely formal criteria such as registration. If we take *Medvedyev* as an example, it

[152] On extraterritorial *refoulement* in the context of refugee law, see J. Hathaway, *The Rights of Refugees under International Law* (Cambridge University Press, 2005), 335–42, as well as UNHCR, *Advisory Opinion on the Extraterritorial Application of Non-Refoulement Obligations under the 1951 Convention relating to the Status of Refugees and its 1967 Protocol*, 26 January 2007, available at <http://www.unhcr.org/refworld/docid/45f17a1a4.html>.

[153] App. No. 27765/09.

[154] Italy has also been criticized by the CPT for its 'push-back' of migrants as constituting a violation of its *non-refoulement* obligations, even if it took place on ships outside Italy's territorial sea—see Report to the Italian Government, CPT/Inf (2010) 14, 28 April 2010, esp. para. 29. In that regard, the CPT seems to have adopted a personal conception of state jurisdiction: 'Italy is bound by the principle of *non-refoulement* wherever it exercises its jurisdiction, which includes via its personnel and vessels engaged in border protection or rescue at sea, even when operating outside its territory. Moreover, all persons coming within Italy's jurisdiction should be afforded an appropriate opportunity and facilities to seek international protection.' Ibid., para. 49.

[155] See, e.g., Convention for the Suppression of Unlawful Seizure of Aircraft [Hijacking Convention], 860 UNTS 105, entered into force 14 October 1971, Art. 4(1).

is the fact that France controlled the individuals aboard the ship that created its jurisdiction, even if the individuals were located on a Cambodian-flagged ship.

If the killing takes place within the territorial sea or the airspace of a country (over which it maintains effective control, or is presumed to do so), whether it is the flag state or a third state, then we could indeed say that the killing took place within that state's jurisdiction. If, on the other hand, the killing takes place on the high seas or in international airspace, then it is to me far from obvious that the mere fact of formal registration, absent any tangible degree of control, should matter at all. Imagine, for example, if the aircraft in question was registered in one state, but was leased to an airline operating in another state half the world away. Could the state of registration still be said to have an obligation to secure the human rights of all persons aboard that airplane?

If, on the other hand, it is only purely factual control over ships and aircraft that should matter—as is the case with territory and places—then we would need to define some of the contours of that control. First, it scarcely needs pointing out that control over such an object can be effected with relative ease by any state. Secondly, the simplest case of such control would be where the state's soldiers actually board a vessel, or control it directly in a similar fashion. Thirdly, even so, control over these objects need not necessarily be so direct. If, say, a migrant ship was interdicted by a state warship, which has every ability to blow it out of the water, or if an aircraft was intercepted by military jets, this would amount to effective control just as if the ship or aircraft was actually boarded by troops.[156]

In short, precisely because control over such objects is relatively easy to achieve, it is the *potential* of such control, rather than the immediate exercise of control in any given situation, that could be said to give rise to state jurisdiction over a vessel.

E. Outlook

Ultimately, what are we to make of the spatial model of jurisdiction? Its application is certainly the clearest when it comes to effective state control over relatively large portions or areas of territory, as in *Loizidou.* Though of course the facts on the ground may vary, these situations resemble most the level of control, and accordingly the capacity to either violate or protect human rights, that the state normally has over its own territory. This model also comports the best with the territorial sense of the word 'jurisdiction' as it is used in state practice. As a matter of policy, its adoption allows us to extend human rights protection to those extraterritorial situations where it is most likely that human rights will be violated and yet could be effectively protected, primarily belligerent occupation.

But the main drawback of the spatial model is precisely that, as a matter of policy, its rigid application will allow numerous extraterritorial cases to slip through and be beyond the scope of application of human rights treaties. There is a profound feeling of injustice and arbitrariness, as measured against the benchmark

[156] See also Guilfoyle, above note 150, at 155. Similarly, see the *Women on Waves* case cited in note 145 above.

of universality, in allowing states to avoid human rights constraints in extraterritorial situations over which they still have complete control. Why should, for example, the US be free to torture persons detained in CIA black sites located in territories under the jurisdiction of other states, when the US is still in control of the acts of its own agents and in control of the individuals subjected to its authority?

This is why the pure or default version of the spatial model, that of effective overall control of (large) areas of territory, needs to be mitigated, and there are several ways of doing so. First, by applying the spatial model to ever smaller areas, with an ever more tenuous degree of control, so that the 'control of an area' test ultimately covers control over places and objects. Secondly, by in some cases discarding the spatial model and opting for the personal model of jurisdiction as control over individuals. There is also a third option—that of discarding the jurisdiction threshold altogether with respect to some obligations—which I personally favour, and which I will turn to shortly.

From the Strasbourg perspective, the virtue of the first approach is that it allows the Court to avoid using the second one. This, in turn, allows it to maintain some rigour to the Article 1 jurisdiction threshold, which would be in danger of collapsing if jurisdiction meant factual control over individuals pure and simple. Even so, however, because the spatial model can only be stretched so far, the Court is, due to the normative pull of universality, in many cases unable to resist conceptualizing jurisdiction as control over individuals, thereby creating a tension with *Bankovic*. As we have seen, the embassy, ship, and aircraft cases that the Grand Chamber in *Bankovic* rationalized on the basis of some supposedly special characteristics of these objects in international law, actually turned on considerations of factual control by state agents over individuals. Even in *Al-Saadoon*, which can be read as embracing a general theory of jurisdiction as control over places, the Court vacillated between control over places and control over individuals as grounds of jurisdiction.[157]

As we have also seen above, it is impossible to clearly define the geographical space that the spatial model of jurisdiction refers to. The more we descend from large areas of territory to places and objects, the more artificial it seems to apply the spatial model at all: it simply collapses to control over individuals, rather than control over territory, however minuscule in size. And because this is so, even if stretched to the utmost extent, the spatial model of jurisdiction will allow numerous extraterritorial situations to slip through and be outside the reach of human rights protection. This, in turn, would in many cases simply be morally intolerable, unjustifiable from the standpoint of universality, and inexcusable by any apparent consideration of effectiveness.

Consider, for example, the assassination of Alexander Litvinenko in London in 2006. Assume that it was, in fact, a Russian agent who killed Mr Litvinenko—say if a KGB/FSB operative openly admitted to the crime. It would nonetheless be impossible to say under any conception of the spatial model of jurisdiction that

[157] See also Guilfoyle, above note 150, at 154.

this killing violated Russia's ECHR obligations, which it would have done if it had taken place on Russian soil. It would surely be absurd to argue that Russia had control over the *Japanese restaurant* in London where Litvinenko was fed the radioactive sushi, and that this somehow established its jurisdiction. It is only if the spatial model was discarded that we could bring Litvinenko within the ECHR's reach. This could be done either by adopting a personal model of jurisdiction as control over Litvinenko himself, or by saying that the jurisdiction threshold should not apply at all to negative obligations, as would be my preference. Or, one could try distinguishing Litvinenko's case from *Bankovic* on the less than principled ground that the killing took place within the ECHR's *espace juridique*.[158]

The same would go, say, for the use of drones by the US in Pakistan or Afghanistan. Extraterritorial torture in a place outside the state's control—for example, if a US agent was to interrogate a detainee in a Pakistani prison—would also slip through the cracks of the spatial model. So would the various complicity scenarios, as with the US or the UK providing intelligence and even the questions to the Pakistani interrogators.[159] Various instances of extraterritorial law enforcement, ranging from searches and seizures to abductions, would likewise be out, at least until the individual concerned is brought into a place or an object under the state's control,[160] as would transboundary environmental harm, as in the example of Colombia's aerial herbicide spraying having effects in Ecuador which we discussed above.[161]

This would be the price to pay were the spatial model, however tweaked and adjusted, to be applied exclusively. None of these individuals would be entitled to any human rights protection. And, of course, what makes this result morally intolerable is that the states in question in all of these cases would still have total control over the acts of their own agents. There would be nothing preventing them from complying with their human rights obligations, if such obligations were to exist. As I have argued above, considerations of effectiveness would need to be taken into account when *applying* these obligations, so that the possibly extraordinary circumstances of extraterritorial application do not produce unrealistic restrictions on states.[162] But should effectiveness truly mean that a state is free to kill and torture outside its borders, and if so why?

[158] See above, Chapter III, Section 6.

[159] See, in that regard, *R. (Hassan) v. Secretary of State for Defence* [2009] EWHC 309 (Admin), where the claimants were the family of an individual detained in the initial stages of the invasion by UK forces in Iraq, who transferred him to the custody of US armed forces pursuant to a memorandum of understanding, under which the UK remained the detaining power under the Third Geneva Convention for the individual in question. Applying the spatial model of jurisdiction, the High Court found that, as the detainee was held on a US military base, he was not within the UK's jurisdiction. As the detaining power, however, and on the basis of its memorandum of understanding with the US, the UK did retain a legal authority over the detainee, which could have sufficed to bring him within its jurisdiction under the personal model. For a critique of *Hassan* in that vein, see H. King, 'Unravelling the Extraterritorial Riddle: An Analysis of *R. (Hassan) v. Secretary of State for Defence*', (2009) 7 *JICJ* 633.

[160] See Section 1.E.

[161] See Section 1.F.

[162] See above, Chapter III, Section 10.

No good answer to this question can be given from the standpoint of universality. This is why the spatial model has in practice frequently been complemented by a personal one, to which I will now turn.

3. The Personal Model: Jurisdiction as Authority and Control over Individuals

A. Introduction

Conceiving of state jurisdiction within the meaning of human rights treaties in personal, rather than spatial terms, would appear to solve most of the policy problems with the spatial model that we have seen above. This is indeed why the personal model of jurisdiction as state authority and control over individuals has been advocated in the literature,[163] and why it has been adopted by various courts or quasi-judicial bodies in numerous cases, to which I will turn below (and many of which we have already seen). Saying, however, that a state has jurisdiction over an individual whenever it controls him poses its own set of problems.

First, it is entirely doubtful that the personal model is reconcilable with the text of at least some of the relevant treaties. The CAT, for instance, specifically refers to jurisdiction over *territory*, rather than individuals, and no amount of interpretation can make these words go away.

More fundamentally, even if the personal model presented a legitimate option in principle, what would actually count as state control over an individual? Just as we had to examine what counts as control over territory, so we would need to answer the same question here. However, it is hard to say whether it is possible to give any meaningful answer to that question. First, we could adopt the broadest possible definition of control, and consequently of jurisdiction over a person—a state would have such control whenever it had the ability to substantively violate an individual's rights. Yet if we did so, the personal model of jurisdiction would essentially collapse. It would serve no useful purpose as a threshold for the application of a human rights treaty, since the treaty would apply whenever the state could actually infringe it.[164] Secondly, to prevent this collapse we could try limiting the notion of personal control, for instance by saying that only physical custody over an individual could satisfy the threshold. As we will see, however, such a limitation is not possible by reference to any non-arbitrary criterion.

[163] Most ably and prominently by R. Lawson, 'Life after Bankovic: On the Extraterritorial Application of the European Convention on Human Rights', in Coomans and Kamminga 83.

[164] Note that this would not mean, as argued by O'Boyle, that the concept of jurisdiction would be conflated with that of attribution, in other words that the inquiries under Art. 2(a) and Art. 2(b) ILC ASR would collapse into each other. Rather, the obligation whose violation is being attributed would not be territorially limited, as is the case for example with the Geneva Conventions. See M. O'Boyle, 'The European Convention on Human Rights and Extraterritorial Jurisdiction: A Comment on "Life After Bankovic"', in Coomans and Kamminga 125, and Chapter II, Section 3 above.

In *Bankovic*, the European Court realized that the personal model cannot be limited, and this is precisely why it rejected it, at least in its more expansive form:

[The applicants] claim that the positive obligation under Article 1 extends to securing the Convention rights in a manner proportionate to the level of control exercised in any given extra-territorial situation. The Governments contend that this amounts to a 'cause-and-effect' notion of jurisdiction not contemplated by or appropriate to Article 1 of the Convention. The Court considers that the applicants' submission is tantamount to arguing that anyone adversely affected by an act imputable to a Contracting State, wherever in the world that act may have been committed or its consequences felt, is thereby brought within the jurisdiction of that State for the purpose of Article 1 of the Convention.

The Court is inclined to agree with the Governments' submission that the text of Article 1 does not accommodate such an approach to 'jurisdiction'. Admittedly, the applicants accept that jurisdiction, and any consequent State Convention responsibility, would be limited in the circumstances to the commission and consequences of that particular act. However, the Court is of the view that the wording of Article 1 does not provide any support for the applicants' suggestion that the positive obligation in Article 1 to secure 'the rights and freedoms defined in Section I of this Convention' can be divided and tailored in accordance with the particular circumstances of the extra-territorial act in question and, it considers its view in this respect supported by the text of Article 19 of the Convention. Indeed the applicants' approach does not explain the application of the words 'within their jurisdiction' in Article 1 and it even goes so far as to render those words superfluous and devoid of any purpose. Had the drafters of the Convention wished to ensure jurisdiction as extensive as that advocated by the applicants, they could have adopted a text the same as or similar to the contemporaneous Articles 1 of the four Geneva Conventions of 1949 (see § 25 above).

Furthermore, the applicants' notion of jurisdiction equates the determination of whether an individual falls within the jurisdiction of a Contracting State with the question of whether that person can be considered to be a victim of a violation of rights guaranteed by the Convention. These are separate and distinct admissibility conditions, each of which has to be satisfied in the afore-mentioned order, before an individual can invoke the Convention provisions against a Contracting State.[165]

Note that the Court here does not explicitly address the applicants' reliance on prior case law which indicated that authority and control over individuals would lead to jurisdiction—presumably because it did not wish to overrule prior jurisprudence, thus maintaining the pretence that *Bankovic* presented no departure from it. The Court does, however, quite clearly reject the idea that anyone whose rights would substantively be violated by a state's extraterritorial act would be within its jurisdiction.

That said, as we will see, the personal conception of jurisdiction has not remained absent from Strasbourg case law, either before or since *Bankovic*, even if the Court's position on it is entirely, and one could say purposefully, unclear. Other human rights institutions, however, have been much less cautious. And in all fairness to the European Court, it must be said that this is not because these other bodies somehow have a better grasp of the whole *problématique* of extraterritorial application. Rather, it is because the stakes are higher in Strasbourg, because its

[165] *Bankovic*, para. 75.

judgments are not only binding formally (which of course is not the case with UN treaty bodies), but, more importantly, have attained a high degree of authority and compliance in practice, that it is much harder for Strasbourg to throw caution to the wind. Let us now take a closer look at the case law.

B. Case law

1. UN treaty bodies

I will start off by examining the jurisprudence of UN treaty bodies. Although it is generally not as rich as that of Strasbourg, nor chronologically the first, it is precisely these bodies that have been the staunchest proponents of a personal conception of jurisdiction.

The Human Rights Committee first addressed the issue of the extraterritorial application of the ICCPR in two nearly identical cases, *Lopez Burgos v. Uruguay*[166] and *Celiberti de Casariego v. Uruguay*.[167] The former concerned an individual resident in Argentina who was kidnapped in Argentina by Uruguayan security forces, held in incommunicado detention for a period of time in Argentina, and then transferred to Uruguay. The second applicant suffered essentially the same fate, except in Brazil rather than Argentina. Both of the applicants complained of abduction and unlawful detention. In *Lopez Burgos*, the Committee held as follows:

> The Human Rights Committee further observes that although the arrest and initial detention and mistreatment of Lopez Burgos allegedly took place on foreign territory, the Committee is not barred either by virtue of article 1 of the Optional Protocol ('... individuals subject to its jurisdiction...') or by virtue of article 2(1) of the Covenant ('... individual within its territory and subject to its jurisdiction...') from considering these allegations, together with the claim of subsequent abduction into Uruguayan territory, inasmuch as these acts were perpetrated by Uruguayan agents acting on foreign soil.
>
> The reference in article 1 of the Optional Protocol to 'individuals subject to its jurisdiction' does not affect the above conclusion because the reference in that article is not to the place where the violation occurred, but rather to the relationship between the individual and the State in relation to a violation of any of the rights set forth in the Covenant, wherever they occurred.
>
> Article 2(1) of the Covenant places an obligation upon a State party to respect and to ensure rights 'to all individuals within its territory and subject to its jurisdiction', but does not imply that the State party concerned cannot be held accountable for violations of rights under the Covenant which its agents commit upon the territory of another State, whether with the acquiescence of the Government of that State or in opposition to it. According to article 5 (1) of the Covenant:
>
> '1. Nothing in the present Covenant may be interpreted as implying for any State, group or person any right to engage in any activity or perform any act aimed at the

[166] *Lopez Burgos v. Uruguay*, Communication No. R.12/52, UN Doc. Supp. No. 40 (A/36/40) at 176 (1981).

[167] *Celiberti de Casariego v. Uruguay*, Communication No. 56/1979, UN Doc. CCPR/C/OP/1 at 92 (1984).

destruction of any of the rights and freedoms recognized herein or at their limitation to a greater extent than is provided for in the present Covenant.'

In line with this, it would be unconscionable to so interpret the responsibility under article 2 of the Covenant as to permit a State party to perpetrate violations of the Covenant on the territory of another State, which violations it could not perpetrate on its own territory.[168]

The Committee adopted identical views in *Celiberti*.[169] Note, first, how the Committee conceives of state jurisdiction in purely personal terms, as a reference 'not to the place where the violation occurred, but rather to the relationship between the individual and the State in relation to a violation of any of the rights set forth in the Covenant, wherever they occurred'. Of course, the Committee somehow had to do away with the explicit mention of territory in Article 2(1) ICCPR, which it did by making an unpersuasive reference to Article 5(1), as well pointed out by Christian Tomuschat in his separate opinion:

I concur in the views expressed by the majority. None the less, the arguments set out in paragraph 12 for affirming the applicability of the Covenant also with regard to those events which have taken place outside Uruguay need to be clarified and expanded. Indeed, the first sentence in paragraph 12.3, according to which article 2(1) of the Covenant does not imply that a State party 'cannot be held accountable for violations of rights under the Covenant which its agents commit upon the territory of another State', is too broadly framed and might therefore give rise to misleading conclusions. In principle, the scope of application of the Covenant is not susceptible to being extended by reference to article 5, a provision designed to cover instances where formally rules under the Covenant seem to legitimize actions which substantially run counter to its purposes and general spirit. Thus, Governments may never use the limitation clauses supplementing the protected rights and freedoms to such an extent that the very substance of those rights and freedoms would be annihilated; individuals are legally barred from availing themselves of the same rights and freedoms with a view to overthrowing the regime of the rule of law which constitutes the basic philosophy of the Covenant. In the present case, however, the Covenant does not even provide the pretext for a 'right' to perpetrate the criminal acts which, according to the Committee's conviction, have been perpetrated by the Uruguayan authorities.

To construe the words 'within its territory' pursuant to their strict literal meaning as excluding any responsibility for conduct occurring beyond the national boundaries would, however, lead to utterly absurd results. The formula was intended to take care of objective difficulties which might impede the implementation of the Covenant in specific situations. Thus, a State party is normally unable to ensure the effective enjoyment of the rights under the Covenant to its citizens abroad, having at its disposal only the tools of diplomatic protection with their limited potential. Instances of occupation of foreign territory offer another example of situations which the drafters of the Covenant had in mind when they confined the obligation of States parties to their own territory. All these factual patterns have in common, however, that they provide plausible grounds for denying the protection of the Covenant. It may be concluded, therefore, that it was the intention of the drafters, whose sovereign decision cannot be challenged, to restrict the territorial scope of the Covenant in view of such situations where enforcing the Covenant would be likely to encounter exceptional obstacles. Never was it envisaged, however, to grant States parties unfettered

[168] *Lopez Burgos*, paras 12.1–12.3.

[169] *Celiberti*, paras 10.1–10.3.

discretionary power to carry out wilful and deliberate attacks against the freedom and personal integrity against their citizens living abroad. Consequently, despite the wording of article 2(1), the events which took place outside Uruguay come within the purview of the Covenant.

The somewhat peculiar reference to Article 5(1) aside,[170] note how both the majority and Professor Tomuschat rely on *universality* in making their argument in favour of extraterritorial application. First, the Committee states that the reason why the ICCPR should apply to Uruguayan actions in Argentina is because 'it would be unconscionable to so interpret the responsibility under article 2 of the Covenant as to permit a State party to perpetrate violations of the Covenant on the territory of another State, which violations it could not perpetrate on its own territory'. The word 'unconscionable' is key—it represents the value judgment that the Committee is making from the standpoint of universality. And so is Tomuschat, when he says that construing 'the words "within its territory" pursuant to their strict literal meaning as excluding any responsibility for conduct occurring beyond the national boundaries would, however, lead to utterly absurd results'. Again, it is in 'utterly absurd results' that we have the universality-speak. This is not an assessment made from pure logic, reason, or law, but rests on the moral and philosophical foundation of human rights—that all human beings should enjoy them by virtue of being human, and that a good reason must be given for denying them.[171]

Secondly, note how effectiveness emerges in Tomuschat's arguably more far-sighted opinion. He thus notes that the ICCPR jurisdiction clause has a specific purpose of taking 'care of objective difficulties which might impede the implementation of the Covenant in specific situations', difficulties which make a state 'normally unable to ensure the effective enjoyment of the rights under the Covenant to its citizens abroad'. And as no such difficulties existed here, since Uruguay was in perfect control of the conduct of its own agents in Argentina, the ICCPR should apply extraterritorially.

Note also how Tomuschat's opinion appears questionable in two elements, at least with the benefit of almost thirty years of hindsight. Firstly, he says that one situation in which there could be 'plausible grounds for denying the protection of the Covenant' is military occupation—precisely the one situation which today most of us would consider as undeniably requiring extraterritorial application, per *Loizidou* and the ICJ's *Wall* and *Congo v. Uganda* decisions. Secondly, Tomuschat relies somewhat problematically on the *nationality* of the victim when he says that it was never envisaged by the states parties that they would have a 'discretionary power to carry out wilful and deliberate attacks against the freedom and personal integrity against their citizens living abroad'. This, I hope, was nothing more than a rhetorical point—the result would surely have been no

[170] See also M. Nowak, *CCPR Commentary* (Engel, 2nd revised edn, 2005), at 860.
[171] See also Chapter III, Section 7 above.

different had Mr Lopez Burgos been an Argentinian, rather than a Uruguayan national.[172]

The Human Rights Committee had few other individual cases dealing with extraterritorial application after *Lopez Burgos* and *Celiberti*.[173] As we have seen, on its consideration of the reports of states parties to the ICCPR it has also affirmed the validity of the spatial model of jurisdiction.[174] Its position, however, remains the same, has solidified due to events surrounding the 'war on terror', and is now stated in its General Comment No. 31:

> States Parties are required by article 2, paragraph 1, to respect and to ensure the Covenant rights to all persons who may be within their territory and to all persons subject to their jurisdiction. This means that a State Party must respect and ensure the rights laid down in the Covenant to *anyone within the power or effective control of that State Party, even if not situated within the territory of the State Party*. As indicated in General Comment 15 adopted at the twenty-seventh session (1986), the enjoyment of Covenant rights is not limited to citizens of States Parties but must also be available to all individuals, regardless of nationality or statelessness, such as asylum seekers, refugees, migrant workers and other persons, who may find themselves in the territory or subject to the jurisdiction of the State Party. This principle also applies to those *within the power or effective control of the forces of a State Party acting outside its territory*, regardless of the circumstances in which such power or effective control was obtained, such as forces constituting a national contingent of a State Party assigned to an international peace-keeping or peace-enforcement operation.[175]

This paragraph is notable for several reasons. First, the Committee now openly espouses the so-called disjunctive–conjunctive interpretation of the jurisdiction clause of Article 2(1) ICCPR—first argued for by Thomas Buergenthal[176]—by saying that the words 'within its territory and subject to its jurisdiction' mean that ICCPR rights must be respected and ensured to all persons who may be within the state's territory *and* to all persons subject to its jurisdiction. Secondly, the Committee again endorses a personal notion of jurisdiction as 'power or effective control' over an individual, regardless of the place where such control is exercised.[177] Thirdly, the Committee's final remark that even state forces in a 'national contingent

[172] Similarly, see M. Scheinin, 'Extraterritorial Effect of the International Covenant on Civil and Political Rights', in Coomans and Kamminga 73, at 75, n. 7; Lawson, above note 163, at 94.

[173] One exception is the so-called passport cases, dealing with the refusal of a state's diplomatic or consular mission abroad to issue a passport to the state's citizen—see, e.g., *Vidal Martins v. Uruguay*, Communication No. R.13/57, U.N. Doc. Supp. No. 40 (A/37/40) at 157 (1982), as well as notes 260 and 261 below and accompanying text. For commentary, see Nowak, above note 170, at 861.

[174] See above note 28.

[175] Human Rights Committee, General Comment No. 31, Nature of the General Legal Obligation on States Parties to the Covenant, UN Doc. CCPR/C/21/Rev.1/Add.13 (2004), para. 10 (emphasis added).

[176] T. Buergenthal, 'To Respect and to Ensure: State Obligations and Permissible Derogations', in L. Henkin (ed.), *The International Bill of Rights: The Covenant on Civil and Political Rights* (Columbia University Press, 1981), at 72–91. See also T. Meron, 'Extraterritoriality of Human Rights Treaties,' (1995) 89 *AJIL* 78, as well as the works cited in Chapter III, above note 11.

[177] The Committee has continued doing so in its examination of state party reports—see, e.g., Concluding Observations of the Human Rights Committee—United Kingdom, UN Doc. CCPR/C/GBR/CO/6, 18 July 2008, para. 14 (requesting of the UK to 'state clearly that the Covenant applies to all individuals who are subject to its jurisdiction and control').

of a State Party assigned to an international peace-keeping or peace-enforcement operation' have ICCPR obligations may be overbroad, in that it neglects potentially very complex issues of attribution. In cases in which the conduct of such a contingent is actually *not* attributable to the sending state, there could also be no violation of the ICCPR by that state arising from that particular conduct.[178] This issue is, of course, beyond the scope of the present study.

Other UN treaty bodies have largely followed the Human Rights Committee's lead with regard to the personal model. We have already seen that the Committee Against Torture interprets the CAT jurisdiction clauses referring to a state's obligations in 'any territory under its jurisdiction' as not only extending to 'all areas where the State party exercises, directly or indirectly, in whole or in part, *de jure* or *de facto* effective control', as well as places or facilities under such control, i.e. within the spatial model, but also that it has adopted a variant of the personal model:

> The Committee considers that the scope of 'territory' under article 2 must also include situations where a State party exercises, directly or indirectly, *de facto* or *de jure* control over persons in detention.[179]

As a purely textual matter, this interpretation makes very little sense. In fact, it is directly contrary to the text of the treaty; to say that the concept of 'territory' includes control over persons is simply mind-boggling. The Committee, of course, had sound reasons of policy to adopt the personal model. While the spatial model could be adapted to cover places rather than large areas of territory, so that, for instance, a Guantanamo detainee would be protected by the CAT since he was, in fact, in an area or place under US control, even the spatial model can only be stretched so far. At the same time, it is quite possible for states to conduct activities that would normally be contrary to the CAT but to do so in places or areas outside their control—this particularly goes for various complicity scenarios. And it is here, to fill this gap, that the personal model becomes especially useful.[180] Otherwise, the activities of, say, a UK agent complicit in the inhumane treatment of a person actually detained by the US or Pakistan, or even interrogating the prisoner himself, would simply fall outside the CAT's scope. Perhaps the Committee's purpose would have been better served by interpreting the CAT as containing an array of implicit negative obligations unlimited by the jurisdiction clauses, an alternative to which I will turn in more detail below. Be that as it may, I cannot see how the Committee's interpretation of the word 'territory' to accommodate the personal model can withstand serious scrutiny.

[178] See also Section 2.C.4 above, as well as Chapter II, Section 3.D.

[179] Committee Against Torture, General Comment No. 2, UN Doc. CAT/C/GC/2, 24 January 2008, para. 16.

[180] Indeed, the Committee seems to have adopted the personal model precisely because of Guantanamo—see Committee Against Torture, Conclusions and Recommendations: United States, UN Doc. CAT/C/USA/CO/2, 25 July 2006, para. 15: 'The State party should recognize and ensure that the provisions of the Convention expressed as applicable to "territory under the State party's jurisdiction" apply to, and are fully enjoyed, by all *persons under the effective control of its authorities*, of whichever type, wherever located in the world' (emphasis added).

2. *Inter-American institutions*

The Inter-American Commission of Human Rights has also championed the personal model, though the Court itself has not yet had the opportunity to rule on matters of extraterritorial application. The Commission's mandate is not only to interpret the American Convention on Human Rights, a binding international treaty, but also the world's first international human rights instrument, the American Declaration on the Rights and Duties of Man. The Declaration is itself not a treaty, but it imposes obligations upon American states indirectly, through their membership in the Organization of American States (OAS). Crucially, while Article 1(1) ACHR contains a jurisdiction clause almost identical to that in the ECHR, the Declaration contains no such clause, nor any other explicit limitation of its territorial scope.[181]

The Commission's first case explicitly dealing with the extraterritorial application of the ACHR was *Saldaño v. Argentina*,[182] filed against Argentina by an Argentinian national incarcerated on death row in the United States, who claimed that Argentina had failed to fulfil its positive obligation to protect him. Relying, *inter alia*, on the European Commission's decision in *Cyprus v. Turkey*, which first espoused the personal model of jurisdiction, the Inter-American Commission considered that 'a state party to the American Convention may be responsible under certain circumstances for the acts and omissions of its agents which produce effects or are undertaken outside that state's own territory'.[183] However, because the conduct complained of was essentially that of the United States, not Argentina, and because 'the petitioner has not adduced any proof whatsoever that tends to establish that the Argentine State has in any way *exercised its authority or control either over the person of Mr Saldaño*, prior or subsequent to his arrest in the United States, or over the local officials in the United States involved in the criminal proceeding taken against him', the petitioner was not within Argentinian jurisdiction.[184] The mere bond of Argentinian nationality did not suffice.[185]

In *Saldaño* the Commission thus clearly adopted the personal model with regard to the ACHR. It did the same several months later with regard to the American Declaration, in *Coard v. United States*, a case dealing with the acts of US military forces in Grenada:

> While the extraterritorial application of the American Declaration has not been placed at issue by the parties, the Commission finds it pertinent to note that, under certain circumstances, the exercise of its jurisdiction over acts with an extraterritorial locus will not only be consistent with but required by the norms which pertain. The fundamental rights of the individual are proclaimed in the Americas on the basis of the principles of equality and non-discrimination—'without distinction as to race, nationality, creed or sex.' Given that

[181] See generally C. Cerna, 'Extraterritorial Application of the Human Rights Instruments of the Inter-American System', in Coomans and Kamminga 141.

[182] *Saldaño v. Argentina*, Report No. 38/99, Annual Report of the IACHR 1998, esp. paras 15–20.

[183] Ibid., para. 17.

[184] Ibid., para. 21 (emphasis added).

[185] Ibid., paras 22–3.

individual rights inhere simply by virtue of a person's humanity, each American State is obliged to uphold the protected rights of *any person subject to its jurisdiction*. While this most commonly refers to persons within a state's territory, it may, under given circumstances, refer to conduct with an extraterritorial locus where the person concerned is present in the territory of one state, but *subject to the control of another state—usually through the acts of the latter's agents abroad*. In principle, the inquiry turns not on the presumed victim's nationality or presence within a particular geographic area, but on whether, under the specific circumstances, the State observed the rights of a person *subject to its authority and control.*[186]

Note how the Commission read into the American Declaration a state jurisdiction requirement that it does not actually contain, only to then adopt the personal model of jurisdiction as authority and control over individuals. The reason it gave for doing so was that 'individual rights inhere simply by virtue of a person's humanity', i.e. universality. The Commission has maintained this approach ever since,[187] for example in a number of (practically rather fruitless) cases against the United States dealing with Guantanamo.[188]

3. *Early Strasbourg case law*

Though in recent years it has been the UN treaty bodies and the Inter-American Commission that have most ardently championed the personal model of jurisdiction, it was in fact first devised by the European Commission of Human Rights. As we have seen, in its first decision on northern Cyprus,[189] the Commission opined that

[i]n Art. 1 of the Convention, the High Contracting Parties undertake to secure the rights and freedoms defined in Section 1 to everyone 'within their jurisdiction' (in the French text: 'relevant de leur juridiction'). The Commission finds that this term is not, as submitted by the respondent Government, equivalent to or limited to the national territory of the High Contracting Party concerned. It is clear from the language, in particular of the French text, and the object of this Article, and from the purpose of the Convention as a whole, that the High Contracting Parties are bound to secure the said rights and freedoms *to all persons under their actual authority and responsibility, whether that authority is exercised within their own territory or abroad*. The Commission refers in this respect to its decision on the admissibility of Application No. 1611/62—X. v Federal Republic of Germany—Yearbook of the European Convention on Human Rights, Vol. 8, pp. 158–169 (at pp. 168–169).

[186] *Coard et al. v. United States*, Case No. 10.951, Report No. 109/99, Annual Report of the IACHR 1999, para. 37 (emphasis added).

[187] See *Armando Alejandre Jr and Others v. Cuba ('Brothers to the Rescue')*, Case No. 11.589, Report No. 86/99, 29 September 1999, para. 23. See further J. Cerone, 'The Application of Regional Human Rights Law Beyond Regional Frontiers: The Inter-American Commission on Human Rights and US Activities in Iraq', *ASIL Insights*, 25 October 2005, available at <http://www.asil.org/insights051025.cfm>.

[188] See, e.g., Decision on Request for Precautionary Measures (Detainees at Guantánamo Bay), 12 March 2002, (2002) 41 ILM 532. For more extensive discussion, see Cerna, above note 181, at 159 *et seq*; Gondek, at 217–19.

[189] *Cyprus v. Turkey* (dec.), App. Nos 6780/74 and 6950/75, 26 May 1975.

The Commission further observes that nationals of a State, including registered ships and aircrafts, are partly within its jurisdiction wherever they may be, *and that authorised agents of a State, including diplomatic and consular agents and armed forces, not only remain under its jurisdiction when abroad but bring any other persons or property 'within the jurisdiction' of that State, to the extent that they exercise authority over such persons or property*. Insofar as, by their acts or omissions, they affect such persons or property, the responsibility of the State is engaged.[190]

The Commission has affirmed the personal model time and again, and did not limit its scope either to specific places or objects, such as embassies, ships, or aircraft, or to specific agents, such as diplomats, but applied it to *all* state agents.[191] Again, all of the cases that the European Court in *Bankovic* tried to rationalize (or even ignore) as supposedly being exceptional, tied to the 'special' nature of either the agents or the premises on which they acted, were decided on the basis of a general conception of jurisdiction as state authority and control over individuals. The Commission's approach to Article 1 jurisdiction was, in fact, limitless—whenever there was in substance a potential violation of the ECHR, the Commission found that state jurisdiction existed.[192]

All of this changed, of course, with *Bankovic*, where the Court explicitly rejected 'a "cause-and-effect" notion of jurisdiction',[193] and gave short shrift to the contrary practice of other human rights institutions, on which the applicants relied:

Fourthly, the Court does not find it necessary to pronounce on the specific meaning to be attributed in various contexts to the allegedly similar jurisdiction provisions in the international instruments to which the applicants refer because it is not convinced by the applicants' specific submissions in these respects (see § 48 above). It notes that Article 2 of the American Declaration on the Rights and Duties of Man 1948 referred to in the above-cited *Coard* Report of the Inter-American Commission of Human Rights (§ 23 above), contains no explicit limitation of jurisdiction. In addition, and as to Article 2 § 1 the CCPR 1966 (§ 26 above), as early as 1950 the drafters had definitively and specifically confined its territorial scope and it is difficult to suggest that exceptional recognition by the

[190] *Cyprus v. Turkey* (dec.), App. Nos 6780/74 and 6950/75, 26 May 1975, at 136, para. 8.

[191] See, e.g., *Stocké v. Germany*, App. No. 11755/85, Commission Report, 12 October 1989, para. 166. See also Gondek, at 126.

[192] Sarah Miller suggests that the first Cyprus case can somehow be interpreted as being about control over territory, rather than individuals—see S. Miller, 'Revisiting Extraterritorial Jurisdiction: A Territorial Justification for Extraterritorial Jurisdiction under the European Convention,' (2010) 20 *EJIL* 1223, at 1237, arguing that 'the essential predicate was that Turkey had established so great a presence in Cyprus that it was in a position effectively to control administration in the region. Likewise, *Banković* emphatically suggests that one cannot conflate questions of effective control of a region with subsequent attribution of acts to state officials'. With respect, Miller is simply mistaken. There is nothing in the Commission's first decision on Cyprus, nor in its progeny, that suggests that the Commission saw 'jurisdiction' as being about control over territory, rather than individuals. That strand of Strasbourg jurisprudence begins in the Court's *Loizidou* judgment many years later. It is likewise simply anachronistic to read the Commission's case law in light of *Bankovic*. While it is true that, under the Commission's approach, ECHR states parties would be responsible for violating the ECHR whenever state agents exercise authority and control over individuals, this does not mean that the issues of attribution and the existence of breach of obligation are *conflated*, just that the obligation is not limited territorially.

[193] *Bankovic*, para. 75.

Human Rights Committee of certain instances of extra-territorial jurisdiction (and the applicants give one example only) displaces in any way the territorial jurisdiction expressly conferred by that Article of the CCPR 1966 or explains the precise meaning of 'jurisdiction' in Article 1 of its Optional Protocol 1966 (§ 27 above). While the text of Article 1 of the American Convention on Human Rights 1978 (§ 24 above) contains a jurisdiction condition similar to Article 1 of the European Convention, no relevant case-law on the former provision was cited before this Court by the applicants.[194]

As we have seen, in recent years the adoption of the personal model has grown in the jurisprudence of other human rights bodies, and the European Court could not today dismiss it as easily. Be that as it may, having adopted a strict spatial, territorial model of jurisdiction in *Bankovic*, the Court soon found itself faced with cases where the application of this model would have led to unacceptable results.

4. *Post-*Bankovic *Strasbourg case law*

The first post-*Bankovic* case to directly contradict it was *Issa*, where the Chamber endorsed the personal model of jurisdiction in addition to the spatial one:

> Moreover, a State may also be held accountable for violation of the Convention rights and freedoms of persons who are in the territory of another State but who are found to be under the former State's authority and control through its agents operating—whether lawfully or unlawfully—in the latter State (see, *mutatis mutandis*, *M. v. Denmark*, application no. 17392/90, Commission decision of 14 October 1992, DR 73, p. 193; *Illich Sanchez Ramirez v. France*, application no. 28780/95, Commission decision of 24 June 1996, DR 86, p. 155; *Coard et al. v. the United States*, the Inter-American Commission of Human Rights decision of 29 September 1999, Report No. 109/99, case No. 10.951, §§ 37, 39, 41 and 43; and the views adopted by the Human Rights Committee on 29 July 1981 in the cases of *Lopez Burgos v. Uruguay* and *Celiberti de Casariego v. Uruguay*, nos. 52/1979 and 56/1979, at §§ 12.3 and 10.3 respectively). Accountability in such situations stems from the fact that Article 1 of the Convention cannot be interpreted so as to allow a State party to perpetrate violations of the Convention on the territory of another State, which it could not perpetrate on its own territory (*ibid.*).[195]

Of interest is not only that the Chamber considered 'jurisdiction' to mean state authority and control over individuals, but *how* and *why* the Chamber reached this conclusion. As for the *how*, note that the Chamber relies precisely on all those cases so brusquely rejected by the Grand Chamber in *Bankovic*: the old European Commission's case law; the *Coard* decision of the Inter-American Commission; and the *Lopez Burgos* and *Celiberti* cases of the Human Rights Committee. As for the *why*, the Chamber made a direct appeal to universality, and did so in the exact language of *Lopez Burgos* and *Celiberti*—'the Convention cannot be interpreted so as to allow a State party to perpetrate violations of the Convention on the territory of another State, which it could not perpetrate on its own territory'.

[194] *Bankovic*, para. 78. [195] *Issa*, para. 71.

It is also worth mentioning that three of the judges sitting in the Chamber that decided *Issa* in 2004—Judges Costa, Baka, and Thomassen—also sat on the Grand Chamber that decided *Bankovic* in 2001. While the Grand Chamber was unanimous in *Bankovic*, the Court's (quite unfortunate) practice is to disallow separate opinions of judges in admissibility decisions, which only state whether they were decided unanimously or by a majority.[196] Thus, though all of the judges in *Bankovic* were in agreement that the case should be dismissed, there could have been serious disagreement on the bench on the exact reasoning for doing so—and that disagreement might have re-emerged in *Issa* three years later. This, of course, is purely speculative. What is not speculative, however, is that no amount of (honest) distinguishing can reconcile *Issa* with *Bankovic*, as the former relied on a personal model of jurisdiction which the latter discarded.

Now let us take a look at how the *Issa* approach fared in subsequent Strasbourg jurisprudence. Its first mention was in that ill-fated case, filed (without any hint of self-irony) by that great friend of human rights Saddam Hussein against several European states, claiming that his continued detention in Iraq was unlawful. A Chamber (Fourth Section) declared his complaint inadmissible since he was being held by US troops, and was therefore outside the respondent states' jurisdiction.[197] Although the Chamber cited *Issa*, it did not definitively embrace its approach that control over an individual may establish state jurisdiction, though it did allow for the possibility that state involvement in detention could be a basis for jurisdiction.

In several subsequent cases the *Issa* approach was cited with explicit approval. In *Pad and Others v. Turkey*,[198] the applicants were Iranian nationals, living close to the Turkish border. They were killed by a Turkish helicopter, as Turkey claimed when they attempted to cross the border illegally. Though it was undisputed that they were killed by Turkish forces, it was disputed whether the killing took place on the Turkish or the Iranian side of the border. Relying on *Issa*, the Chamber (Third Section) held that

> ... a State may be held accountable for violations of the Convention rights and freedoms of persons who are in the territory of another State which does not necessarily fall within the legal space of the Contracting States, but who are found to be under the former State's authority and control through its agents operating—whether lawfully or unlawfully—in the latter State.
>
> However, in the instant case, it was not disputed by the parties that the victims of the alleged events came within the jurisdiction of Turkey. While the applicants attached great importance to the prior establishment of the exercise by Turkey of extraterritorial jurisdiction with a view to proving their allegations on the merits, the Court considers that it is not required to determine the exact location of the impugned events, given that the Government

[196] This was also most notably the situation in *Behrami*—see further Milanovic and Papic, above note 96, at 293–4.

[197] *Saddam Hussein v. Albania and others* (dec.), App. No. 23276/04, 14 March 2006. One could question whether the United Kingdom was in a different position than the other respondent states, due to its status as a joint occupying power in Iraq. See above Section 2.C.4.

[198] *Pad and Others v. Turkey* (dec.), App. No. 60167/00, 28 June 2007.

had already admitted that the fire discharged from the helicopters had caused the killing of the applicants' relatives, who had been suspected of being terrorists.[199]

In other words, the Chamber in *Pad* seems to have accepted that because the applicants were killed by Turkish agents, they were brought within Turkish jurisdiction, regardless of whether the killing took place in Turkey or in Iran. This is, of course, again in direct contradiction to *Bankovic*—even the killing itself took place by missile fire from an aircraft! In *Isaak and others v. Turkey*[200] the same Chamber declared admissible the complaint by a family of an individual beaten to death by TRNC police in northern Cyprus *in a UN buffer zone*, and therefore not in an area over which Turkey had effective overall control under *Loizidou*. The Chamber stated, *inter alia*, that

> [i]n view of the above, even if the acts complained of took place in the neutral UN buffer zone, the Court considers that the deceased was under the authority and/or effective control of the respondent State through its agents (see *Issa and Others*, cited above). It concludes, accordingly, that the matters complained of in the present application fall within the 'jurisdiction' of Turkey within the meaning of Article 1 of the Convention and therefore entail the respondent State's responsibility under the Convention.

Two similar cases, *Solomou and Others v. Turkey*[201] and *Andreou v. Turkey*,[202] were decided by a Chamber (Fourth Section)—the same Chamber which later decided *Al-Saadoon*, though with a somewhat different composition. The former case simply repeated *Isaak*, while in the latter case, the Chamber declared admissible a complaint by an individual who was shot by Turkish or TRNC soldiers while in the UN buffer zone:

> The Government argued that there was no indication that at the time when she was struck by the bullet the applicant had been in the territory of Turkey or of the 'TRNC'. It was contended that Turkey had no 'jurisdiction' and/or control over the UN-controlled buffer zone or the Greek-Cypriot National Guard ceasefire line where the act complained of by the applicant had occurred.
>
> The Court reiterates that, in exceptional circumstances, the acts of Contracting States which produce effects outside their territory and over which they exercise no control or authority may amount to the exercise by them of jurisdiction within the meaning of Article 1 of the Convention. The Court notes that, according to UNFICYP's press release and the UN Secretary-General's report on the events of 14 August 1996, the applicant's injuries were caused by Turkish and/or Turkish Cypriot uniformed personnel, who fired some 25 to 30 rounds into the crowd. These agents of the State were at the time of opening fire in the territory of the 'TRNC'. The Court further notes that, when she was hit by the bullet, the applicant was standing outside the neutral UN buffer zone and in close vicinity to the Greek-Cypriot National Guard checkpoint. Unlike the applicants in the *Bankovic and Others* case (cited above) she was accordingly within territory covered by the Convention.

199 Ibid., paras 53–4.
200 *Isaak and Others v. Turkey* (dec.), App. No. 44587/98, 28 September 2006.
201 *Solomou and Others v. Turkey*, Judgment, App. No. 36832/97, 24 June 2008, paras 44–5, 51.
202 *Andreou v. Turkey* (dec.), App. No. 45653/99, 3 June 2008.

In these circumstances, even though the applicant sustained her injuries in territory over which Turkey exercised no control, the opening of fire on the crowd from close range, which was the direct and immediate cause of those injuries, was such that the applicant must be regarded as 'within [the] jurisdiction' of Turkey within the meaning of Article 1 and that the responsibility of the respondent State under the Convention is in consequence engaged.[203]

It is notable that in this particular case, though the Chamber approvingly cited *Issa*, it seemed to have found that Turkey had jurisdiction either because the bullet was fired at the victim from an area controlled by Turkey, or because when the applicant was hit by the bullet she was located in an area controlled by Cyprus—a situation in the Chamber's view distinguishable from that in *Bankovic* possibly because it was in the ECHR's *espace juridique*. It is unclear, at least to me, why the Chamber in *Andreou* did not apply the *Issa* control over individuals' basis of jurisdiction, as in *Isaak* and *Solomou*.

In *Ben El Mahi and Others v. Denmark*,[204] a Chamber (Fifth Section) declared inadmissible an application lodged against Denmark by Moroccan residents who alleged that Denmark had discriminated against them and violated their freedom of religion by permitting the publication of cartoons disrespectful towards the prophet Mohammed in a private Danish newspaper, which provoked an uproar in the Muslim world. The Chamber found that the applicants were not within Danish jurisdiction. It did, however, cite *Issa* and did accept in principle that jurisdiction may arise from control over individuals.

In sum, three of the Court's five sections cited *Issa* with approval for the proposition that Article 1 ECHR jurisdiction may arise from state control over an individual, rather than territory. This is both contrary to *Bankovic* and directly undermines it. In none of these cases, of course, did the Chamber in question openly acknowledge the conflict with the *Bankovic* Grand Chamber, choosing rather to pretend that the Court's case law is perfectly consistent and coherent.

In that regard, *Issa* and its progeny notwithstanding, any rumours about *Bankovic*'s imminent demise would be greatly exaggerated. Yes, it has been chipped away at bit by bit, but this happened mostly in politically relatively unimportant, low-profile cases. It is when the judges in these cases realized that the spatial model failed them—that, for example, they could not establish the effective overall control of Turkey over the UN buffer zone in Cyprus—and that this would lead to manifestly unjust or arbitrary results, that they decided to instead rely on the personal model of jurisdiction. The Grand Chamber has, however, affirmed *Bankovic* as recently as in its 2010 *Medvedyev* judgment discussed above:

In keeping with the essentially territorial notion of jurisdiction, the Court has accepted only in exceptional cases that acts of the Contracting States performed, or producing effects, outside their territories can constitute an exercise of jurisdiction by them for the purposes of Article 1 of the Convention (see *Banković*, cited above, § 67, and *Ilaşcu and Others v. Moldova and Russia* [GC], no. 48787/99, § 314, ECHR 2004-VII). In its first *Loizidou*

[203] *Andreou v. Turkey* (dec.), App. No. 45653/99, 3 June 2008.
[204] *Ben El Mahi and Others v. Denmark* (dec.), App. No. 5853/06, 11 December 2006.

judgment (*preliminary objections*), for example, the Court found that bearing in mind the object and purpose of the Convention, the responsibility of a Contracting Party might also arise when as a consequence of military action—whether lawful or unlawful—it exercised effective control of an area outside its national territory (see *Loizidou v. Turkey* (*preliminary objections*) [GC], 23 March 1995, § 62, Series A no. 310). This excluded situations, however, where—as in the Banković case—what was at issue was an instantaneous extraterritorial act, as the provisions of Article 1 did not admit of a 'cause-and-effect' notion of 'jurisdiction' (Banković, § 75).[205]

The Court not only reaffirmed the *Bankovic* rejection of a 'cause-and-effect' notion of jurisdiction, but added a new gloss to *Bankovic*—that it dealt with 'an instantaneous extraterritorial act', i.e. a killing. The Court may thus be opening the doors to the application of the personal model to, well, *non*-instantaneous acts, however defined, most notably detention. This, of course, brings us to the single most crucial problem of the personal model—what does, in fact, amount to state 'authority and control' over an individual? In other words, can the personal model be limited so as to serve a useful purpose as a threshold for the application of human rights treaties?

In my view, the answer to these questions is a resounding 'no'—but this does not mean that attempts have not been made. Let us look at some of them.

C. What amounts to authority and control? Can the personal model be limited?

1. *Physical custody*

One possible way of attempting to limit the personal conception of jurisdiction is to say that it applies only when a state has *physical custody* of an individual, and not to any other form of control. Deprivation of liberty is by definition a 'non-instantaneous act', to paraphrase *Medvedyev*, and many European cases applying the personal model, such as *Öcalan* and *Sanchez Ramirez*, dealt precisely with detention. By so bounding the personal model, 'instantaneous acts' such as an extraterritorial deprivation of life would fall outside the state's jurisdiction, unless they were committed in the context of detention. *Bankovic* is, of course, undoubtedly an authority at least for the former proposition.

Though, of course, the European Court could always limit by fiat the personal conception of jurisdiction to physical custody, the real question is whether this would be justified by any consideration of principle. I submit that it would not, and there is no better case to illustrate this, as well as all the perils of the personal model, than *Al-Skeini*.

Recall that all of the English courts dealing with the case found that the five applicants killed by UK troops on patrol were not within the UK's jurisdiction, while Mr Baha Mousa, the sixth applicant, who was killed in British custody, was within the UK's jurisdiction. That outcome can, in my view, only be explained by

[205] *Medvedyev* [GC], para. 64.

the fact that the jurisdiction issue in that case was reverse-engineered, as it were.[206] From the standpoint of the English courts, while the killing of five Iraqis by UK troops on patrol may have been regrettable, it was perhaps perfectly justified in some cases, and could be rationalized as an accident of war in others. However, this could not have been done with the killing of Baha Mousa, the sixth applicant. Nothing could possibly have justified the beating to death of a defenceless prisoner in UK custody, other than an unpalatable, overt adoption of double standards in the name of either relativism or *realpolitik*, or both. Baha Mousa therefore *had* to be protected by the ECHR—but what of the other five applicants?

One option, and the only option that would in my view truly have been principled, would have been to engage in a substantive Article 2 ECHR analysis regarding the lawfulness of the deprivation of life of all six applicants. While such an analysis would have assuredly resulted in a violation with regard to Baha Mousa, the result need not have been the same in respect of the applicants killed on patrol, depending mainly on each particular set of facts, and on the degree to which the courts would have been willing to attenuate their Article 2 analysis to better accommodate the extraordinary circumstances of an occupied southern Iraq in the throes of an insurgency.

But this, of course, was not the option favoured by the English courts. It was simply too fraught with peril, political as well as legal. As it seemed to them impossible to reconcile universality with effectiveness, all of the English courts opted to dismiss the cases of the five applicants killed on patrol on the preliminary grounds of lack of UK jurisdiction, and consequent lack of extraterritorial application of the ECHR, and hence had to come up with a way of distinguishing their cases from that of Baha Mousa. The High Court and the House of Lords chose to do so by tweaking the spatial model so that it would cover UK military prisons abroad, by making the extremely dubious analogy to embassies. The Court of Appeal per Lord Justice Brooke thought, however, that such an analogy made little sense, and quite rightly so. It therefore opted for a personal model of jurisdiction as control over individuals,[207] but it first had to consider *Bankovic*, which seemed to have squarely rejected it:

> I would therefore be more cautious than the Divisional Court in my approach to the *Bankovic* judgment. It seems to me that it left open both the ECA [effective control of an area, i.e. the spatial model] and SAA [state agent authority, i.e. the personal model] approaches to extra-territorial jurisdiction, while at the same time emphasizing (in para 60) that because a SAA approach might constitute a violation of another state's sovereignty (for example, when someone is kidnapped by the agents of a state on the territory of another state without that state's invitation or consent), this route to any recognition that extra-territorial jurisdiction has been exercised within the meaning of an international treaty should be approached with caution.

[206] See above, Chapter III, Sections 4.A and 6.

[207] See *Al-Skeini CA*, para. 49 (per Brooke LJ), distinguishing between the spatial and personal models, with the latter being referred to as 'state agent authority', or SAA.

Viewed in this light, the Court's conclusion in *Bankovic* was fairly inevitable. It was not a case in which it could possibly be said that the NATO forces were in effective control of the relevant area when they bombed the television station in Belgrade. What was in issue on the applicants' arguments was whether, because the bombing was NATO's direct act, any of the states involved in the bombing could be said to have exercised Article 1 jurisdiction over any of the citizens of the FRY who were affected by it, and the court ruled out this possibility in para 71 of its judgment. *I do not read the judgment as excluding the possibility that if a person is in the custody or control of the agents of a member state (whether they be military personnel or police officers) that state may not be fixed with having exercised jurisdiction over that person ratione personae, particularly if, like the British army in Iraq, those agents form part of an occupying force in the eyes of international law.* This question did not arise for decision in *Bankovic*. On the other hand, I consider that the attempt by the claimants to rely on some of the conclusions reached by the Commission on SAA grounds in its 1976 *Cyprus v Turkey* report (see para 58 above) as authority for the proposition that the citizens of Basrah City came within the jurisdiction of the UK simply because they were affected by the activities of the street patrols, without more, as being authoritatively refuted by *Bankovic*.[208]

The Court thus considered *Bankovic* not to exclude the personal model in principle, but to do so with regard to a deprivation of life without physical custody. It then analysed the various cases applying the personal model that I have examined above, including *Öcalan*, *Freda*, and *Sanchez Ramirez*,[209] and quite correctly concluded that

[t]hese cases have nothing to do with the principle of public international law relating to activities within aircraft registered with a state when the aircraft is airborne. They reflect examples of the SAA doctrine applying when someone is within the control and authority of agents of a contracting state, notwithstanding that he comes within that control and authority outside the *espace juridique* of the Council of Europe, and apparently whether or not the host state has consented to this exercise of control and authority on its soil.[210]

For Lord Justice Brooke, the validity of the personal model was therefore clear, and in his view it covered Baha Mousa:

Throughout the case law to which I have referred in paras 98–106 above there has been the constant refrain that a state may be fixed with having exercised extra-territorial jurisdiction if it has exercised control and authority over a complainant. The court's analysis will then be centred on the particular complaint that is made. If the complaint concerns a breach of Article 2 or Article 3 rights, the court will not only consider whether those rights have been violated but also whether the state was in breach of the positive obligations imposed on it in connection with the duty to secure those rights. The Secretary of State now concedes that the UK had jurisdiction in relation to Baha Mousa throughout the period that led up to his death, because he was being held in a British military prison that was operating in Iraq with the consent of the Iraqi sovereign authorities and contained arrested suspects (see DC 287). In my judgment, Mr Mousa came within the control and authority of the UK from the time he was arrested at the hotel and thereby lost his freedom at the hands of British troops.[211]

208 Ibid., paras 80–1 (per Brooke LJ), emphasis added.
209 Ibid., paras 98–106 (per Brooke LJ).
210 Ibid., para. 107 (per Brooke LJ).
211 Ibid., para. 108 (per Brooke LJ).

Thus, from the moment that Baha Mousa was in British custody, he was within the UK's jurisdiction under the personal model. As for the other five applicants:

> None of them were under the control and authority of British troops at the time when they were killed. This is one of the main points that was decided in *Bankovic* (see para 81 above). The only case which might give rise to an argument to the contrary is that of Muhammed Salim (on the basis that British troops assumed control of the house when they burst in), but it would in my judgment be thoroughly undesirable for questions about the applicability of the ECHR to turn for their resolution on sophisticated arguments of this kind. The soldiers, for instance, might have found they were by no means in control of the house if they had all been shot dead by hostile gunfire after they had broken in. It is essential, in my judgment, to set rules which are readily intelligible. If troops deliberately and effectively restrict someone's liberty he is under their control. This did not happen in any of these five cases.[212]

This paragraph is absolutely fascinating, for two reasons. First, the Court of Appeal found that Baha Mousa was under the control and authority of British troops, and thus under UK jurisdiction, but that the five other applicants were not. In other words, it thought that a 'mere' killing did not amount to control over an individual for Article 1 ECHR purposes, while the arrest of an individual did. Secondly, the Court briefly considered applying the spatial model as control by British troops over the *house* where one of the applicants was killed, but decided that this would be stretching it too far—it would be 'thoroughly undesirable for questions about the applicability of the ECHR to turn for their resolution on sophisticated arguments of this kind'. This is exactly right, and this collapse of the spatial model is exactly why not just the Court of Appeal in *Al-Skeini*, but other courts as well, have thought it necessary to apply the personal model of jurisdiction.

Now, as for the distinction between physical custody, which does amount to the control over an individual, and killing, which does not, note that the *only* reason the Court gives for this distinction is that *Bankovic said so*. While *Bankovic* did not even openly consider the personal model, the Court was right in interpreting it as precluding a result whereby the power to kill alone could constitute jurisdiction for Article 1 ECHR purposes.

Bankovic aside, of course, the question is whether this distinction makes any sense—and it simply does not. Why, after all, should the state's power to kill an individual not be considered as an exercise of the state's control over that individual?[213] Is not killing, in fact, the ultimate form of control?[214] It is certainly true that if a state has an individual in custody, it can do *more* to him than just kill him, by for example subjecting him to torture or inhuman treatment. But should this matter? Aside from killing, one can easily imagine a situation where, for instance,

[212] See *Al-Skeini CA*, para. 110 (per Brooke LJ).

[213] See also Scheinin, above note 172, at 77–8; Lubell, above note 50, at 221–4.

[214] In that regard, one could perhaps draw a helpful analogy to the interpretation of the Fourth Amendment to the US Constitution, which prohibits 'unreasonable searches and seizures'. The US Supreme Court has thus held not only that '[w]henever an officer restrains the freedom of a person to walk away, he has seized that person, but also that 'there can be no question that apprehension by the use of deadly force is a seizure subject to the reasonableness requirement of the Fourth Amendment'. *Tennessee v. Garner*, 471 U.S. 1, 7 (1985). See also ibid., at 25 (O'Connor J., dissenting).

one state has custody, but another does the torture, or is at least complicit in it—as with the alleged involvement of the US and UK intelligence services in interrogations conducted in Syria, Egypt, or Pakistan. If, say, a CIA agent was to prod a detainee held in Pakistani custody with a hot poker, why should Pakistan be the only state to blame for the torture?

Indeed, if we take a look at some of the case law, we can see that the personal model has been applied even in cases which did *not* involve physical custody, as, for example, in *Pad, Isaak*, and *Solomou*.[215] All of them involved a killing that took place outside detention. After all, the whole universality rationale of *Issa* was that 'the Convention cannot be interpreted so as to allow a State party to perpetrate violations of the Convention on the territory of another State, which it could not perpetrate on its own territory'.[216] From the standpoint of universality, it makes no sense to say that a state acting extraterritorially may not *detain* arbitrarily, but that it may, in fact, *kill* arbitrarily. If anything, such a position would precisely create an incentive for states to kill rather than capture,[217] as can indeed be seen from the escalation of drone attacks against suspected terrorists in Pakistan and Afghanistan, and a massive decrease in the number of those who are actually detained.[218]

In short, saying that physical custody is the *only* form of state authority and control over an individual that is acceptable for the personal model of jurisdiction is precisely the kind of 'sophisticated legal argument' that the Court of Appeal rightly decided to disregard in respect of the spatial model. It is, indeed, sophistry rather than sophistication, merely a device to exclude from the ambit of judicial review those cases which present extreme political or policy difficulties, and yet are morally much more blurry than is the killing of a defenceless prisoner—as with the unfortunate, but perhaps justifiable killing of civilians on patrol, or as collateral damage in airstrikes, or what have you. It is to avoid dealing with such cases that the personal model is artificially limited, but such unprincipled limitations by their very nature cannot persist indefinitely.

As of the time of writing, *Al-Skeini* is pending before the Grand Chamber of the European Court. Compared to the House of Lords' judgment, the Court could do better or worse, and much depends on just how many issues the Court will reach.

Thus, from the standpoint of the spatial model, the Court could do the following:

(1) Affirm that the effective overall control of an area conception of jurisdiction applies outside the ECHR's *espace juridique*, as the Chamber had done in *Issa*, or, conversely, agree with the House of Lords that the spatial model

[215] See above, Section 3.B.4.
[216] *Issa*, para. 71.
[217] See also Lubell, above note 50, at 224.
[218] With this particular example, the incentive to kill rather than capture was created in great part by the cautious inroads of US courts into the Executive's detention policies abroad, as with Guantanamo, and pending litigation regarding Bagram—see, e.g., 'Special Report: How the White House learned to love the drone', *Reuters*, 18 May 2010, available at <http://www.reuters.com/article/idUSTRE64H5SL20100518>.

extends only to the territories of the ECHR states parties—certainly the worst possible result;[219]

(2) Assuming that the spatial model does apply, the question then would be whether Basra was under the UK's effective overall control. The Court could either say that the UK's status as an occupying power necessarily meant that it was in control of Basra, despite all the difficulties that it had encountered, and that hence all six applicants were within the UK's jurisdiction, or it could agree with the English courts that Basra was not under the UK's effective control;[220]

(3) A finding that Basra, was, in fact, an area under the UK's jurisdiction would dispose entirely of the preliminary question of extraterritorial application. If, however, the Court were to find that Basra was not under the UK's control, it then may decide to apply the attenuated version of the spatial model as control over *places* in order to bring Baha Mousa under the UK's jurisdiction upon his detention, as in *Al-Saadoon*.[221] Doing so, however, would leave outside the ECHR's scope any events that took place between Baha Mousa's arrest and his ultimate transfer to the UK detention facility.[222]

The possibilities of the spatial model would thus be exhausted, and the Court would then have to examine the case from the standpoint of the personal model:

(1) It could reject the validity of the personal model altogether, or narrow its application to some arbitrarily selected exceptional circumstances, and thus exclude all of the applicants from the UK's jurisdiction, unless it opted to fiddle with the spatial model to protect Baha Mousa, as above;

(2) In a similar vein, it could adopt the approach of the Court of Appeal, saying that the personal model requires physical custody, and that accordingly only Baha Mousa was within the UK's jurisdiction;

(3) Finally, it could radically depart from *Bankovic*, and say that the five applicants killed by UK troops on patrol were also under the UK's control, and accordingly within its jurisdiction. This would be a welcome, but not very likely development.

[219] See above, Chapter III, Section 6.C. [220] See above, Section 2.C.3.

[221] See above, Section 2.D.

[222] This issue was manifest in the parties' arguments in *Al-Skeini*—see *Al-Skeini CA*, para. 183 (per Sedley LJ):

> [H]ow, Mr Rabinder Singh QC asks, can one draw a rational line at the prison wall, so that if Mr Mousa had been beaten to death by troops in the hotel where he was arrested rather than in the prison to which he was taken, there would have been no violation of his Convention rights? The logic of Mr Greenwood's submission is that, bar the special status of the military prison in international law, there can be no extra-territorial reach to the ECHR. The logic of Mr Singh's submission is that, given that the Convention reaches the prison, by parity of reasoning it reaches everywhere in Iraq where British forces are the only functioning form of government. That at least is his narrower submission, based on what he calls effective control authority (ECA). His broader and preferred submission is that the Convention reaches everywhere that agents of a state signatory operate (SAA).

The Court may of course very well surprise us. Whatever it does will inevitably be a consequence of how it perceives the tension between universality and effectiveness on the facts of the case, and on their broader policy implications, especially with regard to the personal model.

2. *Control over an individual in a specific place, or by specific agents*

Another possible way of limiting the personal model would be to mix it with the spatial one so that it would apply only in a certain kind of place, or to limit it only to a certain kind of state agent. For example, one could read cases like *Öcalan* as implying that custody over the individual was insufficient of itself to create jurisdiction, but that it had to take place in a context where general international law somehow recognizes that states may exceptionally exercise their jurisdiction extraterritorially, e.g. aboard a registered aircraft or in an embassy. This is, in essence, a variation of the special approach to diplomatic agents or premises, or to ships and aircraft, that we have already examined above, and is no more persuasive. None of the relevant cases actually say that the nature of the place in which the control over an individual is exercised matters, nor is it obvious that it should matter. Nor would this, again, be warranted by any consideration of general international law. As I have argued above, it is perfectly irrelevant for the question of the extraterritorial application of human rights treaties that customary law recognizes that the state may regulate conduct aboard registered ships and airplanes, or that diplomatic and consular agents and premises have a distinct status protected by jurisdictional immunities.[223]

3. *Nationality and membership in the armed forces*

Yet another potential avenue of limiting the personal model would be to do so on the basis of the victim's nationality or some other status, most notably membership in the state's armed forces. So far we have generally considered examples where a state agent affects the rights of a national of another state when acting extraterritorially—for instance, when UK soldiers killed the Iraqi applicants in *Al-Skeini*. But what of the scenario where one UK soldier kills another, for instance in a friendly fire incident, or in a plain, old-fashioned murder? Would the UK have an Article 2 ECHR obligation to investigate?

Such a scenario is obviously not hypothetical. UK courts first encountered it in *Gentle*, in which the applicants were the mothers of two UK soldiers who both died while serving in Iraq.[224] One was killed by a roadside bomb, the other in a friendly fire incident. Inquests were held regarding both deaths, and the relevant facts were established beyond any doubt. However, the applicants contended that Article 2 ECHR required the UK not only to fully investigate the circumstances of the deaths themselves, but also to establish an inquiry into the 2003 Iraq war as a

[223] See above, Section 2.D. [224] *R. (Gentle) v. Prime Minister* [2008] 1 AC 1356.

whole, particularly into the process and the legal advice given before the troops were committed to an armed conflict. In effect, they were trying to inject *jus ad bellum*-type arguments into the framework of Article 2. The House of Lords found unanimously that such a far-reaching inquiry fell outside the scope of state positive obligations imposed by Article 2. It therefore avoided ruling on whether the two soldiers were protected by the ECHR in the first place, i.e. whether it applied to them extraterritorially. That issue was, however, briefly addressed by two of the law lords. Lord Bingham thus noted that

> [s]ubject to limited exceptions and specific extensions, the application of the Convention is territorial: the rights and freedoms are ordinarily to be secured to those within the borders of the state and not outside. Here, the deaths of Fusilier Gentle and Trooper Clarke occurred in Iraq and although they were subject to the authority of the respondents they were clearly not within the jurisdiction of the UK as that expression in the Convention has been interpreted... But I think there is a more fundamental objection: that the appellants' argument, necessary to meet the objection of extra-territoriality, highlights the remoteness of their complaints from the true purview of article 2.[225]

He thus opted for the spatial model, and considered that the two soldiers were clearly not covered by it. Baroness Hale, on the other hand, stated the following:

> Nor have I much difficulty with the proposition that these soldiers were within the jurisdiction of the United Kingdom when they met their deaths. If Mr Baha Mousa, detained in a military detention facility in Basra, was within the jurisdiction, then a soldier serving under the command and control of his superiors must also be within the jurisdiction: see *R (Al-Skeini) v Secretary of State for Defence* [2007] UKHL 26; [2007] 3 WLR 33. The United Kingdom is in a better position to secure to him all his Convention rights, modified as their content is by the exigencies of military service, than it is to secure those rights to its detainees.[226]

Thus, for Baroness Hale, UK soldiers were within the UK's jurisdiction simply because they were under the command and control of their superiors—precisely the variation of the personal model that I am interested in here.

The question of extraterritorial application could not have been avoided in *Smith*, which also concerned a UK soldier who died in Iraq. An inquest was held, which established that he died from heatstroke. The soldier's mother, however, contended that her son's death warranted an Article 2 ECHR inquiry, which would not be limited only to the immediate cause of his death, but would also examine any possible systemic failures by the UK government—such as, for instance, not providing the soldiers with appropriate equipment or facilities—that could have led to her son's death. The question of extraterritorial application thus inevitably arose: was Private Smith protected by the ECHR while in Iraq, or not?

Now, crucially, Private Smith actually died on a UK military base. Per the UK government's concession in *Al-Skeini*, the House of Lords' quite dubious analogy between a military prison or base and an embassy, and the European Court's

[225] Ibid., para. 8(3) (per Lord Bingham). [226] Ibid., para. 60 (per Baroness Hale).

admissibility decision in *Al-Saadoon,* that fact alone would have brought Private Smith within the UK's jurisdiction. In other words, under the spatial model as applicable to *places* Private Smith would have been within the UK's jurisdiction, and therefore entitled to protection under Article 2 ECHR. However, issue was raised in the lower courts as to whether he would have been within the UK's jurisdiction even if he did *not* die on the base, or in any area under the UK's effective control, but essentially under the same circumstances.

In the High Court, Mr Justice Collins thought that the answer to this question could only be in the affirmative:

At the material time, the Coalition Provisional Authority in Iraq (following the ouster of the government of Saddam Hussain) had issued an order whereby the Multinational Force (of which British troops formed a part) should be 'immune from Iraqi legal process' and that all personnel should be 'subject to the exclusive jurisdiction of their sending States': see section 2 1) and 3). Thus the U.K.'s jurisdiction over its own nationals was clearly maintained. In any event, members of the armed forces remain at all times subject to the jurisdiction of the U.K. It would obviously be wholly artificial to regard a soldier sent to fight in the territory of another state as subject to the jurisdiction of that state.[227]

The judge accordingly applied a *personal* conception of jurisdiction, as applying to UK nationals and/or soldiers. Now, obviously, this position is of great practical relevance for UK military operations abroad. Many soldiers have lost their lives outside areas under UK effective control, and they would under Mr Justice Collins' ruling all at least in principle be entitled to the protections of Article 2 ECHR. The government therefore appealed his decision, arguing that Article 1 jurisdiction is exclusively territorial, and would apply only if a soldier died on a UK military base or a UK military hospital (as was in fact the case in *Smith*)—in other words, in a *place* under UK jurisdiction. The government relied, *inter alia*, on Lord Bingham's opinion in *Gentle*, which Mr Justice Collins considered to have been dicta.[228] The Court of Appeal affirmed the lower court's ruling in the following terms:

In our judgment, if it is permissible to answer the question posed by Lord Rodger, namely whether there was a sufficient link between Private Smith and the UK when he died, on the assumption for this purpose that he died outside the base or a hospital, in a broad and commonsense way, the answer is in our opinion plainly yes. As the judge put it at [9] of his addendum, there is a degree of artificiality in saying that a soldier is protected so long as he remains in the base or military hospital but that he is not protected as soon as he steps outside.

As the Commission succinctly puts it in its skeleton argument, there is no question but that members of the British armed forces are subject to UK jurisdiction wherever they are. They remain subject to UK military law without territorial limit and may be tried by court martial whether the offence is committed in England or elsewhere. They are also subject to the general criminal and civil law. Soldiers serve abroad as a result of and pursuant to the exercise of UK jurisdiction over them. Thus the legality of their presence and of their actions depends on their being subject to UK jurisdiction and complying with UK law. As a matter

[227] *R. (Smith) v. Secretary of State for Defence* [2008] EWHC 694 (Admin), para. 12.
[228] *R. (Smith) v. Secretary of State for Defence* [2009] EWCA Civ 441, paras 13–15.

of international law, no infringement of the sovereignty of the host state is involved in the UK exercising jurisdiction over its soldiers serving abroad.

It is not in dispute that the army owes soldiers a duty of care while they are in Iraq, as elsewhere. However, it does not follow from this that a soldier in Iraq is within the jurisdiction of the UK under the Convention. We stress that we are not saying that such a soldier is within the jurisdiction merely because the army may owe soldiers a duty of care. We recognise that that is a different question. However, it is accepted that a British soldier is protected by the HRA and the Convention when he is at a military base. In our judgment, it makes no sense to hold that he is not so protected when in an ambulance or in a truck or in the street or in the desert. There is no sensible reason for not holding that there is a sufficient link between the soldier as victim and the UK whether he is at a base or not. So too, if he is court-martialled for an act committed in Iraq, he should be entitled to the protection of article 6 of the Convention wherever the court martial takes place: see in this regard per Lord Brown in *Al-Skeini* at [140]. In these circumstances we accept the submission made by the Commission that there would have to be compelling reasons of principle for drawing a distinction for the purposes of jurisdiction under article 1 of the Convention between the soldier while at his base and the soldier when he steps outside it, at any rate so long as he is acting as a soldier and not, in the old phrase, on a frolic of his own. In our judgment, no such compelling reasons have been advanced on behalf of the Secretary of State.[229]

Like the High Court, the Court of Appeal adopted a personal conception of jurisdiction, based on the victim's status as a member of the armed forces. The Court quite rightly pointed out that it would be artificial and defy common sense to say that a UK soldier was protected by the ECHR while on a UK base, but would lose all protection once he stepped outside it.

The Supreme Court disagreed.[230] By a majority of 6 to 3 (Lady Hale and Lords Mance and Kerr dissenting), the justices found that mere membership in the armed forces was insufficient to establish a jurisdictional link for the purposes of Article 1 ECHR. This result is in my view ultimately correct. I am personally less than happy, however, with the reasoning of either the majority or the minority in the process of reaching this result. Because, of necessity, they had to start from *Bankovic*, with all its methodological flaws, the judgments of both the majority and the minority suffer from a great deal of otherwise needless conceptual confusion. In other words, because they are national judges applying a treaty primarily supervised by an international court, it is natural that the justices followed the European Court's approach, however flawed, and whatever their misgivings about the mess that Strasbourg has made.

Thus, in his judgment for the majority, Lord Phillips mainly emphasized considerations of effectiveness, by saying that it is for him simply inconceivable that ECHR states parties assumed upon themselves the obligation to conduct an Article 2-compliant investigation into the death of any of their soldiers abroad, nor have they behaved accordingly.[231] So did Lord Collins,[232] who added that there were no 'policy grounds for extending the scope of the Convention to armed forces

[229] *R. (Smith) v. Secretary of State for Defence* [2009] EWCA Civ 441, paras 27–9.
[230] *R. (Smith) v. Secretary of State for Defence* [2010] UKSC 29.
[231] Ibid., paras 54–61. [232] Ibid., paras 303–9.

abroad. On the contrary, to extend the Convention in this way would ultimately involve the courts in issues relating to the conduct of armed hostilities which are essentially non-justiciable.' Both justices stressed that there was no ECHR case law directly supporting the proposition that membership in the armed forces could qualify as 'authority and control' over an individual, and form a basis for Article 1 jurisdiction.

But, Bankovic and its lamentable offspring aside, is this explanation truly satisfactory? Is the majority really correct that Private Smith was not within the UK's jurisdiction under the personal model? Is it not true, as Lord Mance pointed out in his dissenting judgment, that it is 'commonsense' that UK soldiers are within the UK's jurisdiction?[233] Is it not true, as Lord Kerr thought,[234] that 'the United Kingdom brought its soldiers into Iraq; it not only asserted complete authority over them while they remained there, it explicitly excluded the exercise of authority over those soldiers by any other agency or state; and it has always been clear that soldiers remain subject to the laws of the UK during their service abroad'? And is it not true that, as noted by a commentator, '[t]here is a simple, intuitive point—if a soldier is always subject to UK law, wherever he may be, should he not thereby be also entitled to the protection of UK law (including the Human Rights Act)?'[235]

I would submit that this intuition is wrong, however commonsense it may seem at first glance. It not only assumes that the spatial model can be supplemented by a variant of the personal model based on legal links such as nationality or membership in the armed forces, but it rests on a confusion—confusion that we have seen time and again between the various meanings that the word 'jurisdiction' can have in general international law. We can recall that in its first northern Cyprus decision, the European Commission made the same category error, when it remarked that

> [t]he Commission further observes that nationals of a State, including registered ships and aircraft, are partly within its jurisdiction wherever they may be, and that authorised agents of a State, including diplomatic and consular agents and armed forces, *not only remain under its jurisdiction when abroad* but bring any other persons or property 'within the jurisdiction' of that State, to the extent that they exercise authority over such persons or property. Insofar as, by their acts or omissions, they affect such persons or property, the responsibility of the State is engaged.[236]

As I have explained above,[237] it is true that a state may legislate for its nationals and members of its armed forces even when they are abroad, and in fact all states do that to a greater or a lesser extent. It is also true that a state's soldiers are its *de jure* organs, and that their conduct performed in their official capacity is attributable to the state. Neither of these premises can, however, lead to the conclusion that a

[233] Ibid., para. 192. [234] Ibid., para. 322.

[235] See A. Bailin, 'Case preview—R (Smith) v Secretary of State for Defence—on appeal from [2009] EWCA Civ 441', *UKSCblog*, 13 March 2010, available at <http://ukscblog.com/case-preview-%E2%80%93-r-smith-v-secretary-of-state-for-defence-on-appeal-from-2009-ewca-civ-441>.

[236] *Cyprus v. Turkey* (dec.), App. Nos 6780/74 and 6950/75, 26 May 1975, at 136, para. 8 (emphasis added). Similarly, see *Saldaño v. Argentina*, paras 20–2.

[237] See Chapter II, Section 2.

state's soldiers, or diplomats, any other agents or even nationals are, regardless of their location, *always* within that state's jurisdiction within the meaning of the jurisdiction clauses of human rights treaties. This 'jurisdiction' is simply not the jurisdiction to prescribe in general international law. It is *not* about the application and extent of UK domestic law, exclusion from Iraqi law, or even some sort of mutuality of obligations.

Likewise, the Court of Appeal's statement that '[a]s a matter of international law, no infringement of the sovereignty of the host state is involved in the UK exercising jurisdiction over its soldiers serving abroad' is perfectly true. It is also perfectly beside the point. There is equally no infringement on the sovereignty of the host state if a UK soldier is prohibited by UK law or by human rights law from shooting an Iraqi national.[238]

If the personal model of jurisdiction is valid, its acceptance of nationality or membership in the armed forces as grounds of state jurisdiction would lead to an open embrace of double standards—and *there*, I may say, is where the intuitive appeal actually lies. Let us, for example, simply transplant *Smith* to the facts of *Al-Skeini*—imagine that one of the five applicant Iraqi citizens killed by UK troops on patrol was actually a dual UK/Iraqi national. Imagine further that the sixth applicant was not Baha Mousa, but a UK soldier killed by friendly fire by UK troops on patrol. The reasoning of the lower courts or the minority of the Supreme Court in *Smith* would lead to the result that the dual national and the UK soldier were within the UK's jurisdiction and thus entitled to the protection of Article 2 ECHR, while the other applicants were not, *even though all of them were killed in exactly the same way, in exactly the same place.*

In other words, the appeal of the minority approach lies precisely in the fact that we would feel it unjust if the UK government did not owe human rights obligations to its own soldiers in Iraq,[239] while our feelings may be more mixed when it comes to Iraqi civilians shot on patrol. We simply *care* more about 'our boys', than about their enemies or even their victims, and we thus feel them more deserving of protection. Starting from the assumption that *Bankovic* and *Al-Skeini* were correctly decided, as they had to, the minority's approach is ultimately about 'us' versus 'them', if not openly so, and is antithetical to the whole idea of universality. This is not to say that nationality or membership in the armed forces is totally irrelevant for the substantive application of human rights norms. This, however, only depends on the *content* of a very limited set of rights, but the application of the ECHR as such cannot depend on nationality or any other personal status.

[238] See above, Chapter III, Section 4.

[239] The irony is that before the progress of internalization of human rights has progressed to the extent that it has today, soldiers were considered to have waived the rights that they would have had as civilians, and had no recourse to judicial review merely because of their membership in the armed forces. Similarly, it has been a point of some dispute within UK academic literature whether 'public authorities' within the meaning of s. 6 of the Human Rights Act 1998 could themselves enjoy rights under the Act against *other* public authorities. See further D. Oliver, 'The Frontiers of the State: Public Authorities and Public Functions under the Human Rights Act', (2000) *PL* 476, at 491–2; H. Davis, 'Public Authorities as 'Victims' under the Human Rights Act', (2005) 64 *CLJ* 315.

In short, while it may be artificial to say that a UK soldier would be protected by the ECHR while on a UK base, but not off it, this is no more artificial than saying that an Iraqi national is protected by the ECHR while on a UK base or while in UK custody, but not on the streets of Basra,[240] as was the result in *Al-Skeini*. Thus, though Lord Phillips' judgment for the majority in *Smith* mainly rests on considerations of practicality or effectiveness to deny the extraterritorial applicability of the ECHR to UK soldiers abroad on account of their status, I would submit that, more fundamentally (and perhaps with a degree of irony), this is the only result consistent with universality as the normative foundation of the ECHR and all other human rights treaties—unless, of course, it is *Bankovic* and *Al-Skeini* themselves which were wrongly decided.

4. *Exercise of a legal power*

A more promising method of limiting the personal model of jurisdiction as authority and control over individuals may be to rely on the 'authority' part by requiring the exercise of some sort of purported legal power over the victim of the alleged human rights violation. One could possibly deduce this limitation from a number of cases dealing with what I will broadly term extraterritorial law enforcement, in which no issue of extraterritorial application of human rights treaties was raised, or the applicability of the relevant treaty was assumed. In my view, this limitation would also not rest on any grounds of principle, but let us first examine it in more detail.

Note, first, that as with control over territory, which the European Court in *Loizidou* thought could be either lawful or unlawful, so too would the personal model have to apply to both lawful and unlawful exercises of state power over an individual. Not only would any other solution open the door to abuse, but it would actually deny human rights protection precisely in those situations where they are being affected by an exorbitant exercise of state enforcement jurisdiction or power pure and simple as a matter of international law. Thus, for example, whether the state of Israel had a *right* in international law to abduct Adolf Eichmann from Argentina (it clearly did not) should have no bearing on whether Eichmann had any rights vis-à-vis Israel.

Indeed, if we take a look at the European Court cases that applied the personal model, we will see that they explicitly say that jurisdiction arises even when the state concerned acts unlawfully: 'a State may also be held accountable for violation of the Convention rights and freedoms of persons who are in the territory of another State but who are found to be under the former State's authority and control through its agents operating—*whether lawfully or unlawfully*—in the latter State'.[241] Likewise, in *Lopez Burgos* the Human Rights Committee noted that

[240] See also B. Silverstone, 'R. (on the application of Smith) v Oxfordshire Assistant Deputy Coroner: Human Rights and the Armed Forces', (2009) 4 *EHRLR* 566, esp. at 576.

[241] *Issa*, para. 71 (emphasis added). See also *Pad and others v. Turkey* (dec.), para. 53; *Isaak* (dec.), at A.2; *Solomou*, para. 45; *Andreou* (dec.), at A.3; *Ben El Mahi* (dec.), section 'The Law'.

> Article 2(1) of the Covenant places an obligation upon a State party to respect and to ensure rights 'to all individuals within its territory and subject to its jurisdiction', but does not imply that the State party concerned cannot be held accountable for violations of rights under the Covenant which its agents commit upon the territory of another State, *whether with the acquiescence of the Government of that State or in opposition to it.*[242]

This of course makes perfect sense. As I have explained above,[243] the sovereignty of the territorial state may or may not have already been violated by the extraterritorial act of another state, depending on whether valid consent was given or whether the latter state had some other authority for its actions, such as self-defence or a Security Council resolution. But the territorial state's sovereignty is in any event immaterial for the question of whether the individuals in that state are entitled to human rights protection against a third state.

Therefore, the personal model could not be limited only to *lawful* exercises of state power, as between two states and as a matter of general international law. But could it, perhaps, be limited to those extraterritorial acts which could be unlawful, or not, as a matter of international law, but would still be an exercise of a *legal* power? In other words, could the personal model of jurisdiction apply only when the state acts extraterritorially under colour of law—say if it arrests or abducts an individual in another state on the basis of a domestic judicial warrant?[244]

If we examined cases dealing with such instances of extraterritorial law enforcement, or law enforcement which is intraterritorial even though the victim of the alleged violation is outside the state's territory, we would see that human rights treaties were generally considered to be applicable. To my mind, the best such examples are judicial proceedings, whether as in absentia trials of a person located in another country, or indeed as extraterritorial trials by courts of one state sitting in another. Surely we would say in such cases that the state concerned would be bound by fair trial guarantees? When, for instance, the special Scottish criminal court convened at the neutral venue of Camp Zeist in the Netherlands to try those accused of the Lockerbie bombing, surely the defendants were entitled to Article 6 ECHR rights, even though they were not in the UK? After all, is there anything more jurisdiction-y, as it were, then putting someone on trial?

And, sure enough, if we looked, for example, at *Sejdovic v. Italy*,[245] we would see that neither the *Bankovic*-aware Grand Chamber of the European Court, nor any of the parties, thought that there was an Article 1 ECHR issue with respect to an applicant, a Montenegrin national, who was tried in Italy in absentia for murder *while he had absconded to Germany*. *Of course* he was entitled to Article 6 ECHR fair

[242] *Lopez Burgos*, para. 12.3 (emphasis added).

[243] See above, Chapter III, Section 4.

[244] This is, for example, what was argued by the respondent governments in *Bankovic*, para. 36:

> As to the precise meaning of 'jurisdiction', they suggest that it should be interpreted in accordance with the ordinary and well-established meaning of that term in public international law. The exercise of 'jurisdiction' therefore involves *the assertion or exercise of legal authority, actual or purported*, over persons owing some form of allegiance to that State or who have been brought within that State's control (emphasis added).

[245] See, e.g., *Sejdovic v. Italy* [GC], App. No. 56581/00, Judgment, 1 March 2006.

trial rights, even though he was at the time certainly *not* within Italy's 'essentially territorial' jurisdiction. Note that in this example the possible violation of the applicant's rights would not have been extraterritorial, as the trial itself took place in Italy, but that it is the *victim* who needs to be within the state's jurisdiction at the time of the violation for the ECHR to apply—at least under either the spatial or the personal model.[246]

Similarly, in *Martin v. UK*[247] the applicant was the son a UK soldier serving in Germany who was accused of murder, and who was under UK law subject to military law as a family member residing with a member of the armed forces, and was tried by a British court-martial sitting in Germany. Again, neither the parties nor the Court raised any questions about whether he would be protected by the ECHR vis-à-vis the UK, *even though the trial itself took place in Germany*, and indeed the Court did find a violation of Article 6, on the grounds of a lack of independence and impartiality of the tribunal.[248]

To give some more relatively random examples, in *Mullai and Others v. Albania*,[249] the applicants were seven Albanian nationals, as well as an Albanian company, who were embroiled in a dispute regarding a building permit. In particular, the applicants alleged that Albanian authorities failed to enforce a final court judgment which recognized the validity of their building permit, and this was in violation of Article 6 and Article 1 of Protocol 1 to the ECHR. Crucially for our purposes, some of the individual applicants were actually not living in Albania, but in Italy and the United States. Again, however, nobody thought that these particular applicants were not within Albania's jurisdiction. Likewise, in *Vrbica v. Croatia*,[250] the applicant was a Montenegrin national living in Montenegro who complained of the refusal of Croatian courts to enforce a judgment given by a Montenegrin court and subsequently recognized in Croatia, which awarded him damages against two Croatian companies for injuries he sustained in a traffic accident. Again, even though the applicant was outside Croatia at the relevant time, neither the Court nor the parties thought that there was an Article 1 issue to be considered.

246 See also above, Chapter I, Section 2 on the definition of extraterritorial application.

247 *Martin v. United Kingdom*, App. No. 40426/98, Judgment, 24 October 2006.

248 Of interest may be a similar US case, where a US court sitting in West Berlin was trying an East German national for hostage-taking and the hijacking of an airplane which he used to defect to West Germany. In this case, however, the extraterritoriality issue *was* explicitly raised, with the US prosecutors arguing that the US Constitution and its guarantees of due process and jury trial simply did not apply to a US court sitting in West Berlin, which was even obliged to follow instructions issued by the State Department. The presiding judge was quite understandably appalled, and promptly went on to empanel a jury composed of West Berliners. The judge later wrote a book about the trial, which was then made into a decent, if slightly preachy low-budget movie (*Judgment in Berlin*, 1988), remarkably starring Martin Sheen and Sean Penn. See also K. Raustiala, *Does the Constitution Follow the Flag? The Evolution of Territoriality in American Law* (Oxford University Press, 2009), at 151–4.

249 *Mullai and Others v. Albania*, App. No. 9074/07, Judgment, 23 March 2010.

250 *Vrbica v. Croatia*, App. No. 32540/05, Judgment, 1 April 2010.

Nor are such cases limited to Strasbourg. Consider, for example, the well-known *Gueye et al. v. France* case before the Human Rights Committee,[251] in which the applicants were more than 700 Senegalese nationals, all of them residing in Senegal, who had served in the French army prior to Senegal's independence. They alleged that France discriminated against them on the basis of race or nationality because it had enacted a law reducing their pension benefits when compared to retired French soldiers of French nationality. The Committee duly found a violation of Article 26 ICCPR, but again neither it nor France thought that the applicants were not protected by the ICCPR, even though they resided in Senegal when the relevant law was enacted.[252]

Common to all of these cases is that even though the applicants were all located outside the relevant state's territory, they were subjected by that state to some form of legal process. None of these cases could be regarded as being within the relevant state's jurisdiction pursuant to any variant of the spatial model, however attenuated, as the victims of the human rights violations were not located in any area or place under that state's control. Likewise, in all these cases we would feel that it would be manifestly unjust to deny the applicability of human rights treaties. It would simply be inconceivable for the ECHR not to apply to in absentia trials in which the accused has fled to another country, or, say, to the seizure of assets by the UK of a tax-evading UK national residing in Monaco. It is *obvious* that the Convention should apply, and this is why the issue of extraterritorial applicability of human rights treaties is practically never raised in such situations.[253]

But, other than our intuition, on what grounds could we justify the extraterritorial applicability of human rights treaties in such situations, where the victim is not

[251] *Ibrahima Gueye et al. v. France*, Communication No. 196/1983, U.N. Doc. Supp. No. 40 (A/44/40) at 189 (1989). While examining the merits, the Committee only cryptically observed that 'the authors are not generally subject to French jurisdiction, except that they rely on French legislation in relation to the amount of their pension rights'. Ibid., para. 9.4.

[252] Similarly, see *El Orabi v. France* (dec.), App. No. 20672/05, 20 April 2010, where the European Court declared inadmissible the application of an Algerian national living in Algeria who was married to an Algerian former member of the French armed forces, whose request for a survivor's pension was denied. The Court declared the application manifestly ill-founded, but it did not say that there was any Article 1 issue in the case, even though both the applicant and her husband were residing in Algeria at the time of her husband's death. See also *Carson and Others v. the United Kingdom* [GC], App. No. 42184/05, Judgment, 16 March 2010, similarly dealing with UK pensions of persons residing outside the UK, where likewise no jurisdiction issue was raised.

[253] We could also consider the situation where it is not the *state* which initiates legal proceedings against an individual outside its territory, but when it is rather that individual who does so. Thus, in *Markovic and Others v. Italy* [GC], App. No. 1398/03, Judgment, 14 December 2006, dealing with domestic proceedings initiated in Italy by the victims of the NATO bombing of the Belgrade TV station at issue in *Bankovic*, the European Court held that (para. 54):

> [e]ven though the extraterritorial nature of the events alleged to have been at the origin of an action may have an effect on the applicability of Article 6 and the final outcome of the proceedings, it cannot under any circumstances affect the jurisdiction *ratione loci* and *ratione personae* of the State concerned. If civil proceedings are brought in the domestic courts, the State is required by Article 1 of the Convention to secure in those proceedings respect for the rights protected by Article 6. The Court considers that once a person brings a civil action in the courts or tribunals of a State, there indisputably exists, without prejudice to the outcome of the proceedings, a 'jurisdictional link' for the purposes of Article 1.

within the state's jurisdiction conceived of territorially? One answer, which I prefer, would be to simply say that the state jurisdiction threshold does not apply to potential violations of states parties' negative obligation to respect human rights. Otherwise, we would have to accept a personal conception of jurisdiction as authority and control over individuals, in these particular cases exercised as a *legal* authority or power by the state over the individual concerned.

Consider, in that regard, *Stephens v. Malta*, an absolutely fascinating case.[254] The applicant was a UK national living in Spain, who was suspected by Maltese authorities of having conspired with other persons in Spain to transport drugs to Malta. A warrant for his arrest was issued by a Maltese court, and he was detained in Spain following a request for his extradition. While still awaiting extradition in Spain, the applicant retained counsel in Malta who challenged the lawfulness of the arrest warrant before Maltese courts, *inter alia* on the grounds that the court which issued the warrant lacked jurisdiction over a non-Maltese national accused of committing a crime outside Malta. After several judicial instances, the original arrest warrant was found to have been procedurally defective, while in Spain the applicant was released on bail. However, after a new request for extradition was filed, the applicant was rearrested, extradited to Malta, and ultimately convicted on the criminal charges against him. Before the European Court, he challenged the lawfulness of his detention *in Spain* pending extradition—but he did so by claiming that *Malta* had violated Article 5(1) ECHR.

Of course, the merits of his case aside, the issue is precisely that until his extradition the applicant did not set foot on Malta. In purely territorial terms, as a UK national residing and then detained in Spain by Spanish authorities, he could not possibly have been within Malta's jurisdiction. While, as in other cases we have seen above, the parties did not raise the question of the ECHR's extraterritorial application, the Court apparently realized the problem and did so *proprio motu*. The Court first set out some general principles:

> The exercise of jurisdiction is a necessary condition for a Contracting State to be able to be held responsible for acts or omissions imputable to it which give rise to an allegation of the infringement of rights and freedoms set forth in the Convention (see *Ilaşcu and Others v. Moldova and Russia*, [GC], no. 48787/99, § 311, ECHR 2004 - . . .).
>
> According to established case-law Article 1 of the Convention must be considered to reflect this ordinary and essentially territorial notion of jurisdiction, other bases of jurisdiction being exceptional and requiring special justification in the particular circumstances of each case. The Court refers to its case-ląw on the notion of "jurisdiction" and how that notion has been interpreted and applied in different contexts (see *Ilaşcu and Others* [GC], no. 48787/99, cited above; *Banković and Others v. Belgium and 16 Other Contracting States* (dec.) [GC], no. 52207/99, ECHR 2001-XII; *Assanidzé v. Georgia*, [GC], no. 71503/01, ECHR 2004 - . . . ; *Soering v. the United Kingdom*, 7 July 1989, Series A no. 161; *Cruz Varas and Others v. Sweden*, 20 March 1991, Series A no. 201; *Vilvarajah and Others v. the United Kingdom*, 30 October 1991, Series A no. 215; *Loizidou v. Turkey* (preliminary objections), 23 March 1995, Series A no. 310; *Loizidou v. Turkey*, 18 December 1996, *Reports of*

[254] *Stephens v. Malta (No. 1)*, App. No. 11956/07, Judgment, 31 March 2009.

Judgments and Decisions 1996-VI; *Issa and Others v. Turkey*, no. 31821/96, 16 November 2004; *Behrami and Behrami v. France and Saramati v. France, Germany and Norway* (dec.) [GC], nos. 71412/01 and 78166/01, 2 May 2007; *Drozd and Janousek v. France and Spain*, 26 June 1992, Series A no. 240, and *Hess v. the United Kingdom*, no. 6231/73, 28 May 1975, Decisions and Reports (DR) no. 2, p. 72).[255]

It then applied these general principles—to the extent that it actually identified any of them—to the facts of the case:

The Court considers that for the purposes of the examination of this application and in view of its conclusions in respect of the applicant's various complaints, it suffices to consider the matter of Malta's jurisdiction solely in respect of the Article 5 complaints.

The Court notes that the applicant was under the control and authority of the Spanish authorities in the period between his arrest and detention in Spain on 5 August 2004 and his release on bail on 22 November 2004. In so far as the alleged unlawfulness of his arrest and detention is concerned, it cannot be overlooked that the applicant's deprivation of liberty had its sole origin in the measures taken exclusively by the Maltese authorities pursuant to the arrangements agreed on by both Malta and Spain under the European Convention on Extradition.

By setting in motion a request for the applicant's detention pending extradition, the responsibility lay with Malta to ensure that the arrest warrant and extradition request were valid as a matter of Maltese law, both substantive and procedural. In the context of an extradition procedure, a requested State should be able to presume the validity of the legal documents issued by the requesting State and on the basis of which a deprivation of liberty is requested. It is to be noted that in the instant case the arrest warrant had been issued by a court which did not have the authority to do so, a technical irregularity which the Spanish court could not have been expected to notice when examining the request for the applicant's arrest and detention. Accordingly, the act complained of by Mr Stephens, having been instigated by Malta on the basis of its own domestic law and followed-up by Spain in response to its treaty obligations, must be attributed to Malta notwithstanding that the act was executed in Spain.[256]

First, note how the Court's reasoning in these three paragraphs has absolutely nothing to do with the general principles that it had considered before. Indeed, the Court seemed to have felt that there was jurisdiction on the facts of the case since the Convention simply *had* to apply[257]—but then it realized that whatever it did, this case simply does not fit any of its prior case law on extraterritoriality. Therefore, the Court decided to cite and throw *everything* in, the kitchen sink and all, from *Bankovic* and *Behrami* to *Soering* and *Issa*. Secondly, note also how the Court says that 'it suffices to consider the matter of Malta's jurisdiction solely in respect of the Article 5 complaints', thus seemingly being at odds with the Grand Chamber's regime integrity-inspired holding in *Bankovic* that 'the wording of Article 1 does not provide any support for the applicants' suggestion that the positive obligation in Article 1 to secure "the rights and freedoms defined in Section I of this

[255] *Stephens v. Malta (No. 1)*, App. No. 11956/07, Judgment, paras 48 and 49.
[256] Ibid., paras 50–2.
[257] This of course rests on the assumption that the applicability of *all* of the Convention rests on a single threshold, that of state jurisdiction, which need not be the case.

Convention" can be divided and tailored in accordance with the particular circumstances of the extra-territorial act in question'.[258] Indeed, if anything, the Court applied precisely the kind of 'cause-and-effect' notion of jurisdiction that *Bankovic* so pointedly rejected—Malta's warrant was the cause of the applicant's arrest in Spain, and therefore the applicant was within Malta's jurisdiction.

Thirdly, the Court pins its analysis on quite sensible grounds of policy—that Spanish courts could reasonably have relied on the validity of the Maltese arrest warrant, and could not have been expected to know of its procedural defects under Maltese law. That said, the Court arrives at the remarkable conclusion that the detention of the applicant, 'having been instigated by Malta on the basis of its own domestic law and followed-up by Spain in response to its treaty obligations, must be attributed to Malta notwithstanding that the act was executed in Spain'.

By so deciding, the Court was presumably trying to explain why it was Malta, and not Spain, that was the proper respondent in the case, even though the applicant was at all times during his detention under Spain's 'control and authority'. And though one cannot say that the Court's approach is unreasonable as a matter of policy, it is entirely at odds with the general rules on state responsibility to say that the applicant's detention was not attributable to Spain, but to Malta. The detention was carried out by the *de jure* organs of Spain, who were not placed at Malta's disposal nor were acting under Malta's control.[259] What *was* attributable to Malta was the defective arrest warrant issued by its court, and the request for extradition that it transmitted to Spain. Spain then may have had the legal *obligation* to comply with that request, but it could just as easily have refused to do so; the applicant's subsequent detention was still most certainly attributable to it.

Again, the ultimate result that the Court reached was perfectly sensible. Had it been otherwise, Malta would have had human rights obligations towards the applicant only once his extradition was completed, thus putting the period of unlawful detention outside the scope of the ECHR. This still does not mean, however, that the Court produced a satisfactory explanation as to *why* the ECHR protected the applicant extraterritorially.

As I have said above, I see only two possible explanations. First, the Court could have said, *contra Bankovic*, that the ECHR could apply even without Malta having jurisdiction over the applicant—but that was not going to happen, for obvious reasons. Secondly, the Court could have openly adopted the personal model, and said that the applicant was within the jurisdiction of both Malta and Spain, with regard to the former because Malta exercised a *legal* authority over him.[260]

[258] *Bankovic*, para. 75. [259] See Arts 4, 6, and 8 ILC ASR.

[260] See, in that regard, *Lichtensztejn v. Uruguay*, Communication No. 77/1980, UN Doc. CCPR/C/OP/2 at 102 (1990), which concerned the refusal by Uruguayan authorities to issue a passport to an Uruguayan national residing in Mexico. The question, therefore, is whether the individual was within Uruguay's jurisdiction, and the Human Rights Committee found that he was (para. 6.1):

> When considering the admissibility of the communication, the Human Rights Committee did not accept the State party's contention that it was not competent to deal with the communication because the author did not fulfil the requirements of article 1 of the Optional Protocol. In that connection, the Committee made the following observations:

In other words, if we operate within the personal model, that is, if we accept that 'jurisdiction' in human rights treaties means not only effective control over territory, but also authority and control over individuals, then we must also accept that such authority over individuals can be exercised by a state simply by putting in motion a legal process, without having physical control of the person in question.[261] The bigger question, however, is whether the personal model could be *limited* to such legal authority or power.

In my view, it clearly cannot, for the same reason that it could not be limited to the *lawful* exercise of state power under general international law—doing so would open the door to abuse. Any of the possible violations that we have just seen that the state has committed under colour of law it could have done through brute, unadulterated force. Instead of putting someone on trial in absentia, or initiating extradition proceedings, it could simply have sent its agents to abduct him, as the Eichmanns, Öcalans, and Carloses of this world can attest. Instead of taking somebody's house through a legal procedure, it could have done so simply by burning it down. And instead of France reducing the pensions of Senegalese soldiers by passing a law, it could have simply stopped the payments by fiat.

Or, consider the following example: an ECHR state party organizes parliamentary elections and allows its citizens living abroad to vote by post or by going to designated polling stations in their host state—something that states are not obliged to do in principle,[262] but many states in fact do. It then decides, however, to

article 1 applies to individuals subject to the jurisdiction of the State concerned who claim to be victims of a violation by that State of any of the Covenant rights. The issue of a passport to a Uruguayan citizen is clearly a matter within the jurisdiction of the Uruguayan authorities and he is 'subject to the jurisdiction' of Uruguay for that purpose. Moreover, a passport is a means of enabling him 'to leave any country, including his own', as required by article 12 (2) of the Covenant. Consequently, the Committee found that it followed from the very nature of that right that, in the case of a citizen resident abroad, article 12 (2) imposed obligations both on the State of residence and on the State of nationality and that, therefore, article 2 (1)of the Covenant could not be interpreted as limiting the obligations. of Uruguay under article 12 (2)to citizens within its own territory.

261 A commentator has suggested that factual conceptions of the word 'jurisdiction' in human rights treaties, such as the one that I advocate in this study, cannot explain cases in which jurisdiction was found to exist because of the exercise of a legal power by a state, such as the issuance of a passport as in the *Lichtensztejn* case cited in n. 260 above. In his view, such cases can only be explained if the word 'jurisdiction' referred both to a state's factual control over territory and individuals and to its lawful competence in international law—see H. King, 'The Extraterritorial Human Rights Obligations of States', (2009) 9 *HRLR* 521, at 537 *et seq*. To my mind, such an approach is unsatisfactory for two simple reasons. First, it fails to explain how *semantically*, as a matter of treaty interpretation, a single word—'jurisdiction'—can mean two radically different things, both a state's act affecting an individual and its *right* to so act. Secondly, if the personal model of jurisdiction is accepted as valid, it is entirely unnecessary for its application to require that a state exercise its *lawful* competence in international law to prescribe legal rules and enforce them. It is enough, for example, that a state has subjected an individual to a trial in absentia. Whether it had a *right* to do so in international law—in other words, whether it violated, or not, the sovereignty of another state when it did so—is to my mind perfectly irrelevant. After all, why should the applicability of fair trial guarantees depend on whether the in absentia trial was an exorbitant exercise of the state's jurisdiction to prescribe or enforce? And if it does not, then the enquiry into the lawfulness, *vel non*, of the state's act vis-à-vis a third state becomes entirely beside the point.

262 See, e.g., *X. and Association Y. v. Italy* (dec.), App. No. 8987/80, 6 May 1981.

blatantly discriminate among its citizens living abroad by saying, for example, that only those of a certain race, ethnicity, or gender may vote, and passes a law to that effect. Would these people residing extraterritorially be protected by the ECHR and its equality guarantee? Well *of course* they would, either because we consider the jurisdiction threshold not to apply to a state's violation of its negative obligations, or because we adopt the personal model of jurisdiction, and say that the state has exercised its authority over the individuals concerned, even though it had not laid a hand on them physically.

But what if the state in question discriminated in exactly the same way, yet was smart enough not to pass a *law* to that effect and *in fact* prevented the undesirables from voting while having the most pristine electoral legislation imaginable. Would the result be any different? Would the ECHR not apply just because the state has creatively managed to avoid acting under the colour of law, but accomplished the same invidious ends through other means?

Asking this question, I think, is to answer it. To the extent that the personal model of jurisdiction is both valid and necessary, an exercise of a legal power or authority by a state over an individual outside its territory would *suffice* to satisfy the jurisdictional threshold. *Limiting* the personal model to such purported exercises of legal power, however, would be entirely arbitrary, and would only serve to undermine the rule of law by creating a perverse incentive for states to act outside their own legal system if they wish to violate human rights.

D. The personal model collapses

What, then, have we learned about the personal model? It cannot be limited to physical custody. It cannot be limited on the basis of nationality or some special status of the victim, or indeed of the perpetrator. It cannot be limited only to lawful exercises of state power over individuals, nor to extraterritorial acts to which the host state consents, nor indeed to acts committed under the colour of law. It cannot, in short, be limited on the basis of *any* non-arbitrary criterion. 'Authority and control over individuals' as a basis for state jurisdiction simply boils down to the proposition that a state has obligations under human rights treaties towards all individuals whose human rights it is able to violate.[263]

In other words, the main feature of the personal model of jurisdiction—its ability to cover individuals who would be unprotected by the spatial model—is also its main fault. It quite literally collapses, and serves no useful purpose as a threshold for the application of human rights treaties. Unless the personal model is limited artificially on the basis of some essentially arbitrary criteria, there *is* no threshold.

[263] See, in that regard, Scheinin, above note 172, at 76–8. This does not mean that a state's authority and control would exist whenever a state could *affect* an individual through its actions in a more general way (e.g. by cutting exports or development aid). Rather, the act in question would need to be capable of directly violating the treaty—see Lawson, above note 172, at 95, 104.

This, I think, was well realized by the Grand Chamber in *Bankovic* when it rejected a 'cause-and-effect' notion of jurisdiction, which ultimately is what the personal model boils down to. The same sentiment was shared by Lord Brown in *Al-Skeini*, when he remarked that

> [a]lternatively it would stretch to breaking point the concept of jurisdiction extending extra-territorially to those subject to a state's 'authority and control'. It is one thing to recognise as exceptional the specific narrow categories of cases I have sought to summarise above; it would be quite another to accept that whenever a contracting state acts (militarily or otherwise) through its agents abroad, those affected by such activities fall within its article 1 jurisdiction. Such a contention would prove altogether too much. It would make a nonsense of much that was said in *Bankovic*, not least as to the Convention being 'a constitutional instrument of European public order', operating 'in an essentially regional context', 'not designed to be applied throughout the world, even in respect of the conduct of contracting states' (para 80). It would, indeed, make redundant the principle of effective control of an area: what need for that if jurisdiction arises in any event under a general principle of 'authority and control' irrespective of whether the area is (a) effectively controlled or (b) within the Council of Europe?[264]

Put aside, for one moment, the understandable fear of judges that the application of the personal model up to its logical conclusion would render them unable to avoid dealing with the merits of legally and factually complex and politically extremely controversial cases. Even if that fear, based as it is on considerations of effectiveness, could be addressed in a satisfactory way—as I think it can be—under the personal model jurisdiction clauses in human rights treaties would essentially be rendered meaningless. Any act capable of violating a person's rights as a substantive matter would also be capable of bringing that person within the state's jurisdiction, and no jurisdiction issue would therefore actually arise. And, as Lord Brown well notes, if the personal model is applied consequentially, what would be the use for the spatial model?

A further difficulty with the personal model is its lack of an adequate textual basis in at least some of the treaties. While the words individuals 'within its jurisdiction' and 'subject to its jurisdiction' can reasonably be interpreted as denoting a power-based relationship between the state and the individual, as we have seen such an interpretation is textually simply impossible for some jurisdiction clauses, such as the CAT's 'any territory under its jurisdiction'. These words unambiguously denote a relationship between the state and the territory and only *indirectly* between the state and the individuals located in that territory. No amount of creative interpretation can plausibly change this result.

But again, more important than the textual difficulties is the fact that the personal model collapses. Because it does, it is unlikely to be endorsed wholeheartedly by courts, especially those in which the stakes are the highest, as in Strasbourg. Rather, they tend to use it in a wholly unpredictable and unprincipled fashion, to carve limited exceptions out of *Bankovic* and its territorial model of jurisdiction one

264 *Al-Skeini HL*, para. 127 (per Lord Brown).

sliver at a time, in cases in which it would be morally difficult not to do so, and yet not too inconvenient politically. Do they have some other choice, other than running headlong into utopia? Is there a better way?

4. A Third Model: Territorial Jurisdiction and the Distinction Between Positive and Negative Obligations

A. Universality unbound

The driving force behind the personal model is easy to spot—how could it be justified against the normative baseline of universality that a state which is in full control of its own agents is dispensed from respecting the human rights of persons whose lives its agents affect, merely on account of their location? Yet it is precisely this same driving force which ultimately leads to the collapse of the personal model of jurisdiction. Why then not simply say that states have to comply with their negative obligation to respect human rights in all circumstances, regardless of whether they exercise jurisdiction over a particular territory or area?

There is truly a fundamental distinction between a state's obligation to respect human rights, which requires it only to *refrain* from infringing upon an individual's rights without adequate justification, and its duty to *secure* or *ensure* human rights to the inhabitants of a certain territory, which, in certain circumstances, compels the state to prevent the violations of human rights even by private parties.[265] According to the Human Rights Committee,

> The article 2, paragraph 1, obligations are binding on States [Parties] and do not, as such, have direct horizontal effect as a matter of international law. The Covenant cannot be viewed as a substitute for domestic criminal or civil law. However the positive obligations on States Parties to ensure Covenant rights will only be fully discharged if individuals are protected by the State, not just against violations of Covenant rights by its agents, but also against acts committed by private persons or entities that would impair the enjoyment of Covenant rights in so far as they are amenable to application between private persons or entities. There may be circumstances in which a failure to ensure Covenant rights as required by article 2 would give rise to violations by States Parties of those rights, as a result of States Parties' permitting or failing to take appropriate measures or to exercise due diligence to prevent, punish, investigate or redress the harm caused by such acts by private persons or entities.[266]

[265] See, e.g., T. Buergenthal, 'To Respect and to Ensure: State Obligations and Permissible Derogations', in L. Henkin (ed.), *The International Bill of Rights: The Covenant on Civil and Political Rights* (Columbia University Press, 1981), at 72–91. P. van Dijk et al., *Theory and Practice of the European Convention on Human Rights* (Intersentia, 4th edn, 2006), at 13; Nowak, above note 170, at 37–41. In the context of socio-economic rights, Eide has in particular developed a more detailed typology of rights, separating the broader obligation to ensure into the obligations to protect and fulfil—see generally 'The Right to Adequate Food as a Human Right', Final Report submitted by Asbjørn Eide, Special Raporteur, UN Doc. E/CN.4/Sub.2/1987/23 (1987). I do not believe that such a disaggregated typology would be of direct relevance for the issue of extraterritorial application, so I will not address it any further.

[266] Human Rights Committee, General Comment No. 31, para. 8.

The positive obligation of a state to ensure the human rights of persons within its jurisdiction from violations by private parties is not absolute, as states are neither omniscient nor omnipotent. What they must do is to exercise due diligence, i.e. to take all measures reasonably within their power in order to prevent violations of human rights. As held by the Inter-American Court,

> ... in principle, any violation of rights recognized by the Convention carried out by an act of public authority or by persons who use their position of authority is imputable to the State. However, this does not define all the circumstances in which a State is obligated to prevent, investigate and punish human rights violations, nor all the cases in which the State might be found responsible for an infringement of those rights. An illegal act which violates human rights and which is initially not directly imputable to a State (for example, because it is the act of a private person or because the person responsible has not been identified) can lead to international responsibility of the State, not because of the act itself, but because of the lack of due diligence to prevent the violation or to respond to it as required by the Convention.[267]

In order to be realistically complied with, the obligation to respect human rights requires the state to have nothing more than control over the conduct of its own agents. It is the positive obligation to secure or ensure human rights which requires a far greater degree of control over the area in question, control which allows the state to create institutions and mechanisms of government, to impose its laws, and punish violations thereof accordingly.

This is then what my proposed third model would amount to: the notion of jurisdiction in human rights treaties would be conceived of only territorially, as *de facto* effective overall control of areas and places. Having now looked at the text of the relevant treaties and the treaty practice of states generally, as well as at the case law, this is indeed the most natural way of interpreting the term 'jurisdiction'. This threshold would, however, apply only to the state's obligation to secure or ensure human rights, but *not* to its obligation to respect human rights, which would be territorially unbound.[268]

I should not be taken as arguing that the state is in exactly the same position with respect to its negative obligations when it acts inside its territory in conditions of normalcy, as when it acts outside it, particularly in wartime or other extraordinary situations. What I *am* arguing is that the best way to address these difficulties, and to fully take into account all considerations of effectiveness, is *not* in artificially imposing a threshold for the state's negative obligations, but in applying the *substance* of these obligations to the facts at hand with a greater degree of flexibility.[269] As Lord Justice Sedley aptly put it in *Al-Skeini*:

[267] *Velasquez Rodriguez Case*, Judgment of 29 July 1988, Inter-Am.Ct.H.R. (Ser. C) No. 4 (1988), para. 172.

[268] Similarly, see E. Roxstrom, M. Gibney, and T. Einarsen, 'The NATO Bombing Case (*Bankovic et al. v. Belgium et al.*) and the Limits of Western Human Rights Protection', (2005) 23 *B.U. Int'l L.J.* 55, at 72 *et seq*; J. Cerone, 'Jurisdiction and Power: The Intersection of Human Rights Law & the Law of Non-International Armed Conflict in an Extraterritorial Context', (2007) 40 *Israel L Rev* 72, at 122–3.

[269] See above, Chapter III, Section 10.

> [I]t is not an answer to say that the UK, because it is unable to guarantee everything, is required to guarantee nothing. The question is whether our armed forces' effectiveness on the streets [of Basra] in 2003–4 was so exiguous that despite their assumption of power as an occupying force they lacked any real control of what happened from hour to hour in the Basra region. My own answer would be that the one thing British troops did have control over, even in the labile situation described in the evidence, was their own use of lethal force. Whether they were justified in using it in the situations they encountered, of which at least four of the cases before us are examples, is precisely the subject of the inquiry which the appellants seek. It is in such an inquiry that the low ratio of troops to civilians, the widespread availability of weapons and the prevalence of insurgency would fall to be evaluated.[270]

He did not, however, think that this approach was open to him under existing Strasbourg jurisprudence. And he was right—it is not. Adopting the third model would require a radical rethink of Strasbourg's approach, and to a lesser extent also that of other human rights bodies. I am well aware that this makes the third model less attractive. But having now extensively examined the European Court's convoluted case law on extraterritoriality, would it really be such a bad thing to put it on some sensible, principled foundation? I think not. It bears emphasizing, however, that even in regular peacetime conditions the European Court has been prepared to approach positive obligations flexibly.[271] Thus, in the context of the right to life it remarked that

> [f]or the Court, and bearing in mind the difficulties involved in policing modern societies, the unpredictability of human conduct and the operational choices which must be made in terms of priorities and resources, such an obligation must be interpreted in a way *which does not impose an impossible or disproportionate burden on the authorities*. Accordingly, not every claimed risk to life can entail for the authorities a Convention requirement to take operational measures to prevent that risk from materialising. . . . In the opinion of the Court where there is an allegation that the authorities have violated their positive obligation to protect the right to life in the context of their above-mentioned duty to prevent and suppress offences against the person (see paragraph 115 above), it must be established to its satisfaction that the authorities knew or ought to have known at the time of the existence of a real and immediate risk to the life of an identified individual or individuals from the criminal acts of a third party and that they *failed to take measures within the scope of their powers which, judged reasonably, might have been expected to avoid that risk.*[272]

Let me now make my argument in more detail. I will first try to establish whether the text of the relevant treaties could accommodate the third model. While I am not saying that my argument is free of all difficulties, I believe that such difficulties can be overcome, particularly if the negative obligation to respect human rights is read into some of the treaties implicitly. Further, as I will explain, I am not advocating a crude distinction between negative and *all* positive obligations.

[270] *Al-Skeini CA*, para. 197 (per Sedley LJ).
[271] See Lawson, above note 172, at 106.
[272] *Osman v. United Kingdom* [GC], App. No. 23452/94, Judgment, 28 October 1998, para. 116 (emphasis added).

Rather, some positive obligations are procedural or prophylactic in nature, tied solely to the state's compliance with its negative obligation to respect human rights. In my view, it is only the wide-ranging obligation to *secure* or *ensure* human rights, particularly from acts of third parties, that requires a jurisdictional threshold. Finally, I will try to prove that, when compared to its rivals, this third model actually provides the best balance between universality and effectiveness, if at a cost.

B. Textual interpretation and implicit negative obligations

Is the third model even possible under the text of the relevant human rights treaties? Let us again look at what they say. Article 1 ECHR thus provides that '[t]he High Contracting Parties shall *secure to everyone within their jurisdiction* the rights and freedoms defined in Section I of this Convention' (emphasis added). All this article says, as I read it, is that the positive obligation to *secure* human rights is contingent on state jurisdiction—something that I certainly do not dispute. Except in the descriptive heading of the article, it says nothing about the negative obligation to *respect* human rights, which does not mean that this obligation does not exist. Indeed, many human rights treaties with jurisdiction clauses explicitly refer only to the positive obligations of the states parties.

For example, all of the nine jurisdiction clauses in the CAT denote the scope of a positive obligation. Article 2(1) provides that '[e]ach State Party shall take effective legislative, administrative, judicial or other measures *to prevent* acts of torture in any territory under its jurisdiction' (emphasis added); Articles 5(1)(a), 5(2), and 7(1) require states to criminalize and prosecute torture on the basis of the territoriality and universality principles; Articles 11, 12, and 13 require states to review interrogation rules for the purpose of preventing torture, ensure an effective investigation, and provide remedies to individuals; while Article 16 requires states to prevent cruel, inhuman, and degrading treatment, all in 'any territory under its jurisdiction'. Notably, these provisions do not *expressly* say that the state as such has the negative obligation not to torture individuals or treat them inhumanely, probably because it was obvious that such an obligation existed, yet it would have been somewhat impolitic to spell it out. Similarly, Article 3 CERD provides that 'States Parties particularly condemn racial segregation and apartheid and undertake *to prevent, prohibit and eradicate* all practices of this nature in territories under their jurisdiction' (emphasis added).

The treaties which do explicitly mention the obligation to respect are those whose jurisdiction clauses are to a greater or lesser extent based on Article 2(1) ICCPR, which provides that '[e]ach State Party to the present Covenant undertakes to *respect and to ensure* to all individuals *within its territory and subject to its jurisdiction* the rights recognized in the present Covenant'. The grammatically more natural interpretation of this wording would be to say that the jurisdiction threshold applies to both the obligation to respect and to the obligation to ensure. This interpretation is however not the only plausible one—Article 2(1) could also be read as limiting the jurisdiction threshold only to the obligation to ensure, which would be consistent with its object and purpose.[273] Article 2(1) CRC and Article 7 of the Migrant Workers Convention are similar in this respect. Notably, however, Article 1(1) ACHR stipulates that the 'States Parties to this Convention undertake to respect the rights and freedoms recognized herein *and* to ensure to all persons subject to their jurisdiction the free and full exercise of those rights and freedoms' (emphasis added). This formulation could quite comfortably be seen as imposing a jurisdictional threshold for the obligation to ensure alone.

Coming back to the treaties which do not mention the obligation to respect explicitly, most importantly the ECHR, that particular obligation is either spelled out in provisions which guarantee specific rights, or can if necessary be *implied* into the content of the treaty in question. Crucially, if the negative obligation to respect human rights generally, or for example the obligation not to torture specifically, is implicitly read into the treaties, there is no reason why it should depend on the same jurisdictional threshold of application as the positive obligation to secure or ensure.

The best authority for these points is the ICJ's merits judgment in the *Bosnian Genocide* case.[274] Just like the CAT, which does not say in so many words that states will be responsible if their organs or agents commit torture, so the Genocide Convention does not explicitly provide for state responsibility for the commission of genocide. Thus, Article I of the Convention obliges states to prevent and punish genocide, while other provisions elaborate on these requirements. Both at the preliminary objections and at the merits stage of the case the FRY/Serbia disputed the existence of a separate state obligation under the Convention not to commit genocide, asserting that the Convention was a classical international criminal law treaty, dealing with crimes committed by individuals, not states. All the Convention did, in Serbia's view, was to require states parties to criminalize in their domestic law the crimes that it defines, and then prosecute the perpetrators of these crimes. Article IX of the Convention, which does mention the responsibility of states for genocide, was, in Serbia's argument, merely a compromissory clause which did not impose any additional substantive obligations upon states parties.

The Court first reasoned that Article I of the Convention is not merely hortatory or purposive, and that 'in particular its undertaking to prevent, creates obligations distinct from those which appear in the subsequent Articles [of the Convention]' so that the 'the Contracting Parties have a direct obligation to prevent genocide'.[275] It then held as follows:

> The Court next considers whether the Parties are also under an obligation, by virtue of the Convention, not to commit genocide themselves. It must be observed at the outset that such an obligation is not expressly imposed by the actual terms of the Convention. The Applicant

[273] See, e.g., Cerone, above note 268, at 124; R. Künnemann, 'Extraterritorial Application of the International Covenant on Economic, Social and Cultural Rights', in Coomans and Kamminga 201, at 227–99.

[274] *Application of the Convention on the Prevention and Punishment of the Crime of Genocide*, Judgment, 27 February 2007 (hereinafter *Bosnian Genocide* merits judgment).

[275] Ibid., paras 162, 165.

has however advanced as its main argument that such an obligation is imposed by Article IX, which confers on the Court jurisdiction over disputes 'including those relating to the responsibility of a State for genocide or any of the other acts enumerated in Article III'. Since Article IX is essentially a jurisdictional provision, the Court considers that it should first ascertain whether the substantive obligation on States not to commit genocide may flow from the other provisions of the Convention. Under Article I the States parties are bound to prevent such an act, which it describes as 'a crime under international law', being committed.

The Article does not *expressis verbis* require States to refrain from themselves committing genocide. However, in the view of the Court, taking into account the established purpose of the Convention, the effect of Article I is to prohibit States from themselves committing genocide. Such a prohibition follows, first, from the fact that the Article categorizes genocide as 'a crime under international law': by agreeing to such a categorization, the States parties must logically be undertaking not to commit the act so described. Secondly, it follows from the expressly stated obligation to prevent the commission of acts of genocide. That obligation requires the States parties, inter alia, to employ the means at their disposal, in circumstances to be described more specifically later in this Judgment, to prevent persons or groups not directly under their authority from committing an act of genocide or any of the other acts mentioned in Article III.

It would be paradoxical if States were thus under an obligation to prevent, so far as within their power, commission of genocide by persons over whom they have a certain influence, but were not forbidden to commit such acts through their own organs, or persons over whom they have such firm control that their conduct is attributable to the State concerned under international law. In short, the obligation to prevent genocide necessarily implies the prohibition of the commission of genocide.[276]

The Court thus concluded that the 'Contracting Parties to the Convention are bound not to commit genocide, through the actions of their organs or persons or groups whose acts are attributable to them'.[277] It then extended the same reasoning to crimes ancillary to genocide defined by Article III of the Convention, that is conspiracy to commit genocide, direct and public incitement to genocide, attempt to commit genocide, and complicity in genocide.[278]

Turning back to human rights treaties, to my mind the Article 2(1) CAT obligation of the state 'to prevent acts of torture in any territory under its jurisdiction' necessarily implies the state's obligation to itself refrain from torture, just as with the Genocide Convention. More generally, the Article 1 ECHR obligation to secure human rights implies the obligation to respect them. That implication does not, however, necessarily require the same threshold for its existence, that of state jurisdiction over territory.

Thus, for example, in the *Bosnian Genocide* case the ICJ thought that the obligation not to commit genocide had no territorial limitation. States were *implicitly* prohibited from committing genocide anywhere in the world. The Court had the same view with regard to the positive obligation to prevent, an

[276] *Application of the Convention on the Prevention and Punishment of the Crime of Genocide*, Judgment, 27 February 2007 (hereinafter *Bosnian Genocide* merits judgment), para. 166 (paragraph breaks added).

[277] Ibid., para. 167. [278] Ibid.

obligation of states to exercise due diligence and 'employ all means reasonably available to them, so as to prevent genocide so far as possible',[279] saying that '[t]he substantive obligations arising from Articles I and III are not on their face limited by territory. They apply to a State wherever it may be acting or may be able to act in ways appropriate to meeting the obligations in question.'[280]

However, in his separate opinion Judge Tomka forcefully argued that the positive obligation to prevent should be limited territorially. According to him,

> [U]nder Article I of the Genocide Convention the State does have an obligation to prevent genocide outside its territory to the extent that it *exercises jurisdiction outside its territory*, or exercises control over certain persons in their activities abroad. This obligation exists in addition to the unequivocal duty to prevent the commission of genocide within its territory.[281]

In essence, Judge Tomka argued for the imposition of a threshold criterion on the obligation to prevent—state jurisdiction over a territory—the same threshold as in human rights treaties. Under his approach, a state would have to exercise effective overall control of an area in which there is a serious risk of genocide being committed by some other actor in order for its obligation to prevent genocide to arise. His argument was in essence one of policy—that a state should be expected to prevent genocide only when it had territorial jurisdiction, because that obligation was much more onerous then the simple obligation to refrain from committing genocide.[282]

My point is simply this—there is no inherent contradiction in implying, where necessary, the negative obligation to respect human rights into the relevant treaties and that obligation having a *broader*, territorially unlimited scope of application than the positive duty to secure or ensure human rights, or prevent violations thereof. Thus, under the CAT for example, we could say that the state has the duty to prevent torture only in territories under its jurisdiction, but that it has the obligation to itself refrain from torturing in all circumstances. Likewise, under the ECHR, the state's obligation to secure human rights would be limited to areas under the state's effective overall control, but its duty to respect human rights would apply everywhere, without any territorial limitation. For example, the ECHR would apply to the taking of property by the UK within the UK of a person residing in Monaco, whether in law or merely in fact.

C. Prophylactic and procedural positive obligations

Let me again emphasize that I am not advocating a strict separation between negative and positive obligations. Rather, I am arguing for a separation between those positive obligations which require control over territory in order to be effective, such as the obligation to prevent inhuman treatment or secure human rights generally even from

[279] Ibid., para. 430. [280] *Bosnian Genocide* merits judgment, para. 183.

[281] *Bosnian Genocide* merits judgment, Separate Opinion of Judge Tomka, para. 67.

[282] For more on the territorial scope of application of the Genocide Convention, see M. Milanovic, 'Territorial Application of the Convention and State Succession', in P. Gaeta, ed., *The UN Genocide Convention: A Commentary* (Oxford University Press, 2009), 473.

third parties, and those obligations whose effectiveness depends only on the state's control over its own agents. The practice of the European Court and other human rights bodies has long recognized that some positive obligations are of a procedural or prophylactic nature.[283] Thus, for example, in the context of the right to life, the state has the negative obligation not to take life unjustifiably, but also the positive obligation to conduct an independent and effective investigation into a possible taking of life by its own agents, e.g. the police or the armed forces. As explained by the House of Lords,

> The European Court has also interpreted article 2 as imposing on member states a procedural obligation to initiate an effective public investigation by an independent official body into any death occurring in circumstances in which it appears that one or other of the foregoing substantive obligations has been, or may have been, violated *and it appears that agents of the state are, or may be, in some way implicated.*[284]

In Lord Bingham's view, '[t]his procedural duty does not derive from the express terms of article 2, but was no doubt implied in order to make sure that the substantive right was effective in practice'.[285] This procedural obligation is, however, *not* the same as the other positive aspect of Article 2, which flows from the state's obligation to *secure* human rights, namely 'to establish a framework of laws, precautions, procedures and means of enforcement which will, to the greatest extent reasonably practicable, protect life'.[286] That obligation is much more far-reaching, as it requires the state to prevent and investigate even acts of purely *private* violence, in which the agents of the state are not implicated at all.

Therefore, in my view, the first category of positive obligations, which exist solely to make the state's *negative* obligations truly effective, should apply co-extensively with the negative obligations themselves. On the other hand, those positive obligations which flow from the state's duty to secure or ensure human rights or prevent violations thereof—say prevent private violence or discrimination—require a threshold that sets out the limits of realistic compliance. And that threshold is precisely state jurisdiction, i.e. control over territory.

On the other hand, the *intensity* of these obligations also differs; a state has an absolute duty to investigate violations of human rights committed by its own agents, or committed against persons in its custody.[287] It also has to provide persons whom it deprives of liberty with food, clothing, or health care, even though it might not have such obligations towards the general population.[288] However,

[283] See generally Mowbray, above note 124.

[284] *R. (Middleton) v. West Somerset Coroner* [2004] 2 AC 182, para. 3 (citing Strasbourg case law).

[285] *Gentle*, para. 5 (per Lord Bingham).

[286] *Middleton*, para. 2 (citing Strasbourg case law).

[287] See, e.g., *Selmouni v. France* [GC], App. No. 25803/94, Judgment, 28 July 1999.

[288] Even a system which has an extremely limited acceptance of positive obligations generally, such as the one in the United States—see e.g. *DeSheaney v. Winnebago*, 489 U.S. 189 (1989)—recognizes that the state may accrue positive obligations towards an individual if it acts beforehand so as to restrict or affect the rights or liberty of the individual concerned, for example by incarceration. In the words of the US Supreme Court, 'elementary principles establish the government's obligation to provide medical care for those whom it is punishing by incarceration' as 'it is but just that the public be required to care for the prisoner, who cannot by reason of the deprivation of his liberty, care for

when it comes to its positive obligations with regard to purely private conduct, they are indeed more far-reaching in scope but are also more flexible in content. In the words of the European Court in *Osman*, they must not 'impose *an impossible or disproportionate burden* on the authorities'.[289]

We can thus distinguish between several kinds of positive obligations under human rights treaties, and this distinction should have a bearing on their extraterritorial application. To make this discussion somewhat less abstract, let us go back to *Al-Skeini* and the five applicants killed by British troops on patrol. Assume, for the sake of the argument, that even though the killings took place in British-occupied Basra, because of the strength of the insurgency Basra could not be qualified as an area under the UK's effective overall control,[290] and was hence outside its jurisdiction. Even so, in my view, the UK would still have not only the negative obligation to refrain from depriving the five applicants of life unjustifiably, but would also have the positive procedural obligation to conduct an effective investigation into their killing. Its existence depends solely on the UK's own involvement in the killing, and in order to comply with it the UK need not do anything more than investigate the conduct of its own troops, which it is in principle perfectly able to do. However, were the killings actually done by third parties, be they insurgents or indeed the soldiers of an allied country, the UK would have had no obligation to investigate or prevent the deaths, since they took place in an area outside its jurisdiction, and it would in fact be exceedingly difficult, if not impossible, for the UK to conduct an effective investigation without actually having control over the territory.[291]

Or, take again the example of the assassination of Alexander Litvinenko in London, ostensibly with Russian involvement. The killing clearly took place outside Russia's jurisdiction, if jurisdiction is conceived of territorially. Nonetheless, Russia would still have the obligation to investigate it, to the extent that a credible case can be made that its own agents were involved in Litvinenko's death, for example by providing the killers with polonium, the radioactive substance with which Litvinenko was poisoned. If, however, such a credible case could not be made, and on the evidence available there was no reasonable suspicion that Litvinenko's death was anything other than a purely private act, then it would

himself'. *Estelle v. Gamble*, 429 U.S. 97, 102–4 (1976). See also *Farmer v. Brennan*, 511 U.S. 825 (1994).

[289] *Osman*, para. 116 (emphasis added).

[290] See above, Section 2.C.3.

[291] I am not denying that the UK might be faced with serious evidentiary and forensic difficulties even when investigating the conduct of its own troops in a territory outside its control, e.g. because it does not have access to the crime scene. In such a situation the prophylactic positive obligation to investigate should again be interpreted flexibly, so that the UK is obliged to do only what it can in fact do. Above all, however, the UK would be expected to put in place reasonable safeguards and procedures before it mounts a military operation that would allow it to investigate allegations of misconduct by its own troops.

only be the UK which would have the obligation to investigate the murder since it took place within its jurisdiction.[292]

Similarly, on the facts of *Bankovic*, the respondent states should have been asked by the Court to justify on the merits their killing of individuals who were not within their jurisdiction territorially conceived, as the killing implicates the states' negative obligation to which some positive obligations may attach. In doing so, the Court should have taken into account the relevant rules of international humanitarian law and the extraordinary circumstances of armed conflict, and adopted a more flexible approach to Article 2 than in a situation of normalcy. As I have argued above, however, there would be limits to that flexibility, motivated by the need to preserve the integrity of the ECHR regime.[293] The Court would also have encountered serious evidentiary difficulties on the merits, but it has managed to deal with such difficulties in the past, as with Chechnya, for example by equitably distributing the burden of proof between the parties. Thus, though in my view the correct result in *Bankovic* would probably have been that the killings were unlawful, it is far from obvious that this should have been the case. The respondent states would have a case to answer, but they would also have something to answer the case with.

Or, if we take the example of the pending *Aerial Herbicide Spraying* case before the ICJ while taking the facts alleged by Ecuador as proven,[294] Colombia would be responsible for violating the rights of Ecuadorian residents adversely affected by its spraying operation, as the operation was conduct by its own agents or on its own behalf. Colombia would not, however, have the same obligation under human rights treaties with regard to pollution or herbicides used by purely private actors operating within its territory and having effects in Ecuador. Similarly, if we consider

[292] See, in that regard, *W v. United Kingdom* (dec.), App. No. 9348/81, (1983) 32 DR 190, a Commission admissibility decision. The applicant's husband was killed in the Republic of Ireland, while her brother was killed in Northern Ireland. She complained that the United Kingdom had failed to secure her husband's and brother's right to life. With respect to the husband, who was killed in the Republic of Ireland, the Commission declared the application to be incompatible *ratione loci* (at 199):

> The Commission further considers that, in determining its competence *ratione loci* in relation to the jurisdiction of the United Kingdom, regard must be had to the position, at the relevant time, of the direct victim (i.e. the applicant's husband) and not of the indirect victim (the applicant herself) of the alleged violation of the Convention. It finds that, at the time of his death at G. Sales Yard, in the Republic of Ireland, the applicant's husband was not 'within the jurisdiction' of the United Kingdom in the sense of Article 1 of the Convention. The Commission has also considered whether any active measures by United Kingdom authorities could have contributed to the murder of the applicant's husband in the Republic of Ireland. However, it notes that even the applicant has not alleged any such action by these authorities.

Note how the Commission applied a spatial model of jurisdiction—the victim was not within UK's jurisdiction because the killing took place in Ireland, and therefore the UK had no obligation to investigate it. However, the Commission did entertain the possibility that the result could have been different if the applicant had shown that UK authorities were involved in the murder.

[293] A further problem is that even though Article 15(2) ECHR permits derogations from Article 2 'in respect of deaths resulting from lawful acts of war', which could serve to add more flexibility to an Article 2 analysis in times of armed conflict, no such derogation was made by ECHR states parties engaging in the bombing of Serbia. See also Chapter V below.

[294] See Section 1.F above.

the scenarios of extraterritorial complicity that we examined above, with say a UK intelligence officer feeding questions and data to a Pakistani torturer interrogating a terrorist suspect in Pakistan,[295] the UK would in my view have a territorially unlimited negative obligation under Article 3 ECHR not to be complicit in the torture of any person anywhere in the world. Accordingly, it would also have the positive obligation to investigate complicity in torture by its own agents wherever it might occur. It would not, however, have such an obligation with regard to an extraterritorial human rights violation in which its own agents did not participate at all, unless the act was committed in a territory under its effective overall control.

D. Reconciling universality and effectiveness

To my mind, this third model provides us with the best balance between universality and effectiveness with regard to the extraterritorial application of human rights treaties. Instead of being artificially limited, universality is brought to its logical (and moral) conclusion. States would have the same obligation to respect human rights both within and outside their territories. Whether they use drones for the targeted killings of suspected terrorists, use force in more conventional military operations, abduct or detain combatants or civilians or put them on trial, poison the crops of innocent farmers, or enforce their laws, states would still have to abide by the restrictions that human rights law places on the arbitrary exercise of their power, and do so regardless of territorial boundaries. When, however, states are expected to do more than just refrain from adversely affecting the lives of others, when they need to take positive steps, from preventing domestic violence[296] and safeguarding private property to protecting lawful public assemblies[297] and the free exercise of religion,[298] they cannot fulfil such obligations effectively without having the tools to do so. Such obligations should, therefore, be territorially limited to areas and places under the state's jurisdiction.

This model would, in principle, be able to accommodate all of the effectiveness concerns that generally militate against the extraterritorial application of human rights instruments, which I have examined above in some detail: flexibility, impact, regime integrity, clarity, and predictability.[299] Flexibility is beyond doubt the most important, and can only be achieved on the merits and substance of each concrete instance of extraterritorial application to a specific set of facts. In the transition from threshold to substantive issues lies both the appeal of this model, and its greatest risk. It would be unreasonable to contend that most human rights obligations can be applied in exactly the same way in peacetime in the state's own territory, and during a military occupation or in an active battlefield abroad. Full account must be

[295] See Section 1.D above.

[296] See, e.g., *Opuz v. Turkey*, App. No. 33401/02, Judgment, 9 June 2009.

[297] See, e.g., *Plattform 'Ärzte für das Leben' v. Austria*, Judgment, 21 June 1988, Ser. A, No. 139, (1991) 13 EHRR 204.

[298] See, e.g., *97 members of the Gldani Congregation of Jehovah's Witnesses and 4 others v. Georgia*, App. No. 71156/01, Judgment, 3 May 2007.

[299] See above, Chapter III, Section 10.

taken of the extraordinary circumstances in which the treaty is being applied extraterritorially, and substantive adjustments need to be made in order for extraterritorial application to truly become realistic and effective—for example, through the greater use of international humanitarian law in interpreting the norms of human rights law.

But, as I have said, though a price must be paid if universality is not to descend into utopia, that price must not be too steep. There still must be a point to extraterritorial application—it must have an actual *impact*—and the integrity of the human rights regime as a whole must not be compromised by its being watered down too much. Again, both of these concerns can only be addressed by examining the numerous substantive issues arising with extraterritorial application, such as targeted killings or preventive detention, which are as such beyond the scope of this study. My only point here is that such concerns *can* be addressed effectively, and that the extraterritorial application of human rights treaties can be both realistic and worthwhile. Obviously, the extension of the negative obligation to respect human rights that I propose can by itself have an enormous impact, as it would at the very least require some sort of justification by states for their acts outside their territory which violate the rights of countless people.

The third model also sufficiently addresses the *Bankovic* concern that human rights treaties would be compromised if state obligations were divided and tailored to match each specific extraterritorial action. The most onerous obligation by far, that to secure or ensure human rights, is strictly confined to areas over which states have effective control. The treaties are not chopped up and divided. Rather, they require states to *respect* human rights when they are in the position to do so.

Likewise, by allowing us to have those substantive disputes which are truly important, on matters like targeted killings or security detention, and by detaching these disputes from the preliminary threshold question of extraterritorial application, the third model would lead to a great simplification of the case law on extraterritorial application, its grounding in principle, and a rejection of casuistry. In other words, we would be able to tell with some clarity when human rights treaties actually apply extraterritorially—something that we are in all honesty incapable of doing with the present state of the jurisprudence, at least when it comes to the ECHR. That the current lack of certainty makes the lives of judges and academics more difficult is the least of its problems; it directly undermines the authority of the Court, and makes other participants in the system—above all states—unable to adequately factor the applicability of human rights treaties extraterritorially into their own policy-making.[300]

[300] See also Miller, above note 192 at 1230, arguing that

> [t]his uncertainty creates the twin risks that states will either under-estimate the jurisdictional scope of the Convention and violate human rights which might otherwise be protected, or that they will over-estimate the Convention's reach and refrain from actions which are strategically essential. Either way, the Court's doctrinal ambivalence prevents signatory states from accurately weighing the legal liabilities associated with particular extraterritorial actions, to the detriment of both human rights protection and security.

I should also say that I am aware that it is impossible to conclusively test whether my proposed model strikes the best balance between universality and effectiveness without actually trying it out. From the perspective of the European human rights system, whose strength derives precisely from the fact that it is the *European* human rights system, the risk of a flood of litigation doing it irreparable damage may be seen by some—particularly judges—as being too great. That risk, it needs to be said, is a realistic one.[301] I maintain, however, that the European Convention is sufficiently flexible, and that the tools given to the European Court are sufficiently powerful, to minimize that risk to a great extent. The floodgates will not open, and the European system will not be overwhelmed—at least not any more than it already is—just because the Court would no longer be able to use the preliminary question of extraterritorial application to avoid a number of vexing and controversial substantive issues. Nothing would prevent the Court, for example, from declaring inadmissible those cases which it considers manifestly ill-founded, after conducting even a deferential substantive analysis. And though the risk may be real, it is one worth taking, if universality of human rights is to have any meaning. Whether this will actually happen will depend on those who sit on the European Court of Human Rights, and on whether it is, in their view, the 'European' or the 'Human Rights' bit of their Court and Convention that matters most.

Finally, to the extent that my textual argument in favour of the third model is not considered to be persuasive, or that its adoption would require too radical a departure from existing jurisprudence, I would (purely as a pragmatic matter) not be opposed to the adoption as a substitute of the personal model of jurisdiction as state authority and control over individuals jointly with the spatial model. If the personal model is considered to be textually necessary, then it should apply to the negative obligation to respect, while the spatial model would apply to the positive obligation to secure or ensure human rights.[302] Indeed, if we took another look at the various cases applying the personal model, we would see that they generally deal only with negative obligations of states, or with procedural or prophylactic positive obligations. So long as the personal model is not in any way limited, its application would lead to the same result as my third model, just with the added pretence of applying a jurisdictional threshold to negative obligations when no such threshold in fact exists. This would be an adequate solution for interpreting the jurisdiction clauses in some treaties, such as in the First Optional Protocol to the ICCPR, which limit the right to individual petition only to those persons subject to the state's jurisdiction. This would also be an adequate answer to Lord Brown's challenge in *Al-Skeini* that the spatial model would become redundant if the personal one were to be adopted, since each would apply to different types of state obligations.

[301] See, in that regard, Miller, above note 192, at 1235.

[302] For a similar contextual approach, see Lawson, above note 172, at 120, as well as R. Lawson, 'Really out of Sight? Issues of Jurisdiction and Control in Situations of Armed Conflict under the ECHR', in A. Buyse (ed.), *Margins of Conflict: The ECHR and Transitions to and from Armed Conflict* (Intersentia, 2010), 57; Lubell, above note 50, at 227 *et seq*; S. Skogly, *Beyond National Borders: States' Human Rights Obligations in International Cooperation* (Intersentia, 2006), at 66 *et seq*, 206.

Having thus outlined the model of extraterritorial application that I would prefer, I again do not wish to be taken as arguing that this model is free of all weaknesses. In some cases at least, the distinction between positive and negative obligations is not as clear-cut as I have made it seem. The same goes for the distinction between positive obligations arising from the duty to secure or ensure human rights, which I would subject to a threshold of territorial jurisdiction, and those which are prophylactic or procedural in nature and thus appurtenant to negative obligations, which I would not. As with most things, there would be grey areas in which my model would not provide a clear answer. And there probably would be areas in which my model would give answers that would run counter to our legal or moral intuitions regarding the territorial scope of human rights treaties. Even so, in my view it provides a principled foundation that allows us to move from the question of *when* human rights treaties apply extraterritorially to the far more important and difficult question of *how* they should do so.

5. The Special Problem of the ICCPR

Having discussed the available models of extraterritorial application, I will briefly turn back to an issue that I have mentioned but until now have more or less studiously avoided—the proper interpretation of Article 2(1) ICCPR, providing that '[e]ach State Party to the present Covenant undertakes to respect and to ensure to all individuals within its territory and subject to its jurisdiction the rights recognized in the present Covenant'. As we have seen, this clause presents two interpretative difficulties: first, whether the obligation to respect human rights is modified and limited by the remainder of the clause, or whether it is only the obligation to ensure which is subject to limitation; and secondly, whether that limitation is conjunctive or disjunctive, i.e. whether individuals have to be *both* within a state's territory *and* subject to its jurisdiction to have ICCPR rights, or is it rather that states have to guarantee these rights to all individuals within their territories *and* to those subject to their jurisdiction? As for the former issue, I have just explained why there is in my view no bar to reading Article 2(1) as imposing no limitation on the obligation to respect, similarly to Article 1 ACHR. As for the latter, I have indicated several times how the Human Rights Committee has now espoused the more expansive disjunctive interpretation of Article 2(1),[303] and it will come as no surprise to the reader that this is an interpretation I agree with in principle—but let me now offer my reasons for doing so, other than that this is simply the result that I prefer as a matter of policy.

First, I think it fair to say that the conjunctive reading of Article 2(1) is textually or grammatically more natural, and interpretation under Article 31 VCLT does of course start from the text. This does *not* mean, however, that this is the *only*

[303] See Buergenthal, above note 265; Human Rights Committee, General Comment No. 31, para. 10: 'States Parties are required by article 2, paragraph 1, to respect and to ensure the Covenant rights to all persons who may be within their territory and to all persons subject to their jurisdiction.'

'ordinary' or 'plain' meaning of the text, as e.g. argued by the United States in its appearances before the Human Rights Committee,[304] or by Michael Dennis in an academic setting.[305] The use of grammatical particles such as 'and' or 'or' frequently leads to ambiguity, and accordingly to more than one available 'plain' or 'ordinary' meaning of the text. That ambiguity may be resolved by the (grammatical) context in which the words appear,[306] but there is nothing in Article 2(1) which assists us in that regard. Recourse must thus be had to the object and purpose of the treaty, which as we have seen is founded upon the inherent dignity of all human beings and the universal nature of human rights. Although the conjunctive reading may be grammatically the more natural, the disjunctive reading is certainly the one more consonant with the treaty's object and purpose.

Again, there is nothing truly novel in regarding words such as 'and' or 'or' as ambiguous and interchangeable in certain respects, if not generally. Not only do such ambiguities frequently arise in everyday linguistic usage,[307] but courts have a long history of addressing them. Almost 150 years ago the US Supreme Court declared that

> [i]n the construction of statutes, it is the duty of the court to ascertain the clear intention of the legislature. In order to do this, courts are often compelled to construe '*or*' as meaning '*and*,' and again '*and*' as meaning '*or*.'[308]

It is precisely in the search for the intent of the legislator (as problematic as the whole idea of legislative intent is in the context of interpretation) and the purpose that a legal enactment is supposed to achieve that the '[d]isjunctive "or" and the conjunctive "and" may be interpreted as substitutes'.[309] And in the international context, for example, it is the very same US administration which rejected a disjunctive reading of the 'and' in Article 2(1) ICCPR which employed a similar disjunctive reading of an 'and' in a rather important Security Council resolution.[310]

[304] This argument was developed to the fullest in the US second and third periodic reports to the Human Rights Committee, submitted in one document, as UN Doc. CCPR/C/USA/3, 28 November 2005, Annex I.

[305] See M. Dennis, 'Application of Human Rights Treaties Extraterritorially in Times of Armed Conflict and Military Occupation', (2005) 99 *AJIL* 119, at 122 *et seq.*

[306] See R. Gardiner, *Treaty Interpretation* (Oxford University Press, 2008), at 178 *et seq.*

[307] If, for example, I were to ask a guest of mine: 'would you like some milk and sugar in your coffee?' we would not normally interpret that question as presenting my guest only with the options of taking both the milk *and* the sugar or neither, even though 'and' is conjunctive in nature. Similarly, if we replaced the 'and' in the question with an 'or', we would not interpret it as precluding my guest from taking *both* milk *and* sugar with his coffee, though the disjunction through an 'or' is normally regarded as being exclusive. (I imagine that not even lawyers and logicians ask their guests whether they would like milk 'and/or' sugar with their tea or coffee.) The ambiguity of this sentence is not resolved by the grammatical context in which the words were uttered, but by their *social* context—i.e. by the fact that plenty of people drink their coffee only with milk, or only with sugar, and that it would be quite impolite not to offer these options to one's guest. Likewise, in treaty interpretation it is the *wider* context, not merely the grammatical one, taken together with other considerations such as object and purpose, which can help us resolve such ambiguities.

[308] *United States v. Fisk*, 70 U.S. (3 Wall.) 445, 447 (1865).

[309] N.J. Singer, 1A *Statutes and Statutory Construction* (West, 6th edn, 2002), § 21:14.

[310] See OLC, Memorandum Opinion for the Counsel to the Vice President: Whether False Statements or Omissions in Iraq's Weapons of Mass Destruction Declaration Would Constitute a

This transformation of an 'and' into an 'or' might be regarded as a transparent lawyer's trick, enabling him or her to get to the policy result that he or she wants. Depending on the circumstances, that might even be true. But while interpretation is not a process that should in principle be result-driven, it is equally true that the object and purpose of a treaty are considerations that *must* be taken into account when interpreting it, and that they may tend to prefer one result over another. In other words, an interpretation of the ICCPR which favours universality and human dignity is within the limits of textual vagueness or ambiguity by definition preferable to an interpretation which runs against the grain of the treaty.

This now brings me to the ICCPR's *travaux préparatoires*, on which the United States has relied in particular in support of its argument for a strictly territorial application of the treaty.[311] In the US view, the *travaux* clearly demonstrate that the ICCPR was to have no extraterritorial application. It was in fact on a US proposal that the reference to territory was introduced into Article 2(1), which had previously contained only a jurisdiction clause, and subsequent attempts by some parties to have this language deleted were rejected. The US representative during the drafting of the then single Covenant, Eleanor Roosevelt, explained this wording in the following terms:

> The purpose of the proposed addition [is] to make it clear that the draft Covenant would apply only to persons within the territory and subject to the jurisdiction of the contracting states. The United States [is] afraid that without such an addition the draft Covenant might be construed as obliging the contracting states to enact legislation concerning persons, who although outside its territory were technically within its jurisdiction for certain purposes. An illustration would be the occupied territories of Germany, Austria and Japan: persons within those countries were subject to the jurisdiction of the occupying states in certain respects, but were outside the scope of legislation of those states. Another illustration would be leased territories; some countries leased certain territories from others for limited purposes, and there might be questions of conflicting authority between the lessor nation and the lessee nation.[312]

In the US argument, this explanation not only conclusively excludes the extraterritorial application of the ICCPR to occupied territories, but also with regard to *leased* territories, foremost among them Guantanamo.

A comprehensive reading of the *travaux*, however, paints a different picture.[313] Rather than giving us clarity, it provides us with an ample dose of confusion and doubt, as is so often the case when having recourse to preparatory work. Even the passage quoted above shows us that the US negotiators at the time were not concerned so much with the application as such of the ICCPR to occupied

'Further Material Breach' under U.N. Security Council Resolution 1441, 7 December 2002, available at <http://www.justice.gov/olc/2002/materialbreach.pdf>.

311 See above note 304.

312 See UN Doc. E/CN.4/SR.138, at 10.

313 See generally N. Rodley, 'The Extraterritorial Reach and Applicability in Armed Conflict of the International Covenant on Civil and Political Rights: a Rejoinder to Dennis and Surena', (2009) *EHRLR* 628; N. Lubell, above note 50, at 195 *et seq*; M. Satterthwaite, 'Rendered Meaningless: Extraordinary Rendition and the Rule of Law,' (2007) 75 *Geo. Wash. L. Rev.* 1333, at 1361 *et seq.*

territories, but with a purported obligation to *legislate* for such territories. This is the result that they wanted to avoid. In truth, bearing in mind the case law we examined, no such general obligation would exist even under the most expansive reading of the obligation to ensure human rights—indeed, as we have seen, Article 43 of the Hague Regulations is designed precisely to prevent occupants from modifying the laws of the occupied territories, except when absolutely necessary to do so.[314] In some cases, however, the obligation to ensure might require the occupant to alter the legal system in order to effectively maintain peace and security in the occupied territory, or to fulfil other human rights commitments, and in exceptional circumstances this obligation might lead to an unresolvable norm conflict. Similarly, the US concern with leased territories was that the extension of human rights obligations might lead to questions of 'conflicting authority between the lessor nation and the lessee nation', i.e. to norm conflict. The US desires aside, however, whether the language inserted into the treaty actually successfully manages to avoid all such norm conflicts is a different matter—I think it does not.

Other sections of the *travaux* tell us that the drafters wanted to avoid states parties assuming the obligation to ensure the rights *of their own nationals* in a foreign country.[315] They did not want the states parties to have to do the impossible, for example to ensure the right to a fair trial of their national before a foreign court.[316] As the ICJ itself said in the *Wall* case,

> The *travaux préparatoires* of the Covenant confirm the [Human Rights] Committee's interpretation of Article 2 of that instrument. These show that, in adopting the wording chosen, the drafters of the Covenant did not intend to allow States to escape from their obligations when they exercise jurisdiction outside their national territory. They only intended to prevent persons residing abroad from asserting, vis-à-vis their State of origin, rights that do not fall within the competence of that State, but of that of the State of residence (see the discussion of the preliminary draft in the Commission on Human Rights, E/CN.4/SR.194, para. 46; and United Nations, *Official Records of the General Assembly, Tenth Session, Annexes*, A/2929, Part II, Chap. V, para. 4 (1955)).[317]

The *travaux* emphatically do not offer us what we most need—a set of first principles as to the proper interpretation of Article 2(1). They simply do not provide conclusive evidence that the drafters wanted a strictly conjunctive reading that would have limited the application of the treaty solely to those territories over which the states parties had title, and that they did so contrary to the basic purpose of the treaty that they themselves professed, that of universality. Nor do the *travaux* tell us, for example, whether the word 'jurisdiction' refers to effective overall control

[314] See above, Chapter III, Section 9, and below, Chapter V.

[315] See M. Bossyut, *Guide to the 'travaux préparatoires' of the International Covenant on Civil and Political Rights* (Nijhoff, 1987), at 53–5, and the sources cited therein.

[316] See Nowak, above note 170, at 43: 'the adoption of these words was intended to avoid obligating States parties to protect persons under their jurisdictional authority but outside their sovereign territory. This would be possible only through diplomatic channels'; Lubell, above note 50, at 201: 'the drafters . . . wished to avoid the risk that the Covenant would create positive duties outside the scope of a state's authority and ability to execute such obligations'.

[317] *Wall*, para. 109.

over a territory, or to authority and control over an individual, or to something else entirely, such as the exercise of legal authority or power or the right to exercise such power. They only give us glimpses of the particular *results* or *applications* of the treaty that (some) of the states parties wanted to avoid.

But it is not such original expected applications of a treaty which form its legally binding meaning. Even if, for example, it was clear from the *travaux* (which it is not) that states parties did not want the ICCPR to apply during occupation, this does not exclude the possibility that the language that they employed, as well as the broader object and purpose of the treaty, do precisely lead to such a result.

Philip Allott famously remarked that a treaty is a 'disagreement reduced to writing'.[318] I would not go so far as to label the ICCPR itself as such, but its *travaux* certainly qualify. I am prepared to concede, however, that if we could by some necromantic art revive Eleanor Roosevelt and the other drafters of the two Covenants, and present them one by one with the various scenarios of extraterritorial application that we are concerned with today, they might be disinclined to opt for an expansive approach. But I would also imagine that if we asked them whether under the treaties they were drafting Auschwitz would *technically* not have been a violation thereof because it was located in occupied Poland, rather than in a territory over which the German Reich had legal title, it would be very doubtful that Roosevelt et al. would have found such an interpretation acceptable. And yet this is *precisely* what the US position on strict territoriality would entail.

If the disjunctive interpretation of Article 2(1) were to be adopted, what would then be the purpose of the words 'within its territory' that were added to the jurisdiction clause? Would they not thus be rendered ineffective? Not necessarily. These words would then make ICCPR obligations more onerous than those under other human rights treaties, as Article 2(1) would require states to ensure the human rights of persons within its territory even when that territory is *not* under its jurisdiction, e.g. if a foreign state or an insurrectional movement was in control thereof. In other words, *à la Ilascu*, states parties would have a positive obligation to ensure the human rights of the population of their sovereign territories even in the absence of territorial control, though this obligation would have to be significantly attenuated.

Alternatively, even if the conjunctive reading of Article 2(1) were to be seen as correct, the application of the ICCPR would not have to be confined only to those territories over which the state has both title and effective control. As we have seen, the negative obligation to respect could comfortably be read as being subject to no territorial limitation. Moreover, though the possessive 'its' in 'within its territory' implies that a state would indeed need to have title or sovereignty over territory, this would not necessarily be the only plausible interpretation. In light of the universality-driven object and purpose of the treaty, the 'within its territory' clause could be read as not requiring title, but control.[319] In other words, the combination of the reference to territory and the reference to state jurisdiction could be taken as setting

[318] P. Allott, 'The Concept of International Law, (1999) 10 *EJIL* 31, 43.

[319] I have argued for a similar reading of Article VI of the Genocide Convention—see Milanovic, above note 282, at 481–2.

out a *spatial* test of jurisdiction over territory that would need to be met in order for the positive obligation to ensure human rights to arise. This interpretation would hence be consonant with my third model of extraterritorial application.

6. Treaties Without Jurisdiction Clauses

This brings me to the problem posed by treaties which contain no provisions on their territorial application such as jurisdiction clauses, or have jurisdiction clauses only for some but not all obligations under the treaty. As we have seen above,[320] other than a brief mention of territorial jurisdiction in its Article 14, the ICESCR contains no jurisdiction clauses. Neither does the CEDAW, while most provisions of the CERD, particularly Articles 2 and 5 which protect a wide range of substantive rights, do not have any kind of territorial limitation. What then is the appropriate model of extraterritorial application of such treaties?

There is in principle no need for a one-size-fits-all approach to these treaties. Each should be evaluated and interpreted on its own merits, with regard to its specificities in text and object and purpose. However, as I see it, three basic options present themselves:

(1) The treaty, or particular obligations within it, could be interpreted as being completely territorially limited. It is hard to see how such a restrictive approach could ever be justified from the standpoint of universality in the absence of reasonably clear text to that effect, particularly because state ability to ensure or secure even the most demanding socio-economic rights does not depend on title, but on control over territory. Such a restrictive approach would hence tend to slide into pure apology for legally unrestrained state power.

(2) The treaty, or particular obligations within it, could be interpreted as being completely territorially unlimited. This would be an approach akin to the ICJ's holding in the *Bosnian Genocide* case that the obligation to prevent genocide was territorially unlimited, and required states to exercise due diligence and 'employ all means reasonably available to them, so as to prevent genocide so far as possible'.[321] However, even in the context of the prevention of genocide such a far-reaching obligation is undermined by its own vagueness. In other contexts, as for instance with regard to socio-economic rights, this obligation would be even vaguer. What would, for example, the US or EU *exactly* need to do to safeguard the right to food of say the starving population of North Korea? With one or two possible exceptions, as for example with the Article 2(1) ICESCR obligation to provide 'international assistance and co-operation' for the realization of socio-economic rights, this approach would tend to slide into utopia.

[320] See Chapter I, Section 4.C.

[321] *Bosnian Genocide* merits judgment, para. 430.

(3) Finally, my third model of extraterritorial application could be read into the relevant treaties so that the negative obligation to respect the rights in question would be territorially unlimited, while the positive obligations arising from such treaties would generally require the exercise of territorial jurisdiction, i.e. effective overall control of an area, as such control is necessary for their effective realization. Thus, if we take as an example the *Aerial Herbicide Spraying* case between Ecuador and Colombia that is currently pending before the ICJ,[322] we could say that Colombia has the obligation towards the people of Ecuador to respect their right to health and food, which the herbicide spraying would in principle be capable of violating. However, Colombia would not have the obligation (other than possibly as reparation for its prior wrongful act) to actually provide food or health care services to the population of Ecuador, since it has no jurisdiction over the relevant parts of Ecuador. Similarly, if we take as another example the facts of the current *Georgia v. Russia* dispute, Russian soldiers engaging in military action on Georgian soil would have had the negative obligation under Article 2(1)(a) CERD 'to engage in no act or practice of racial discrimination against persons, groups of persons or institutions'. However, in the absence of jurisdiction, i.e. territorial control, Russia would not have had the positive obligation under Article 2(1)(d) CERD to 'prohibit and bring to an end, by all appropriate means, including legislation as required by circumstances, racial discrimination by any persons, group or organization', in other words to prevent *private* discrimination.

This discussion is admittedly somewhat simplistic. Each of the treaties in question is deserving of a fuller account, something that I do not wish to engage in at this point.[323] It is particularly the lack of concrete cases that necessarily makes this discussion somewhat abstract, but as we have seen, some relevant cases are already being litigated. As a general matter, however, I believe that as with treaties which contain jurisdiction clauses, the third model provides the best balance between universality and effectiveness and could provide a default position even for treaties without jurisdiction clauses, which could then be modified if the particular circumstances of the treaty in question so require.

322 See above, Section 1.F.

323 For literature dealing specifically with the extraterritorial application of treaties guaranteeing socio-economic rights, see above, Chapter I, note 11.

V

Norm Conflicts, International Humanitarian Law, and Human Rights Law

1. Introduction

The final chapter of this study will be devoted to exploring the relationship between international humanitarian law and international human rights law. Why do so now, and what is the importance of this relationship to our main topic, the extraterritorial application of human rights treaties?

There are several reasons for examining this relationship before providing a synthesis or some general conclusions regarding extraterritorial application. First, as we have already seen, the interaction between IHL and IHRL is a frequent issue in various scenarios of extraterritorial application of human rights treaties. It is particularly the theme of norm conflict which is so often woven into the fabric of the case law on extraterritorial application, as the fear of such conflicts sits heavily on judges' minds.[1] And it is precisely this motive, this idea or perspective of norm conflict that I believe has a major role to play in describing the relationship between IHL and IHRL. Secondly, many of the same policy considerations at play with regard to extraterritorial application are also at play when it comes to the interface between these two bodies of law. Much as the tension between universality and effectiveness determines outcomes with regard to extraterritorial application, so does the tension between humanitarianism and effectiveness explain the interaction of IHL and IHRL. Finally, I have argued several times in this study that the interpretation of human rights treaties in light of IHL can introduce much-needed effectiveness and flexibility in times of armed conflict so as to make the application of IHRL realistic and practicable. In this chapter I will therefore not just try to assess the state of our effort at explaining the relationship between these two bodies of law, but also attempt to determine its prospects and, crucially, its limits.

At the outset it must be said that this examination of the relationship between the two bodies of law is not an abstract academic endeavour, but above all a pragmatic and a practical project that is designed to have a real-life impact. Together with the question of the extraterritorial application of human rights treaties, the IHL/IHRL project serves several purposes. The first, and the broadest,

[1] See above, Chapter III, Section 9.

is the affirmation of an idea: the law applicable in war is no longer solely a law between sovereigns, who agree out of grace and on the basis of reciprocity to limit themselves in their struggles so as to reduce the suffering of innocents. Rather, human beings embroiled in armed conflict still retain those rights that are inherent in their human dignity, which are more, not less, important in wartime than in peacetime, and which apply regardless of considerations of reciprocity between the warring parties.

Consequently, a more radical purpose of the project (perhaps not universally shared even among its adherents) is to shift the balance between effectiveness and humanitarianism, that was struck by states during the drafting of the major IHL treaties, more in the direction of humanitarianism. In other words, what we as participants in this project really want to do by examining the relationship between IHL and IHRL is to further humanize IHL and war generally.[2] We do this partly by using human rights norms to fill the gaps or areas left unregulated or very sparsely regulated by IHL, for example in regard to non-international armed conflicts, and partly by trying to change some outcomes that in fact *are* determined by IHL by introducing human rights rules and arguments into the equation.[3]

Finally, a more down-to-earth purpose of the project is the enforcement of IHL through human rights mechanisms.[4] Even if human rights substantively added nothing to IHL, i.e. if the relationship between IHL and IHRL was such that IHRL in wartime brought no less, but also no more, protection for individuals than IHL, there would still be a point in regarding IHL and IHRL as two complementary bodies of law. IHL, now (jurisdictionally) framed in human rights terms, could be enforced or its enforcement could be attempted before political bodies, such as the Human Rights Council or UN political organs more generally, or through judicial and quasi-judicial mechanisms, such as the ICJ, the European Court of Human Rights, the UN treaty bodies, or domestic courts.[5]

In brief, we wish to (boldly) take human rights to places, be they extraterritorial situations, or those of armed conflict, or both, where, as a matter of practical reality, no human rights have gone before. Saying openly that we are participating in a project with pragmatic and transformative ends in mind, rather than engaging in some sort of abstract discovery of the law, does not mean that we as lawyers are doing so illegitimately, or that we are usurping the legislative prerogatives of states.

[2] See generally T. Meron, 'The Humanization of Humanitarian Law', (2000) 94 *AJIL* 243.

[3] This approach of course rests on the assumption that more human rights in the formal sense would necessarily lead to better outcomes for people on the ground—an assumption which is rarely defended, and is challenged by Modirzadeh in a powerful recent article—see N. Modirzadeh, 'The Dark Sides of Convergence: A Pro-civilian Critique of the Extraterritorial Application of Human Rights Law in Armed Conflict', (2010) 86 *U.S. Naval War College International Law Studies (Blue Book) Series* 349. Whether this assumption actually holds is an empirical question, and, at that, one which is very hard to answer. I do not wish to engage with this issue at this time, beyond the several examples that I raise in the remainder of this chapter.

[4] See, e.g., G. Verdirame, 'Human Rights in Wartime: A Framework for Analysis', (2008) *EHRLR* 689, 691.

[5] A most effective example is the extensive litigation in UK domestic courts under the Human Rights Act 1998 arising from the occupation of Iraq.

It is the states themselves that have affirmed the complementary application of IHL and IHRL, both in the texts of the relevant human rights treaties[6] (that were after all concluded in the aftermath of the most horrible of abuses committed during an *armed conflict*, the Second World War), and in their official pronouncements in international fora. States have also greatly contributed to the humanization of IHL through the adoption of the 1977 Additional Protocols to the 1949 Geneva Conventions, for example through their provisions on fundamental humanitarian guarantees, or through greater limitations on reprisals.[7] As is often the case, states may not have abided in practice by their commitments as they should have, but this does not mean that these commitments were not made.

Having said this, as we will see there are limits to what legitimate methods of interpretation can do to harmonize IHL and IHRL. For human rights-based arguments to be even remotely persuasive in regard to situations of armed conflict, and to avoid the impression of a fluffy, utopian human rightist disregard for the realities of international relations, these arguments must meet certain requirements—the same requirements of effectiveness that we have already examined with regard to extraterritorial application: flexibility, impact, regime integrity, and clarity and predictability.[8]

As with extraterritorial application, there is a price to be paid for the joint application of IHL and IHRL. Though it may be our goal to further humanize IHL, in order to do so we must also be prepared to water down IHRL to make its application possible and practical. To the extent that war and armed conflict are accepted as a reality—as they must be, if this project is to make any sense—human rights norms cannot be applied in a 'business as usual' kind of way. Considerations of effectiveness must always be taken into account.

And as we have also seen above, though human rights have to be watered down to be applied jointly with IHL, they must not be watered down too much. Not only would this defy the whole purpose of the exercise, but it would also potentially compromise the values safeguarded by the human rights regime in peacetime.[9] There must, in other words, be caution in applying IHL together with human

[6] See the derogation clauses in the major human rights treaties, two of which explicitly mention 'war' as a situation in which derogations might be appropriate—Art. 4 ICCPR; Art. 15 ECHR; Art. 27 ACHR. See also Art. 72 of Additional Protocol I to the Geneva Conventions, which refers to 'other applicable rules of international law relating to the protection of fundamental human rights during international armed conflict', as well as the second preambular paragraph of Additional Protocol II, referring to 'international instruments relating to human rights [that] offer a basic protection to the human person'.

[7] For a general overview, see C. Droege, 'Elective Affinities? Human Rights and Humanitarian Law', (2008) 90 *IRRC* 501, at 503 *et seq*.

[8] See above, Chapter III, Section 10.

[9] See also the insightful critique by W. Schabas, '*Lex Specialis?* Belts and Suspenders? The Parallel Operation of Human Rights Law and the Law of Armed Conflict, and the Conundrum of *Jus ad Bellum*', (2007) 40 *Israel L Rev* 592, who argues that if IHL and IHRL are to be reconciled, IHRL must develop an IHL-like indifference to the *jus ad bellum* lawfulness of war, as well as allow the 'collateral' killing of civilians regardless of the overall legality of the use of force, and that this abandonment of pacifism is too high a price to pay for harmony as far as IHRL is concerned.

rights, as it is human rights that might lose something in the process.[10] After all, for all its humanitarian ethos, IHL is still a discipline about killing people, if in a civilized sort of way.[11] In short, at their very core IHL and IHRL are fundamentally incompatible, and we forget this at our peril.[12]

Having taken these considerations into account, we can move on to their implications for the relationship between IHL and IHRL. First, in order to be able to provide judges and other decision-makers with workable rules that they could then apply, and (more importantly) would be *willing* to apply, the focus of our debates must shift from the relationship of the two regimes as such, to the relationship of the particular norms belonging to the two regimes that control specific factual situations. Secondly, since this relationship is of course but one of many that contribute to the wider phenomenon of the fragmentation of international law,[13] we must be prepared to fully employ the toolbox that the doctrine of international law provides us for avoiding or resolving conflicts of norms. Thirdly, we must place the *lex specialis* maxim, whose exact nature remains unclear, within that toolbox. If we do so, we will realize that *lex specialis* must be abandoned as some sort of magical, two-word explanation of the relationship between IHL and IHRL; it confuses far more than it clarifies. Finally, we must be prepared to concede that there are some situations—perhaps not many, but by no means practically irrelevant—where the international lawyer's craft and tools will fail him, and where no legal solution can be provided. Such situations of unresolvable conflict or antinomy can be solved only in the manner in which they were created—through the political process. I will address these issues in turn.

2. A Relationship between Norms, not between Regimes

The relationship between IHL and IHRL, and the fragmentation phenomenon more broadly, are often examined from the high altitude perspective of a relationship

[10] A further risk is that injecting human rights discourse into warfare would actually serve to legitimize war as a phenomenon—see Verdirame, above note 4, at 692. On the other hand, those more devoted to IHL could argue that it is IHL that loses something by being injected with IHRL—for example, with regard to clarity and precision, or practicality and effectiveness of the norms applied—see generally Modirzadeh, above note 3.

[11] This fear, whether entirely conscious or not, that human rights might be compromised by giving too much room to effectiveness and to the realities of armed conflict might provide at least a partial explanation for the European Court's reluctance so far to explicitly take IHL into account in its cases on Chechnya, together with related fears regarding institutional incompetence. See also W. Abresch, 'A Human Rights Law of Internal Armed Conflict: The European Court of Human Rights in Chechnya', (2005) 16 *EJIL* 741. For an overview of the Chechen cases, see P. Leach, 'The Chechen Conflict: Analysing the Oversight of the European Court of Human Rights', (2008) *EHRLR* 732.

[12] See also R. Provost, *International Human Rights and Humanitarian Law* (Cambridge University Press, 2002), at 349–50: 'while there is indeed space for enlightened cross-pollination and better integration of human rights and humanitarian law, each performs a task for which it is better suited than the other, and the fundamentals of each system remain partly incompatible with that of the other'.

[13] See, e.g., A. Cassimatis, 'International Humanitarian Law, International Human Rights Law, and the Fragmentation of International Law', (2007) 56 *ICLQ* 623.

between two or more legal regimes. Though there can be some use in such inquiries, they are unhelpful more often than not. When it comes to the relationship between IHL and IHLR in particular, they can not only be unhelpful—viz. the *lex specialis* mantra—but misleading and even dangerous as well. To see how this can be the case, we need only turn to the most remarked upon examples of such high altitude pronouncements, those of the ICJ.

As is well known, the Court's first foray into the matter occurred in the *Nuclear Weapons* advisory opinion, where it remarked that

> [i]n principle, the right not arbitrarily to be deprived of one's life applies also in hostilities. The test of what is an arbitrary deprivation of life, however, then falls to be determined by the applicable *lex specialis*, namely, the law applicable in armed conflict which is designed to regulate the conduct of hostilities. Thus whether a particular loss of life, through the use of a certain weapon in warfare, is to be considered an arbitrary deprivation of life contrary to Article 6 of the Covenant, can only be decided by reference to the law applicable in armed conflict and not deduced from the terms of the Covenant itself.[14]

Note that the Court in *Nuclear Weapons* did not *really* examine the relationship between IHL and IHRL as regimes. It examined the relationship between one particular IHRL norm, the right to life, and, at that, the right to life as it is formulated in Article 6 ICCPR (which is worded differently than Article 2 ECHR, for instance), and the relevant rules of IHL. It was these *specific rules* that were held to be *lex specialis*, in that they could help interpret the 'arbitrary' part of Article 6 ICCPR in times of armed conflict. Though the Court's pronouncement was thus framed in terms of one particular problem and one particular set of norms, it nonetheless understandably provoked an academic extrapolation to the relationship between IHL and IHRL as a whole. The *Nuclear Weapons* opinion was thus understood as saying that IHL defines what IHRL means in wartime, IHRL guaranteeing no less, but also no more, rights to individuals affected by armed conflict than IHL.[15]

Our discussions have of course evolved since then, and so has the ICJ, whose next pronouncement in the *Wall* advisory opinion was decidedly broader in nature:

> More generally, the Court considers that the protection offered by human rights conventions does not cease in case of armed conflict, save through the effect of provisions for derogation of the kind to be found in Article 4 of the International Covenant on Civil and Political Rights. As regards the relationship between international humanitarian law and human rights law, there are thus three possible situations: some rights may be exclusively matters of international humanitarian law; others may be exclusively matters of human rights law; yet others may be matters of both these branches of international law. In order to answer the question put to it, the Court will have to take into consideration both these

[14] *Legality of the Threat or Use of Nuclear Weapons*, Advisory Opinion, ICJ Reports 1996, 226, para. 25.

[15] See, e.g., L. Doswald-Beck, 'International humanitarian law and the Advisory Opinion of the International Court of Justice on the Legality of the Threat or Use of Nuclear Weapons', (1997) 37 *IRRC* 35.

branches of international law, namely human rights law and, as *lex specialis*, international humanitarian law.[16]

As opposed to *Nuclear Weapons*, the Court here does refer to IHL as *lex specialis* to IHRL, i.e. it considers not only some particular norms, but one regime to be special to the other. The problem with such a characterization is not just that it is overbroad, but also that the concept of *lex specialis* is vague and can mean at least two radically different things, as I will explain below.[17] It should be noted, in that regard, that when the Court in *Congo v. Uganda* quoted its dictum from the *Wall* case,[18] it dropped the reference to *lex specialis*. Whether this omission was intentional is anyone's guess, but let us hope that it was. As a purely practical matter, the three categories outlined by the Court do no work whatsoever, if nothing else then because the Court does not say anything about which rules fall within which category, and on what basis. If you are, say, a British judge asked to rule under the Human Rights Act whether the preventive security detention of persons by British troops in Iraq violates Article 5 ECHR, the ICJ's dicta will provide you with precisely zero guidance.

To say, therefore, that the two 'spheres of law are complementary, not mutually exclusive',[19] as is today's human rights orthodoxy, may be perfectly true, but it is equally unhelpful in providing practical solutions to actual cases. The complementarity claim is only an answer to the equally broad counter-claim that the two regimes are mutually exclusive. To the extent that the latter was for many years the default position—as it admittedly was, as a matter of actual practice—the complementarity claim does have a purpose. But beyond that, it does not solve anything. It is the relationships between specific IHL and IHRL *norms* that need examining.[20]

As I stated above, dwelling too much on the relationship between regimes, instead of the relationship between the relevant norms, is not only unhelpful, but it can even be *dangerous* to refer to IHL as *lex specialis* to IHRL as a whole. Consider, for instance, the Bush administration's legal strategy in its 'war on terror'. It denied that persons detained in Guantanamo where entitled to protection under human rights law on the basis of two separate arguments. First, as we have seen, it contended that the relevant human rights treaties did not apply extraterritorially. Secondly, it claimed that even if they did, they were displaced by IHL, which applied as *lex specialis* to the putative global armed conflict between the US and Al-Qaeda.[21] Even if IHRL did apply formally, it granted no more protection than

[16] *Legal Consequences of the Construction of a Wall in the Occupied Palestinian Territory*, Advisory Opinion of 9 July 2004, ICJ Reports 2004, 136, para. 106.

[17] See also H. Krieger, 'A Conflict of Norms: The Relationship between Humanitarian Law and Human Rights Law in the ICRC Customary Law Study', (2006) 11 *JCSL* 265, 269.

[18] *Case Concerning Armed Activities on the Territory of the Congo (DRC v. Uganda)*, Judgment, 19 December 2005, para. 216.

[19] Human Rights Committee, General Comment No. 31, CCPR/C/21/Rev.1/Add.13 (26 May 2004), para. 11.

[20] See Krieger, above note 17, at 271.

[21] For an overview, see F. Hampson, 'The Relationship between International Humanitarian Law and Human Rights Law from the Perspective of a Treaty Body', (2008) 90 *IRRC* 549, at 550 *et seq.*

IHL. Because the detainees were 'unlawful combatants', they enjoyed no protection under IHL; because the absence of a rule, i.e. an unrestrained freedom for the state to act, is also a rule, IHRL brought nothing to the table, and the Guantanamo detainees had no rights under international law.[22] Now, admittedly, this argument of the departed administration was a succession of non sequiturs, but it was nonetheless made using the same language in which the gentler souls among us argue for a complementary relationship between IHL and IHRL, *lex specialis* and all.[23]

We are past the stage where there is much use in general inquiries as to the relationship between IHL and IHRL as regimes. Moreover, those who continue to oppose the joint application of IHL and IHRL will not be persuaded to the contrary no matter what generalities we or the ICJ might produce. If this is so, then why bother? It is enough to say that the complementary nature of the relationship between IHL and IHRL is confirmed by numerous pronouncements by states and international political bodies,[24] by international courts and tribunals,[25] by the text of derogation clauses of human rights treaties, and above all else, by the treaties' object and purpose.[26] If human rights accrue to human beings solely by virtue of their humanity, why should these rights then evaporate merely because two states, or a state and a non-state actor, engage in an armed conflict? More limited these rights might be, and considerations of effectiveness need to be taken into account, but they cannot be completely extinguished or displaced if their basic universality premise, that they are immanent in the human dignity of every individual, is accepted.[27] And though it is of course quite possible for this premise to be contested on ideological grounds, in *legal* terms that premise is hard-coded into the relevant international human rights instruments. In other words, it is the *law* that human rights are universal and that they accrue to every human being, war or no war. Nothing more needs to be said, and indeed nothing more *can* be said on the matter.[28]

It is to specific practical problems and their solutions that we must turn. Before doing that, however, we must take stock of the tools that are at our disposal to avoid or resolve norm conflicts, which is what I will proceed to do next.

[22] See, in particular, Chapter I above, note 2, and Chapter IV above, note 304.

[23] See further P. Alston, J. Morgan-Foster and W. Abresch, 'The Competence of the UN Human Rights Council and its Special Procedures in relation to Armed Conflicts: Extraterritorial Executions in the "War on Terror"', (2008) 19 *EJIL* 183.

[24] See, e.g., Tehran Conference, General Resolution XXIII, 'Respect for Human Rights in Armed Conflicts', UN Doc. A/Conf.32/41 (13 May 1968); UNGA Res. 2444 (1968); UNGA Res. 2647 (1970); UNSC Res. 237 (1967), UNSC Res. 1649 (2005), UNSC Res. 1882 (2009).

[25] See, e.g., *Nuclear Weapons*, above note 14; *Wall*, above note 16; *Serrano-Cruz Sisters v. El Salvador,* Preliminary Objections, 23 November 2004, Inter-Am. Ct. H.R. Series C, No.118, para. 112.

[26] See also D. McGoldrick, 'Human Rights and Humanitarian Law in the UK Courts', (2007) 40 *Is LR* 527, 531.

[27] See also O. Ben-Naftali and Y. Shany, 'Living in Denial: The Application of Human Rights Treaties in the Occupied Territories', (2003) 37 *Israel L Rev* 17, at 41–2; Verdirame, above note 4, at 691–2.

[28] See Chapter III, Section 2.

3. Norm Conflict Avoidance and Norm Conflict Resolution

A. Defining norm conflict

To provide a norm conflict perspective on the relationship between IHL and IHRL (an expression that I will continue to use merely as convenient shorthand, with all of the caveats stated above), I must first give both a definition of norm conflict and some other conceptual clarifications to the brief account that I have already given above.[29] The notion of conflict will be defined broadly: a relationship of conflict exists between two norms if one norm constitutes, has led to, or may lead to, a breach of the other.[30] With regard to IHL and IHRL norms specifically, the notion of conflict could be defined even more broadly—a norm conflict would exist whenever the application of the two norms would lead to two opposite results, for example if IHL provided that a particular use of force was lawful, while IHRL made it unlawful.

A further distinction must be made between apparent and genuine norm conflicts, and consequently between conflict avoidance on the one hand, and conflict resolution on the other. An apparent conflict is one where the content of the two norms is at first glance contradictory, yet the conflict can be avoided, most often by interpretative means. In instances in which all techniques of conflict avoidance fail, a genuine, as opposed to an apparent, conflict will emerge.[31] These true norm conflicts are those that cannot be avoided, but which it might be possible to resolve. Unlike avoidance, which interprets away any incompatibility, norm conflict resolution requires one conflicting norm to prevail or have priority over the other. Moreover, for a genuine conflict to be truly resolved in a case of conflicting *obligations* it is necessary for the wrongfulness on the part of the state for failing to abide by the displaced norm to be precluded as a matter of state responsibility. It is only if the state bears no legal cost for disregarding one of its commitments in favour of another that a norm conflict has truly been resolved.[32]

An examination of a norm conflict situation will usually proceed in two steps. The first is an inquiry into whether it is possible to avoid a norm conflict by interpreting the two potentially conflicting norms so as to make them compatible. Secondly, if avoidance is impossible, the conflict might be resolved by assigning priority to one norm over the other. However, a third outcome is also possible, and is in fact far from unlikely—that a norm conflict will not only be unavoidable, but also unresolvable. In such situations, which are a consequence of the diffuse and decentralized nature of the international legal system, there can only be a political

[29] See Chapter III, Section 9. I will also refer the reader to my more exhaustive treatment of the subject, in M. Milanovic, 'Norm Conflict in International Law: Whither Human Rights?', (2009) 20 *Duke J. Comp. & Int'l L.* 69, and the authorities cited therein.

[30] J. Pauwelyn, *Conflict of Norms in Public International Law: How WTO Law Relates to other Rules of International Law* (Cambridge University Press, 2003), 176.

[31] See Pauwelyn, above note 30, at 272.

[32] See further Milanovic, above note 29, at 6 *et seq.*

solution to the conflict between two equal norms.[33] Regrettably, despite our best efforts, a number of the potentially conflicting norms of IHL and IHRL might end up in this third category.

But before venturing into this third category, I will examine the utility of the various tools of norm conflict resolution and avoidance. I will start with methods of norm conflict resolution, if nothing else then to show how impracticable they are in addressing IHL and IHRL norm conflicts.

B. Methods of norm conflict resolution

Purely practically speaking, international law is not, and it might never be, a legal system in which a hierarchy based on the *sources* of norms plays an important role. Though, for example, a treaty will usually prevail over a customary rule, this is so only because the customary rule is *jus dispositivum*, and applies only so long as states do not agree otherwise[34]—and the same generally goes for treaties. Of course, IHL and IHRL norms in apparent conflict can emanate both from custom and treaties, the custom usually (and problematically) being derived in large part from the treaties to which many, but not all states concerned may be parties. There are four principal ways of resolving norm conflict, which, as stated above, requires the assigning of priority to one norm over another: (1) *jus cogens*; (2) Article 103 of the UN Charter; (3) conflict clauses in treaties; and (4) *lex posterior*.

There is probably no concept that has attracted so much scholarly attention, yet so little practical application, as *jus cogens*. And by little, I mean zero. There is, to my knowledge, not a single case were *jus cogens* was unambiguously the basis for a court holding that a conflicting rule of international law was null and void.[35] There are several reasons why this is so. First, as a general matter rigid hierarchy does not sit well with the overarching structure of international law, which is still largely the product of consensual law-making between states. Secondly, despite what some of its more fervent adherents might claim, the number of rules that undisputedly belong to the category of *jus cogens* (e.g. the prohibitions of torture, genocide, or slavery) is quite limited.[36] Thirdly, states are not nearly so foolish to conclude treaties or engage in some other sort of international law-making that is openly contrary to *jus cogens* norms.

Finally, *jus cogens* is a blunt instrument, which allows for no balancing or consideration of conflicting interests, but mandates a single result on the basis of

[33] See also Pauwelyn, above note 30, at 418 *et seq*.

[34] See International Law Commission, 'Fragmentation of International Law: Difficulties Arising from the Diversification and Expansion of International Law', UN Doc. A/CN.4/L.682 (13 April 2006), hereinafter 'ILC Fragmentation Study', para. 79 *et seq*. See also A. Verdross, '*Jus Dispositivum* and *Jus Cogens* in International Law', (1966) 60 *AJIL* 55, 58.

[35] This of course does not stop judges from injecting liberal references to *jus cogens*, peremptory or intransgressible norms or whatever in a number of judgments. I am not saying that this is a bad thing—far from it, symbolically at least—but that does not mean that *jus cogens* as a concept has a direct role in the result of the case.

[36] See R. Higgins, 'A Babel of Judicial Voices? Ruminations from the Bench', (2006) 55 *ICLQ* 791, 801.

hierarchy. In cases lacking similar political and moral certainty—and that would be almost all of them—the use of such an instrument would pose an understandably unappealing prospect for most judges. It should likewise be noted in that regard that Article 53 VCLT, which solidified the status of *jus cogens* as positive law, provides not that a particular *norm* that conflicts with a rule of *jus cogens* is void, but that the *treaty* which so conflicts—all of it—would be void.[37] Now, assuming the general validity of Article 53 VCLT, imagine if we were to say that an IHRL rule prevails over an IHL rule contained in one or more of the Geneva Conventions because it has the status of *jus cogens*—this would mean the voidness of the Geneva Conventions as a whole, not just of that particular rule.

Aside from *jus cogens*, there is one other quasi-hierarchical rule of norm conflict resolution that does have practical relevance—Article 103 of the UN Charter. It provides that

> [i]n the event of a conflict between the obligations of the Members of the United Nations under the present Charter and their obligations under any other international agreement, their obligations under the present Charter shall prevail.

Most importantly, this priority rule extends to state obligations under binding resolutions of the UN Security Council.[38] Depending on the situation, the Council can thus inject rules of its own making into the IHL/IHRL calculus, rules that arguably have priority over all others save for *jus cogens*. A good example would be the *Al-Jedda* case before English courts,[39] where a dual Iraqi/British national was held in security detention by British forces in Iraq. He was so held not under any authority of the UK as an occupying power, both because he was a British national and because the occupation had arguably already been terminated at the time. Rather, he was held on the basis of an authorization of preventive detention by the Security Council in its Resolution 1546 (2004), with the House of Lords deciding that this authorization prevailed over Article 5 ECHR, which does not allow for detention on preventive grounds, by virtue of Article 103 of the Charter.[40]

As for conflict clauses, they are used in treaty practice to explicitly give priority to one treaty over another,[41] though not nearly often enough.[42] Obviously, no such clauses exist in the relevant IHL and IHRL treaties. The closest we come to them are the derogation clauses in human rights treaties, which allow for the temporary

[37] See also Art. 64 VCLT.

[38] See *Questions of Interpretation and Application of the 1971 Montreal Convention arising from the Aerial Incident at Lockerbie*, Provisional Measures, Order of 14 April 1992, ICJ Reports 1992, 114, at 126, para. 42.

[39] *R. (Al-Jedda) v. Secretary of State for Defence* [2007] UKHL 58, [2008] 1 AC 332, 12 December 2007 (hereinafter *Al-Jedda*).

[40] See *Al-Jedda* (per Lord Bingham) para. 39. See also *Al-Jedda* (per Baroness Hale) paras 126 and 129, and *Al-Jedda* (per Lord Carswell) para. 136. Note that as of the time of writing *Al-Jedda* is, like *Al-Skeini*, pending before the Grand Chamber of the European Court.

[41] Indeed, Art. 103 of the Charter is in effect a special, proto-constitutional type of prospective conflict clause.

[42] See, e.g., Arts 237(1) and 311 of the UN Convention on the Law of the Sea. On conflict clauses in treaties, see generally C. Borgen, 'Resolving Treaty Conflicts', (2005) 37 *Geo. Wash. Int'l L. Rev.* 573, 584–7; ILC Fragmentation Study, paras 267–71.

suspension of certain rights during an emergency to the extent strictly required by the exigencies of the situation.[43] By limiting the content of applicable human rights law, derogations can potentially avoid or resolve many conflicts with IHL. However, as a practical matter, such derogations are rarely used, particularly in an extraterritorial context, since their use by a state would be interpreted as a concession that the IHRL treaty in question in principle applies extraterritorially to a given situation, and would thus open the state's actions up to judicial scrutiny, even if a curtailed one.[44] Consequently, as we have already seen, the UK has for example not derogated from either the ECHR or the ICCPR in respect of the conflicts in Iraq and Afghanistan, thereby creating, rather than resolving, a host of difficult problems.

Finally, there is that equally magical sibling of *lex specialis* (on which more in a moment), *lex posterior*. In domestic legal systems, which are after all a model for us as to how *any* legal system is supposed to work, if the legislature passes a statute which contradicts a prior statute, the latter in time will with some exceptions normally prevail, even if it does not contain an explicit clause to that effect. We justify this rule basically by inferring the relevant intent on the part of the single, uniform legislature, operating within a unified hierarchical system, which is not supposed to issue contradictory commands to citizens. However, whatever its validity in domestic law, this assumption manifestly does not hold true in the international system.[45] Which is why (what goes for) the international *lex posterior* rule[46] can hardly ever apply to conflicting multilateral law-making treaties, even assuming that they could be said to relate to the same subject-matter which is a prerequisite for the rule's application. When it comes to the IHL and IHRL treaties in particular, not only is there the obvious problem that the law-making in the two areas occurred temporally in several waves, so that it is somewhat absurd to treat any of these treaties as successive to each other in a *lex posterior* sense, but there is also not the slightest hint of state intent that the relationship between the two bodies of law should be governed by this rule.

In conclusion, the only two rules of norm conflict resolution that are of practical relevance for the relationship between IHL and IHRL are Article 103 of the Charter and the derogation clauses of human rights treaties. As is also the case more generally, it is the methods of norm conflict avoidance which are far more useful in practical terms.

C. Methods of norm conflict avoidance

Every legal system, but particularly the international one, can simultaneously tend towards fragmentation, because it tries to accommodate a number of widely

[43] See above note 6.

[44] See also McGoldrick, above note 26, at 555–6.

[45] 'There is no single legislative will behind international law. Treaties and custom come about as a result of conflicting motives and objectives—they are 'bargains' and 'package-deals' and often result from spontaneous reactions to events in the environment.' ILC Fragmentation Study, para. 34.

[46] See Arts 30 and 59 VCLT.

diverging values and interests, and towards harmonization, because without a measure of unity a legal system would soon stop being one, and divide into several particular regimes. The latter tendency is especially evident in the case law, as judges are generally more or less keen on preserving the integrity of the system to which they perceive themselves as belonging. Avoidance of norm conflict hence usually is (and should be) the first recourse.[47] When two norms can be interpreted harmoniously, they generally are—we do norm conflict avoidance all the time.

For the purposes of the present study I propose to (entirely informally) trace the various forms of avoidance on a broad spectrum, ranging from 'consistent' at one end to 'forced' at the other, with 'creative' being somewhere in the middle. Let me explain each more fully.

Consistent avoidance happens when the language, object and purpose, and other structural elements of the two potentially or apparently conflicting norms can be reasonably reconciled without much effort. An example has been given by the ICJ in *Nuclear Weapons*—Article 6 ICCPR prohibits 'arbitrary' deprivations of life, while IHL can tell us what 'arbitrary' means in times of armed conflict. Such a method of interpretation is warranted, *inter alia*, by the principle of systemic integration set out in Article 31(3)(c) VCLT, which requires that other applicable rules of international law be taken into account when interpreting a treaty.

In many cases, the 'fit' between the two norms will be far from perfect. This will usually, but not always, require one of the norms to be 'read down' from what its ordinary meaning would initially suggest, or from how it is ordinarily applied, so as to accommodate the other. For example, I would argue that Security Council resolutions that purport to limit certain human rights have to do so in clear and unambiguous terms.[48] But it is not always other norms of international law that need to be read down to accommodate the growing demands of human rights. In *Al-Adsani*,[49] for instance, the European Court held by 9 votes to 8 that Article 6 ECHR should be read consistently with international rules on state immunity, even in cases of alleged torture.[50] I would now again emphasize the point that there is a price to be paid if the IHL/IHRL project is to work. A large part of human rights law as interpreted in peacetime will have to be read down, to a greater or a lesser extent, in order to be effectively applied in wartime and be realistic rather than utopian. To again take arbitrary deprivations of life as an example, we cannot reasonably judge in the same way situations in which the state has a peacetime monopoly on the use of force and has to plan police operations in such a way so as to absolutely minimize any possible loss of life, and combat situations when a state

[47] To give two domestic law examples, consider the constitutional avoidance canon in US constitutional law, whereby a statute will be interpreted so far as possible not to pose a constitutional question or conflict, thus potentially leading to its unconstitutionality—see, e.g., *INS v. St Cyr*, 533 U.S. 289, 299–300 (2001); *Crowell v. Benson*, 285 U.S. 22, 62 (1932); or s. 3(1) of the UK Human Rights Act 1998, mandating that '[s]o far as it is possible to do so, primary legislation and subordinate legislation must be read and given effect in a way which is compatible with [ECHR] rights'.

[48] See Milanovic, above note 29, at 92 *et seq.*

[49] *Al-Adsani v. United Kingdom* [GC], App. No. 35763/97, Judgment, 21 November 2001.

[50] See also *Jones* v. *Ministry of Interior Al-Mamlaka Al-Arabiya AS Saudiya (the Kingdom of Saudi Arabia)* [2006] UKHL 26, [2007] 1 AC 270.

is embroiled in internal or international armed conflict. In the latter type of situation, it is IHL that would be used to introduce much-needed flexibility into the IHRL regime.

This brings me to *creative avoidance*, which, to put it bluntly, involves a court simply making things up. Such creativity can be employed for good ends or bad, in ways which are legally and logically acceptable, or not. One example is the *Bosphorus* line of cases before the European Court, in which the Court held that the transfer by states of competencies to international organizations that would be immune from judicial process both internally and before the Court itself would not be a violation of the ECHR so long as the organization in question provided an equivalent level of protection of human rights.[51] This was, in effect, a reading down of the ECHR for the sake of enhanced international cooperation through various international organizations, above all the EU. Opinions can differ on whether the Court's approach was justified (I think that it was), but it was indisputably entirely made up—the equivalent protection doctrine flowed from no provision of the ECHR, nor from any general principle of international law.[52]

Another, much less palatable example, would be the *Behrami* case that we have already examined,[53] in which the European Court simply invented on the spot an attribution rule to the effect that the acts of peacekeepers on a UN Security Council-authorized mission were attributable solely to the UN even if it were particular states, rather than the UN, who had operational command and control. The Court did so, contrary to practically unanimous authority, at least partly in order to avoid an apparent norm conflict between Resolution 1244 (1999), whose vague provisions the states concerned interpreted as authorizing preventive detention without judicial review, and Article 5 ECHR, with Article 103 of the Charter and the approach of English courts in *Al-Jedda* looming in the background.[54]

Perhaps the ultimate form of creative avoidance was employed by the European Court of Justice in *Kadi*, where it asserted that the EU legal system was independent of international law, and that UN Security Council resolutions on terrorist sanctions are unable to penetrate this independent system and could thus not prevail over EU fundamental rights guarantees by virtue of Article 103 of the Charter.[55]

Finally, there is what I termed *forced avoidance*, i.e. situations where a court is faced with a result that it deems to be undesirable (for example, a conflict with a *jus*

[51] *M & Co. v. Federal Republic of Germany* (dec.), App. No. 13258/87, 64 DR 138, 9 February 1990; *Waite and Kennedy v. Germany* [GC], App. No. 26083/94, Judgment, 18 February 1999; *Beer and Regan v. Germany* [GC], App. No. 28934/95, Judgment, 18 February 1999; *Bosphorus Hava Yolları Turizm ve Ticaret Anonim Şirketi v. Ireland* [GC], App. No. 45036/98, Judgment, 30 June 2005.

[52] See further above note 29, at 54 *et seq.*

[53] *Behrami and Behrami v. France, Saramati v. France, Germany and Norway* [GC] (dec.), App. Nos 71412/01 and 78166/01, 2 May 2007.

[54] See further M. Milanovic and T. Papic, 'As Bad as it Gets: The European Court of Human Rights' *Behrami and Saramati* Decision and General International Law', (2009) 58 *ICLQ* 267.

[55] Joined Cases C-402/05 P and C-415/05 P, *Kadi & Al Barakaat International Foundation v. Council and Commission*, Judgment, 3 September 2008.

cogens norm, or an unresolvable norm conflict), yet which it cannot avoid through normal means because the text or object and purpose of the relevant provisions simply cannot allow it under any reasonable interpretation. A court then might resort not just to invention or innovation, but to the wholesale rewriting of a particular provision or rule. The limits of judicial authority, and the boundaries between legislation and interpretation, are of course set not so much by the law itself, but by the wider social practices behind it. Just how far a judge can stray from a law-applying to a law-making role is necessarily dependent on a particular political and ideological context. That said, forcible methods of avoidance would by and large be considered as illegitimate, or at the very least improper.[56] To paraphrase Lord Bingham's view of such practices in the context of the Human Rights Act 1998, they do not constitute 'judicial interpretation, but judicial vandalism'.[57] Examples will follow.

D. Unresolvable norm conflicts

Though the international and domestic legal systems have in common both an inbuilt imperative for norm conflict avoidance and many techniques of avoidance, there is a point where the similarities end. In domestic systems, all norm conflicts must eventually be either avoided or resolved, since all forms of legislation at whatever level of governance ultimately fit within a single normative hierarchy. In the international system the avenues of norm conflict resolution are at best rudimentary. As we have seen above,[58] it therefore knows conflicts which are both unavoidable and unresolvable.[59] A more appropriate domestic analogy, if one is needed, would be with private, rather than with public law. Just as I can conclude two equally valid contracts whereby I commit to sell the same thing to two different people, and then have to face a choice as to which obligation to fulfil and which to breach and hence suffer the consequences, so a state can enter into two mutually contradictory, yet equally valid, commitments from which the only escape is a political one. It can do so, moreover, not just bilaterally, but through multilateral law-making treaties.

Take, for example, the *Matthews* case before the European Court, dealing with the elections in Gibraltar for the European Parliament.[60] On the one hand, the ECHR, as interpreted and applied by the European Court, qualified the European Parliament as a legislature in respect of which states had to organize free elections. On the other, a treaty concluded between EU members prohibited the UK from extending the franchise for the European Parliament to Gibraltar. The Court thus found the UK

[56] See Pauwelyn, above note 30, at 244 *et seq.*

[57] *R. (Anderson) v. Secretary of State for the Home Department* [2002] UKHL 46, [2003] 1 AC 837, para. 30.

[58] See Chapter III, Section 9.

[59] But see Droege, above note 6, at 524, (arguing that in cases of genuine norm conflict, one of the norms must prevail). With respect, there is no 'must' here. That it is desirable to resolve conflicts is certainly true—but it is by no means possible in all circumstances.

[60] *Matthews v. United Kingdom* [GC], App. No. 24833/94, Judgment, 18 February 1999.

responsible for violating the ECHR irrespective of its other treaty obligations. This did not mean, however, that the treaty prohibiting the extension of the franchise to the inhabitants of Gibraltar was invalid, or that the ECHR prevailed over it in some hierarchical sense. Both treaties were formally of equal stature, and no norm conflict resolution was possible—the UK could not fulfil its obligations under either treaty without violating the other, thereby incurring state responsibility. Only a *political* solution was possible for this norm conflict, based on a political preference for the ECHR among the other EU member states, and an eventual extension of the franchise.[61]

Or take an even better known case, *Soering*,[62] which we have already mentioned above in the context of effectiveness and norm conflict. There the European Court interpreted Article 3 ECHR as setting out a *non-refoulement* obligation, prohibiting the UK from transferring a person to the US if a real risk of that person being subjected to inhuman or degrading treatment in the US was established. At the other end was a perfectly valid extradition treaty between the UK and the US, which obliged the UK to extradite Soering, and which specified no exception to that obligation like the one devised by the European Court. There was, in other words, a norm conflict between the ECHR and the UK/US extradition treaty, and that was that. Now, one could argue, for instance, that the *non-refoulement* obligation invented by the Court in *Soering* had the status of *jus cogens*, and that it thus invalidated the conflicting extradition treaty—but this would be quite a stretch, to put it mildly.[63] Or, one could forcibly 'read in' a human rights exception into the UK/US treaty, even though there was no such exception—but such a judicial rewriting of a treaty would in my view almost invariably be illegitimate. In reality what we had in *Soering* was an unresolvable norm conflict. The political solution to this conflict was that the US did not press the issue, and that it reached an accommodation with European states generally whereby it would provide assurances that a person whose extradition was being sought would not be tried for a capital offence.

I must emphasize at this point that the fact that the norm conflicts in *Matthews* and *Soering* were unresolvable does not mean that the two *cases* were unresolvable. In both cases the Court was asked the only question that it had the jurisdiction to answer—whether the ECHR was violated. In both cases it said yes. In both cases another treaty would have been violated had the state fulfilled its ECHR obligations, but in both cases this simply did not matter for answering the question that was posed to the Court, as all avenues of norm conflict avoidance and resolution were exhausted. The only way that the Court could have avoided the norm conflict

[61] For a fuller account of *Matthews*, see Milanovic, above note 29.

[62] *Soering v. United Kingdom*, App. No. 14038/88, Judgment, 7 July 1989.

[63] That the prohibition of torture is *jus cogens* does not automatically entail that the prohibition of other forms of ill-treatment is also *jus cogens*. Even if it is, this would not automatically entail that the derived *non-refoulement* obligation would also have such status. Likewise, that all of these norms are absolute—which they are—does not mean that they all necessarily must be *jus cogens*, since peremptory norms must be 'accepted and recognized [as such] by the international community of States as a whole' (Art. 53 VCLT).

in *Soering*, for instance, was by ruling that there was no *non-refoulement* obligation arising under Article 3 ECHR, or by giving this obligation a rather pathetic content. On the other hand, in its interpretation of the ECHR the Court made a variety of legal, policy, and value choices that rendered such a course undesirable, if not impossible. This shows that avoidance of norm conflict is not some sort of absolute priority, but is a value like any other, which will sometimes prevail, and sometimes not. But in any event a court faced with an unresolvable norm conflict is not thereby required to pronounce a *non liquet*.

We can see similar norm conflict considerations in the *R (B and Others)* case[64] before the English Court of Appeal, essentially an extraterritorial *Soering*. The applicants were asylum-seekers in Australia, who were allegedly held in a detention centre in appalling conditions. They escaped and sought refuge in a British consulate, claiming that under Article 3 ECHR the UK could not release them to Australian authorities as they would be subjected to inhuman or degrading treatment. The Court of Appeal held that because the applicants were present on Australian sovereign territory, and the UK had an international legal obligation to surrender them to Australia, the UK's Article 3 *non-refoulement* obligation had to be qualified from what it would normally be under *Soering*:

> In a case such as *Soering* the Contracting State commits no breach of international law by permitting an individual to remain within its territorial jurisdiction rather than removing him to another State. The same is not necessarily true where a State permits an individual to remain within the shelter of consular premises rather than requiring him to leave. It does not seem to us that the Convention can require States to give refuge to fugitives within consular premises if to do so would violate international law. So to hold would be in fundamental conflict with the importance that the Grand Chamber attached in *Bankovic* to principles of international law. Furthermore, there must be an implication that obligations under a Convention are to be interpreted, insofar as possible, in a manner that accords with international law.[65]

Therefore:

> We have concluded that, if the *Soering* approach is to be applied to diplomatic asylum, the duty to provide refuge can only arise under the Convention where this is compatible with public international law. Where a fugitive is facing the risk of death or injury as the result of lawless disorder, no breach of international law will be occasioned by affording him refuge. Where, however, the receiving State requests that the fugitive be handed over the situation is very different. The basic principle is that the authorities of the receiving State can require surrender of a fugitive in respect of whom they wish to exercise the authority that arises from their territorial jurisdiction; see Article 55 of the 1963 Vienna Convention. Where such a request is made the Convention cannot normally require the diplomatic authorities of the sending State to permit the fugitive to remain within the diplomatic premises in defiance of the receiving State. Should it be clear, however, that the receiving State intends to subject the fugitive to treatment so harsh as to constitute a crime against humanity, international

[64] *R. (B and Others) v. Secretary of State for Foreign and Commonwealth Affairs* [2004] EWCA Civ 1344, [2005] QB 643.

[65] Ibid., para. 84.

law must surely permit the officials of the sending state to do all that is reasonably possible, including allowing the fugitive to take refuge in the diplomatic premises, in order to protect him against such treatment. In such circumstances the Convention may well impose a duty on a Contracting State to afford diplomatic asylum.

It may be that there is a lesser level of threatened harm that will justify the assertion of an entitlement under international law to grant diplomatic asylum. This is an area where the law is ill-defined. So far as Australian law was concerned, the applicants had escaped from lawful detention under the provisions of the Migration Act 1958. On the face of it international law entitled the Australian authorities to demand their return. We do not consider that the United Kingdom officials could be required by the Convention and the Human Rights Act to decline to hand over the applicants unless this was clearly necessary in order to protect them from the immediate likelihood of experiencing serious injury.[66]

The Court of Appeal in *B* was faced with an apparent norm conflict. On the one hand, except perhaps in a very limited set of circumstances, customary international law does not allow for diplomatic asylum, and required the UK to respect Australian sovereignty and surrender the applicants to Australia. On the other, Article 3 ECHR prohibited the UK from surrendering the applicants, if there was a serious risk that they would be subjected to ill-treatment. Realizing the existence of the conflict, the Court decided to forcibly avoid it by reading down the *non-refoulement* obligation under Article 3 ECHR. Its reasoning is however problematic in at least two respects.

First, the Court's actual reading down—qualifying *non-refoulement* so that it applies only if 'the receiving State intends to subject the fugitive to treatment so harsh as to constitute a crime against humanity'—is as plainly incorrect as it is forced, since crimes against humanity can never be committed against individuals in isolation, but only in the context of a widespread or systematic attack on a civilian population.[67] This condition would hardly be satisfied even in places like Iraq or Iran, let alone in Australia. Secondly, and more fundamentally, as we have just seen there was in fact a norm conflict even in *Soering*, yet this was not enough for the European Court to read down the ECHR—thus leading to an unresolvable norm conflict.

To bring this discussion of unresolvable norm conflicts a bit closer to home and the relationship between IHL and IHRL, let us turn back to the *Al-Saadoon* case that we have already extensively examined above.[68] Readers will recall the norm conflict in *Al-Saadoon*. Accepting the UK's argument, it had an obligation to transfer the applicants to Iraq.[69] Accepting the applicants' argument, the UK had an ECHR obligation not to do so. In other words, there was a genuine norm conflict *à la Soering* at hand, and no consistent form of avoidance was available, nor could this norm conflict be resolved on the basis of some hierarchical rule. The Court of

[66] Ibid., paras 88–9.

[67] See, e.g., Art. 7 of the Rome Statute of the International Criminal Court.

[68] See Chapter IV, Section 2.B.

[69] This obligation would derive from the customary obligation of all states not to interfere with the sovereignty and jurisdiction of other states—see, *mutatis mutandis*, *Asylum (Colombia v. Peru)*, Judgment, ICJ Reports 1950, 275.

Appeal thus decided to resort to more imaginative forms of avoidance, ranging somewhere from the 'creative' to the 'forced' on my tentative scale. It first held that the ECHR simply did not apply extraterritorially, thus nipping the norm conflict in the bud, because the UK had the applicants in custody solely on the basis of Iraqi legal authority and because it had an obligation to transfer them to Iraq. It then held, relying on its earlier precedent in *B*, that even if the ECHR did apply, and there was in principle an applicable *non-refoulement* obligation, that obligation had to be qualified, because the applicants were held on Iraqi soil and on Iraqi legal authority.[70]

As we have seen, the first of the Court's methods of avoidance is entirely unconvincing—there is no principled reason why the applicability of the ECHR should depend on the existence of other international obligations. The UK undoubtedly had an obligation to extradite in *Soering*, but that had no bearing on the applicability of the ECHR. When *Al-Saadoon* thus came before the European Court, it quite rightly decided that the ECHR was applicable, as its extraterritorial application did not depend on the UK's obligations towards Iraq, but on its *de facto* control over their place of detention.[71]

The Court of Appeal's second method of avoidance is somewhat more credible. It is possible to argue that *Soering* is distinguishable in that the applicant in that case was held on UK territory, while the applicants in *Al-Saadoon* were held on Iraqi territory, and that the respect for Iraqi sovereignty warrants a more attenuated application of the ECHR. Norm conflict avoidance would thus be possible in principle, if the ECHR was read down. This approach remains questionable, however, because of the *values* that the *non-refoulement* rule is meant to protect—the prohibition of all kinds of ill-treatment and the effective implementation of this prohibition—and it is entirely unclear why these values should be balanced away in an extraterritorial setting, but not in an intraterritorial setting.

This is precisely why the Court of Appeal's approach was ultimately rejected by the European Court in its merits judgment in *Al-Saadoon*:

> The Government contended that they were under an obligation under international law to surrender the applicants to the Iraqi authorities. In this connection, the Court recalls that the Convention must be interpreted in the light of the rules set out in the Vienna Convention on the Law of Treaties, 1969, of which Article 31 § 3(c) indicates that account is to be taken of 'any relevant rules of international law applicable in the relations between the parties'. More generally, the Court recalls that the principles underlying the Convention cannot be interpreted and applied in a vacuum. The Convention should be interpreted as far as possible in harmony with other principles of international law of which it forms part (see *Al-Adsani v. the United Kingdom* [GC], no. 35763/97, § 55, ECHR 2001-XI; *Banković*, cited above, §§ 55–57). The Court has also long recognised the importance of international cooperation (see *Al-Adsani*, § 54 and *Bosphorus*, § 150, both cited above).
>
> The Court must in addition have regard to the special character of the Convention as a treaty for the collective enforcement of human rights and fundamental freedoms. Its

[70] *Al-Saadoon CA*, paras 41–51.

[71] *Al-Saadoon and Mufdhi v. United Kingdom* (dec.), App. No. 61498/08, 30 June 2009, paras 87–9.

approach must be guided by the fact that the object and purpose of the Convention as an instrument for the protection of individual human beings requires that its provisions be interpreted and applied so as to make its safeguards practical and effective (see, inter alia, *Soering*, cited above, § 87; *Loizidou v. Turkey* (preliminary objections), cited above, § 72; *McCann and Others*, cited above, § 146).

It has been accepted that a Contracting Party is responsible under Article 1 of the Convention for all acts and omissions of its organs regardless of whether the act or omission in question was a consequence of domestic law or of the necessity to comply with international legal obligations. Article 1 makes no distinction as to the type of rule or measure concerned and does not exclude any part of a Contracting Party's 'jurisdiction' from scrutiny under the Convention (*Bosphorus*, cited above, § 153). The State is considered to retain Convention liability in respect of treaty commitments subsequent to the entry into force of the Convention (see *Bosphorus*, cited above, § 154 and the cases cited therein). For example, in *Soering*, cited above, the obligation under Article 3 of the Convention not to surrender a fugitive to another State where there were substantial grounds for believing that he would be in danger of being subjected to torture or inhuman or degrading treatment or punishment was held to override the United Kingdom's obligations under the Extradition Treaty it had concluded with the United States in 1972.[72]

In short, the Court said that it would go to great lengths to avoid norm conflicts by harmonizing the interpretation of the ECHR and other international obligations, but only *within the limits* set by the ECHR's special character as a treaty with the purpose of protecting human rights. It thus quite correctly pointed out that in *Soering* the Court was unwilling to read down the ECHR to accommodate other international obligations. The Court is mistaken, however, when it says that in *Soering* the ECHR *non-refoulement* obligation 'was held to override' the UK's obligations towards the US. This is simply incorrect if the word 'override' is to be taken as implying a hierarchical relationship. The ECHR is certainly not superior law to a bilateral treaty with a non-state party. Rather, as we have seen, the two treaties were in an unresolvable norm conflict, with the only possible solution to that conflict being political in nature, for example through the renegotiation of either obligation.

The Court then went on to find that it 'is not open to a Contracting State to enter into an agreement with another State which conflicts with its obligations under the Convention. This principle carries all the more force in the present case given the absolute and fundamental nature of the right not to be subjected to the death penalty and the grave and irreversible harm risked by the applicants.'[73] Accordingly, the UK 'was under a paramount obligation to ensure that the arrest and detention did not end in a manner which would breach the applicants' rights'.[74] Because the UK did not attempt to negotiate with Iraqi authorities to obtain binding guarantees from them that the applicants would not be subjected to the death penalty or other ill-treatment, as it could have, the applicants' ECHR

[72] *Al-Saadoon and Mufdhi v. United Kingdom*, App. No. 61498/08, Judgment, 2 March 2010, paras 126–8.
[73] Ibid., para. 138.
[74] Ibid., para. 140.

rights were violated. As for the resulting norm conflict, the Court was of the view that

> . . . the Government have failed to establish that there were no realistic or practicable means available to them by which to safeguard the applicants' fundamental human rights. In these circumstances, the 'objective impediment' claimed by the Government, namely the absence, on 31 December 2008, of any available course of action consistent with respect for Iraqi sovereignty other than the transfer of the applicants, was of the respondent State's own making.[75]

Note how the Court's analysis actually rests on an underlying *value* judgment about the importance of Article 2 and 3 ECHR rights, and the *non-refoulement* obligation which protects them. If it had wanted to, it could have avoided the norm conflict by qualifying this obligation in an extraterritorial setting, but it chose not to, precisely because of the importance it gave to the rights at hand. Note also that whatever the European Court's ruling, this does not change the fact that the UK *did* have an international legal obligation to surrender the applicants to Iraq. As in *Soering*, in *Al-Saadoon* the UK brought itself into a situation of unresolvable norm conflict—a conflict of its 'own making', but a genuine conflict nonetheless.[76] And as in *Soering*, that conflict only had a political solution: the UK could have flatly refused to surrender the applicants to Iraq; it could have tried to negotiate with Iraq and obtain guarantees that the applicants would not be subjected to the death penalty—though Iraq could of course always have refused to negotiate; or, it could have disobeyed its ECHR obligations and surrendered the applicants to Iraq. The UK, as we know, chose the third option.[77] However, the fact that most of us might think that as a matter of policy it should have opted for the second or the first does not mean that this choice was *legally* warranted. The choice was an entirely political one—but it is one for which the UK has incurred state responsibility for violating the ECHR, both with regard to the substance of the *non-refoulement* claim and with regard to the European Court's interim measures order.[78]

[75] Ibid., para. 162.

[76] See also M. Cross and S. Williams, 'Between the Devil and the Deep Blue Sea: Conflicted Thinking in the *Al-Saadoon* Affair', (2009) 58 *ICLQ* 689; N. Bhuta, 'Conflicting International Obligations and the Risk of Torture and Unfair Trial', (2009) 7 *JICJ* 1133.

[77] The norm conflict was even openly acknowledged by the UK government—see Letter dated 26 January 2008 from the Minister of State for the Armed Forced to the Chairman of the Joint Committee for Human Rights, available at <http://www.publications.parliament.uk/pa/jt200910/jtselect/jtrights/85/85we03.htm>. Relying on the Court of Appeal's ruling that the applicants were not within the UK's jurisdiction, and having regard to the expiry of the UK forces' mandate on 31 December 2008, the letter states, *inter alia*, that the government was 'faced with the option of breaching the Rule 39 measure or acting unlawfully in international law'. Of course, the former was no less a violation of international law, particularly in the light of the European Court's decision that the applicants were at all times within the UK's jurisdiction.

[78] For a more extensive discussion of *Al-Saadoon*, see M. Milanovic, 'Al-Saadoon and Mufdhi Merits Judgment', *EJIL: Talk!*, 2 March 2010, available at <http://www.ejiltalk.org/al-saadoon-and-mufdhi-merits-judgment/>.

4. Is *Lex Specialis* a Rule of Conflict Avoidance or of Conflict Resolution?

Having examined the tools that we have at our disposal for avoiding or resolving norm conflicts, I will now turn to finding out the proper place of *lex specialis* in the toolbox. Namely, despite all that has been written on *lex specialis* and the relationship between IHL and IHRL, the meaning of the maxim remains entirely unclear. In the framework that I have given above, we are still unsure whether *lex specialis* is a rule of norm conflict *avoidance* or a rule of norm conflict *resolution*, and this has extraordinary consequences for the practical utility of this rule. In my view, *lex specialis* can only be a tool of conflict avoidance, and not a particularly impressive one at that, as I will now try to explain.

That *lex specialis* can mean two essentially different things has long been recognized,[79] but the importance of this distinction has sometimes been downplayed or overlooked.[80] To see how the distinction between avoidance and resolution remains fundamental, we need only look at a familiar IHL/IHRL example—the lawfulness of preventive detention.

IHL either authorizes or considers as lawful at least two forms of preventive detention during international armed conflicts.[81] First, under Article 21 of the Third Geneva Convention, prisoners of war may be subjected to internment, i.e. they may be detained on purely preventive grounds, so that they do not rejoin the hostilities. IHL does not require the detaining power to prove that it is *necessary* to detain POWs—it simply presumes that this is the case, and POWs may be detained until the cessation of active hostilities. Secondly, under Articles 41–43 and 78 of the Fourth Geneva Convention, civilians can, like combatants, also be subjected to internment 'if the security of the Detaining Power makes it absolutely necessary'. IHL therefore poses a stricter requirement for the internment of civilians than of combatants, as it requires a positive showing of necessity.

When it comes to IHRL, different instruments regulate detention in different ways. Article 9(1) ICCPR provides that

> [e]veryone has the right to liberty and security of person. No one shall be subjected to arbitrary arrest or detention. No one shall be deprived of his liberty except on such grounds and in accordance with such procedure as are established by law.

Unlike Article 9(1) ICCPR, which sets out a standard prohibiting *arbitrary* detention, Article 5(1) ECHR is much more specific:

[79] See, e.g., A. Lindroos, 'Addressing Norm Conflicts in a Fragmented Legal System: The Doctrine of *Lex Specialis*', (2005) 74 *Nord J Int'l L* 27, esp. at 46; N. Prud'homme, '*Lex Specialis*: Oversimplifying a More Complex and Multifaceted Relationship?', (2007) 40 *Israel L Rev* 355, at 366 *et seq*; Droege, above note 7, at 523–4; Krieger, above note 17, at 268–70; ILC Fragmentation Study, paras 56–7, and the authorities cited therein.

[80] See ILC Fragmentation Study, paras 88–97.

[81] See generally J. Pejic, 'Procedural Principles and Safeguards for Internment/Administrative Detention in Armed Conflict and Other Situations of Violence', (2005) 87 *IRRC* 375.

Everyone has the right to liberty and security of person. No one shall be deprived of his liberty save in the following cases and in accordance with a procedure prescribed by law.

The list that elaborates on the 'save in the following cases' clause comprises of six possible situations, none of which can be reasonably interpreted to allow for preventive security detention. Article 5 was quite deliberately drafted exhaustively, in order to prevent overly expansive interpretations, intrusive on personal liberty, of something like the ICCPR arbitrariness standard.

To briefly pose the problem: if somehow, unfathomably, France decided to declare war and invade the UK, would the detention of French POWs on UK soil constitute a violation of Article 5(1) ECHR, in the absence of a UK derogation? Less fancifully, but with the added problem of extraterritoriality, was the detention of thousands of persons by the UK as an occupying power in Iraq after the April 2003 invasion also a violation of Article 5(1), prior to the June 2004 Security Council authorization in Resolution 1546 that could arguably prevail over the ECHR, per *Al-Jedda*?

The answer to these two questions depends on what we think the nature of the *lex specialis* rule is. If it was just a rule of norm conflict avoidance, we could interpret the Article 9(1) ICCPR arbitrariness standard while taking into account the rules of IHL, and we could say that preventive detention in times of international armed conflict in accordance with IHL is not arbitrary. (This, of course, is precisely what the ICJ did in *Nuclear Weapons* in regard to the same standard in relation to deprivation of life.) But, if *lex specialis* was merely a method of avoidance, we would be powerless if the same issue was examined under Article 5(1) ECHR. That provision allows for no 'window' through which IHL could enter; no reasonable interpretation of its text in light of its object and purpose would reach a compatible result. In the absence of a derogation, it is only if the European Court *forcibly* read Article 5 ECHR as if it set out an arbitrariness standard similar to Article 9 ICCPR that conflict avoidance would be feasible. However, not only would such rewriting of the treaty be judicial vandalism, not interpretation, as Lord Bingham so aptly put it, it would also have nothing to do with *lex specialis*. It would in effect amount to a court *disregarding* a clear norm emanating from one treaty in favour of another, on the basis of a *policy* judgment as to the norm's desirability, reasonableness, or effectiveness.

If, on the other hand, *lex specialis* was not a rule of norm conflict avoidance, but of norm conflict resolution, then and only then could we say that IHL *prevails* over the ECHR in some quasi-hierarchical sense (presumably because states intended it to do so), and that it precludes the UK's responsibility for failing to abide by the express requirements of Article 5. But there is *simply no evidence* that *lex specialis* is in fact a rule of conflict resolution. No treaty says so. It has certainly never been used as such in an IHL/IHRL context, least of all by the ICJ, which only used to interpret general IHRL standards such as arbitrariness in light of IHL.[82] Neither

[82] Perhaps the closest to doing so was the Inter-American Commission on Human Rights, which was overruled on that point by the Inter-American Court. See further C. McCarthy, 'Human Rights and the Laws of War under the American Convention on Human Rights', (2008) *EHRLR* 762, esp. at 767 *et seq*.

the texts of the relevant treaties nor the official positions of states support an inference of a state intent to override the express language of human rights treaties by having regard to a principle as nebulous as *lex specialis*,[83] which is for that matter, unlike *lex posterior*, not even mentioned in the VCLT. If anything, the text of the human rights treaties makes it clear that states are supposed to use derogations to avoid conflicts with IHL.[84] Moreover, that states have so far not fully complied with their IHRL obligations in armed conflict hardly counts as an inference of such intent, as they do not fully comply with their IHL obligations either. Perhaps in *some* cases it might be possible to infer an intent by contracting parties to apply the more specific treaty *against the express terms* of a more general treaty, but this is most certainly not the case with IHL and IHRL.[85]

More fundamentally, *lex specialis* as a rule of conflict resolution rests on an unstated assumption—that for any given situation at any given point in time there is one, and can *only* be one, expression of state consent or intent as to how that situation is to be regulated.[86] But that assumption is manifestly unfounded. As we have seen above, states are, like people, perfectly capable of assuming contradictory commitments. And just as *lex specialis* is unable to resolve conflicts such as those in *Matthews*, *Soering*, or *Al-Saadoon*, so it is unable to resolve conflicts between IHL and IHRL. In international law *lex specialis* is nothing more than a sub-species of harmonious interpretation, a method of norm conflict avoidance.[87] All it can do is assist in the interpretation of general terms and standards in either IHL or IHRL by reference to more specific norms from the other branch. It can hence help us determine whether particular deprivations of life or liberty during an armed conflict are 'arbitrary' or not. It can also work both ways—for instance, the more detailed IHRL norms on fair trial and a developed human rights jurisprudence can help us interpret the more general Common Article 3 requirement for trials satisfying 'judicial guarantees which are recognized as indispensable by civilized peoples'.[88]

But it can do no more than that.[89] Above all, it cannot create hierarchies where there are none. And precisely because the use of the term *lex specialis* to describe the

[83] See also ILC Fragmentation Study, para. 104, where the ILC considered that *lex specialis* in the context of the *Nuclear Weapons* case only affected the assessment of the Art. 6 ICCPR arbitrariness standard. The ILC does not consider a scenario where no such malleable standard was available.

[84] See *Wall*, above note 16.

[85] But see Krieger, above note 17, at 272; Droege, above note 7, at 524.

[86] See Pauwelyn, above note 30, at 388.

[87] See M. Akehurst, 'The Hierarchy of the Sources of International Law', (1974–1975) 47 *BYIL* 273, stating that '*lex specialis* is nothing more than a rule of interpretation'; Pauwelyn, at 410. Confusingly, *lex specialis* is sometimes also invoked to describe the relationship between general (customary) international law and treaties, to the effect that treaties usually override custom. But, as stated above, this is only so because most customary (and treaty) rules apply so long as states do not agree differently. See further Lindroos, above note 79, at 49 *et seq*. As we have seen in *Al-Saadoon*, however, a conflict between treaty and custom is nonetheless perfectly possible.

[88] This is in effect what the US Supreme Court did in *Hamdan v. Rumsfeld*, 548 U.S. 557 (2006), n. 66. See further M. Milanovic, 'Lessons for Human Rights and Humanitarian Law in the War on Terror: Comparing *Hamdan* and the Israeli *Targeted Killings* Case', (2007) 89 *IRRC* 373, 389 *et seq*. See also Krieger, above note 17, at 274.

[89] See also T. Thienel, 'The Georgian Conflict, Racial Discrimination and the ICJ: The Order on Provisional Measures of 15 October 2008', (2009) 9 *HRLR* 465, 467–9.

relationship between IHRL and IHL creates a false impression of facility, its use should be avoided.[90] The joint application of two bodies of law indeed requires a great deal of avoidance and creativity, involving a host of different legal, policy, and value judgments. To an extent these judgments will be governed by our impression of which norm is more 'special' in regard to the subject-matter, parties, level of detail, etc.—but that is only one factor among many. And it is because *lex specialis*, like other rules of norm conflict avoidance, must operate within the permissible bounds of interpretation, and may not stray into disregarding perfectly clear yet conflicting rules, that in a number of areas the relationship between IHL and IHRL may actually be that of unresolvable norm conflict.[91] I will now turn to examining some of them.

5. Areas of Potentially Unresolvable Norm Conflict

A. Preventive detention and judicial review of detention

We have just examined one area of potentially unresolvable norm conflict—detention. Let us further compare three situations of detention, all actual situations occurring in Iraq. First, in an *Al-Jedda*-type situation a genuine norm conflict exists between a Security Council authorization and Article 5(1) ECHR, which can be resolved in the former's favour by Article 103 of the Charter. Secondly, in an *Al-Saadoon*-type situation, security detention is transformed into a somewhat unorthodox form of pre-trial detention that is in principle compatible with Article 5(1) ECHR. Thirdly, in the situation of preventive security detention in pre-June 2004 occupied Iraq without any Security Council interference, we have a pure conflict between IHL and IHRL. With regard to the ICCPR, that conflict can be avoided through the harmonious interpretation of the Article 9 ICCPR arbitrariness standard. With regard to Article 5 ECHR, in the absence of a derogation nothing can be done to either avoid or resolve the conflict, short of resorting to probably illegitimate, forceful means of avoidance.

Does the UK's failure to derogate from Article 5 ECHR mean that all of the persons whom it detained as an occupying power in Iraq before the passing of Resolution 1546, or as an occupying power in Afghanistan, were detained unlawfully, IHL notwithstanding? The answer to that question is regrettably a resounding yes as far as the ECHR is concerned. And this is entirely the UK's fault, since it made a conscious political choice not to use the method of avoidance provided for by the ECHR itself, preferring instead to deny that the ECHR applied altogether. There is moreover nothing absurd about such a result, nor does it lead to some sort of *non liquet*. A court dealing with the matter can answer any of the questions asked

[90] Similarly on the ineffectiveness and conceptual vagueness of *lex specialis*, see Prud'homme, above note 79.

[91] Again, where the line between interpretation and legislation actually lies is a difficult and context-specific question, dependent (at least in a positivist framework) on social practices and conceptions of political legitimacy.

of it. Was the detention authorized or at least permitted by IHL? Yes. Was it nonetheless contrary to the ECHR? Again, yes. That the law is in a state of antinomy is simply not the court's problem, after it has weighed all of the options at its disposal to either avoid or resolve that conflict.[92]

In addition to the grounds of detention, another possible area of unresolvable conflict is that of judicial review of the lawfulness of detention. Here the ICCPR and the ECHR are exactly the same: both of them demand such review, and allow for no exceptions from that rule. Thus, Article 5(4) ECHR provides that

> [e]veryone who is deprived of his liberty by arrest or detention shall be entitled to take proceedings by which the lawfulness of his detention shall be decided speedily by a court and his release ordered if the detention is not lawful.

While Article 9(4) ICCPR stipulates that

> [a]nyone who is deprived of his liberty by arrest or detention shall be entitled to take proceedings before a court, in order that that court may decide without delay on the lawfulness of his detention and order his release if the detention is not lawful.

When it comes to IHL, because in respect of POWs it requires no showing of necessity that they be detained, subsequent to a determination of their status under Article 5 of the Third Geneva Convention, it also requires no judicial process of reviewing their detention. In respect of interned civilians, for whose preventive detention a necessity must be shown, Article 43(1) of the Fourth Geneva Convention does require periodic review 'by an appropriate court *or administrative board*'.

The IHRL norm that 'everyone' or 'anyone' must be able to challenge his detention before *a court*, period, cannot be reconciled, absent a derogation, with the IHL norms that the detention of POWs need not be reviewed at all, while the detention of civilians can be reviewed even by a mere administrative board. The only possible accommodation that IHRL can make for IHL in this respect is to read standards such as 'speedily' or 'without delay' more flexibly in armed conflict, but the core of the unambiguously expressed norm cannot just go away. Now, if one takes a more formal or narrow definition of norm conflict, then one could say that there is no conflict here between IHL and IHRL, as nothing prohibited the state from assuming a further set of IHRL obligations that are stricter than its IHL obligations.[93] Substantively, however, and above all *practically* speaking, we are faced with an unresolvable norm conflict, because one branch of the law regards a situation as perfectly lawful, while the other does not.

[92] This is in fact precisely what happened in one of the interstate cases initiated by Cyprus against Turkey before the now defunct European Commission on Human Rights, in which a majority of the Commission held that the preventive detention of POWs by Turkey violated Article 5 ECHR in the absence of a derogation—see *Cyprus v. Turkey*, App. Nos 6780/74 and 6950/75, Report of the Commission, adopted on 10 July 1976. Françoise Hampson finds such a result to be absurd, apparently on policy grounds—see Hampson, above note 21, at 565–6. Legally speaking, however, though it might be quite undesirable to have an unresolvable norm conflict, there is nothing absurd about it, particularly when Turkey could have used a derogation to avoid it.

[93] See Cassimatis, above note 13, at 633.

Moreover, not even a derogation might be able to completely avoid a conflict between IHRL and IHL in respect of detention. Recall that a state may take measures of derogation only 'to the extent strictly required by the exigencies of the situation, provided that such measures are not inconsistent with its other obligations under international law'.[94] Therefore, even if during an armed conflict a state party derogated from Article 5 ECHR or Article 9 ICCPR to allow for preventive detention without judicial review, in a manner completely consistent with IHL, this would not necessarily suffice to make that derogation stand—the measures taken still need to be 'strictly required', as a matter of some sort of objective external assessment.

For example, a derogation restating the blanket IHL position that combatants may *always* be preventively detained may be challenged, say if the armed conflict is quite protracted and the burden of detention consequently grows greater, the security risk posed by the POWs is small, and less restrictive means than internment for preventing their return to hostilities are available. Or, in respect of the internment of civilians in occupied territory, a state which derogates from the ICCPR or the ECHR to allow for the review of their detention by mere administrative boards, but is in reality quite capable of creating independent courts who could do the reviewing, ultimately might not be able to rely on its derogation no matter what IHL might say.[95]

B. Necessity in targeting

The detention example of unresolvable norm conflict is far-reaching enough, but my next example, that of targeting, is no less important. In IHL, the traditional position has been that combatants may always be targeted so long as they are not *hors de combat*,[96] and this position holds true with some temporal limitations even for civilians taking a direct part in hostilities.[97] In other words, one belligerent party does not need to prove any kind of necessity to kill combatants belonging to the other belligerent party, in order to be able to do so lawfully. 'Those who belong to armed forces or armed groups may be attacked at any time.'[98]

When it comes to IHRL, there are significant variations among the relevant treaties when it comes to conditions under which a state may deprive individuals of life. Article 6(1) ICCPR stipulates that '[n]o one shall be arbitrarily deprived of his life', while Article 4(2) thereof prohibits derogations from this prohibition under any circumstances. Article 2 ECHR prohibits any intentional deprivation of life,

[94] See Art. 15 ECHR, Art. 4 ICCPR.

[95] See Human Rights Committee, General Comment No. 29, UN Doc. CCPR/C/21/Rev.1/Add.11 (2001), para. 16, where the Committee asserts, perhaps too broadly, that states may never derogate from the requirement for judicial review of detention.

[96] See Y. Dinstein, *The Conduct of Hostilities under the Law of International Armed Conflict* (Cambridge University Press, 2004), 27–9, 145; N. Lubell, 'Challenges in Applying Human Rights Law to Armed Conflict', (2005) 87 *IRRC* 737, 745–6.

[97] See Art. 51(3) of Additional Protocol I.

[98] Y. Sandoz et al., *Commentary on the Additional Protocols of 8 June 1977 to the Geneva Conventions of 12 August 1949* (1987), 1453.

save 'when it results from the use of force which is no more than absolutely necessary: (a) in defence of any person from unlawful violence; (b) in order to effect a lawful arrest or to prevent the escape of a person lawfully detained; (c) in action lawfully taken for the purpose of quelling a riot or insurrection'. On the other hand, Article 15(2) does allow derogations from the right to life (only) 'in respect of deaths resulting from lawful acts of war', thus opening a window for IHL.

The ICCPR and the ECHR again pose somewhat different interpretative problems. The arbitrariness standard of the ICCPR allows for an easier reconciling with IHL, as the ICJ did in *Nuclear Weapons*. As for the ECHR, Article 2(2)(c) would govern situations of internal armed conflict, yet any deprivation of life would still have to be 'absolutely necessary', while Article 15(2) would apply to situations of international armed conflict.[99] But again, even with Article 15(2) in the picture, the derogation still needs to be 'strictly required by the exigencies of the situation' under Article 15(1), and of course needs to actually be *made*.

This brings me to the potentially unresolvable norm conflict with IHL. Both the ICCPR and the ECHR can easily be interpreted as requiring a showing of necessity before *any* intentional deprivation of life,[100] while, as we have seen, IHL purposefully does not require such a showing in respect of combatants or civilians taking a direct part in hostilities.[101] The IHRL necessity standard may be relaxed somewhat to take into account the fact of armed conflict, but it is hard to see how it can be totally extinguished, as IHL warrants, as soon as an armed conflict takes place and solely on the basis of the person's status.

To see this norm conflict at work we need only to take a look at the *Targeted Killings* judgment of the Supreme Court of Israel.[102] The Court's prime desire in that case was to give clear guidance to the Israeli armed forces and secret services as to under what conditions they can target a suspected terrorist.[103] One of the conditions that it imposed was that a terrorist may not be targeted if the incidental damage caused to nearby civilians is disproportionate to the direct military advantage anticipated from the elimination of the said terrorist.[104] This is of course the IHL proportionality principle, whose purpose is to allow the use of lethal force but to minimize the collateral damage arising from its use.[105]

[99] It is worth noting that no ECHR state party has ever made a derogation under Art. 15(2). For a similar analysis of this example, see Hampson, above note 21, at 564–5.

[100] See, e.g., *McCann and Others v. United Kingdom*, App. No. 18984/91, Judgment, 27 September 1995, paras 149–50.

[101] See also McCarthy, above note 82, at 773 *et seq*; Krieger, above note 17, at 280–1.

[102] *The Public Committee against Torture in Israel et al. v. The Government of Israel et al*, HCJ 769/02, Judgment, 11 December 2006, available at <http://elyon1.court.gov.il/Files_ENG/02/690/007/a34/02007690.a34.HTM> (hereinafter *Targeted Killings*).

[103] Before examining the targeting question, the Court had to resolve two preliminary questions: first, it (somewhat questionably) held that the conflict between Israel and Palestinian groups in the occupied territories should be qualified as an international armed conflict; second, it held that the suspected terrorists designated as targets can be qualified as civilians taking a direct part in hostilities—see *Targeted Killings*, paras 6–18, 33–40.

[104] *Targeted Killings*, para. 40.

[105] See Art. 51(5)(b) of Additional Protocol I.

But the Court imposed a further condition—a terrorist may not be targeted if less harmful means can be employed, i.e. if he can be captured and put on trial. In the Court's own words, '[t]rial is preferable to use of force. A rule-of-law state employs, to the extent possible, procedures of law and not procedures of force.'[106] Now, it is not entirely clear whether the Court derived this rule from IHRL or from domestic constitutional law, but it is clear that it was a human rights norm that it was applying.[107] It used the kernel of a human rights rule—that necessity must be shown for any intentional deprivation of life—to restrict the application of an IHL rule—that in armed conflict no necessity need be shown for the killing of combatants or civilians taking a direct part in hostilities. This does not mean, of course, that when applied to a specific set of facts the legally different IHL and IHRL norms would ultimately produce different results. What it means is that in *every* case some measure of justification will have to be offered as to why the killing of a particular individual was necessary on the facts of that case, and that this justification cannot be exhausted merely by invoking the person's status under IHL.

The Court's holding was not based on *lex specialis* or any other form of mechanical reasoning. It made a policy and value judgment that in the context of the prolonged Israeli occupation of Palestinian territories the traditional IHL answer was no longer satisfactory, and it had a legal basis in human rights law to say so.[108] When one adversary possesses overwhelming strategic and tactical superiority; in the context of an occupation, especially a prolonged one; during a limited insurgency or non-international armed conflict; in situations, in other words, which lend themselves more easily to non-lethal approaches, the imposition of an IHRL necessity requirement becomes more and more attractive.[109]

It is questionable, however, whether this necessity requirement could be effectively applied in a more traditional battlefield setting. There is perhaps no other area of potential norm conflict where the infusion of IHL with IHRL could lead to a greater slide into utopia, and a consequent slide into irrelevance. It might be better to have *some* rules which are effective, than rules which satisfy our moral intuitions but are honoured only in their breach. If the IHRL necessity requirement is to be

[106] *Targeted Killings*, para. 40.

[107] The Court thus cited the *McCann* judgment of the European Court—see above note 100.

[108] 'Arrest, investigation, and trial are not means which can always be used. At times the possibility does not exist whatsoever; at times it involves a risk so great to the lives of the soldiers, that it is not required. However, it is a possibility which should always be considered. It might actually be particularly practical under the conditions of belligerent occupation, in which the army controls the area in which the operation takes place, and in which arrest, investigation, and trial are at times realizable possibilities. Of course, given the circumstances of a certain case, that possibility might not exist. At times, its harm to nearby innocent civilians might be greater than that caused by refraining from it. In that state of affairs, it should not be used.' *Targeted Killings*, para. 40 (citations omitted). See also Milanovic, above note 88, at 390–2.

[109] See in that regard the excellent recent study by M. Sassòli and L. Olson, 'The Relationship between International Humanitarian and Human Rights Law where it Matters: Admissible Killing and Internment of Fighters in Non-International Armed Conflicts' (2008) 90 *IRRC* 599, at 613–14, who come to a similar conclusion when it comes to the dependence of the IHRL rule on the level of state control. Where regretfully I part company with them is in their assertion that the *lex specialis* principle has something to do with this outcome. Similarly, see also L. Doswald-Beck, 'The Right to Life in Armed Conflict: Does International Humanitarian Law Provide All the Answers?', (2006) 88 *IRRC* 881; Prud'homme, above note 79, at 391.

applied at all in such situations, *contra* IHL, it will in any event have to be read down significantly to be effective, flexible, and practicable.[110] But again, even if IHRL is read down so far that it would *never* consider as arbitrary a killing of a combatant who is not *hors de combat*, per IHL, this would not be because of the *lex specialis* principle, but because of a policy judgment that IHRL has to be (forcibly) read down far enough to be effective. This policy judgment on the other hand requires a degree of *legitimacy* to be made, and again it is in my view doubtful that courts possess such legitimacy to fix antinomies left to them by states, the law's creators.

C. (Transformative) occupation

My final example of potentially unresolvable conflict between IHRL and IHL will be that of occupation. Suppose that the UK becomes the belligerent occupant of a territory that has Sharia as part of its domestic law—e.g. Iran. The Penal Code of Iran prescribes stoning as a punishment for adultery, and even has a stunning provision which stipulates that '[t]he size of the stone used in stoning shall not be too large to kill the convict by one or two throws and at the same time shall not be too small to be called a stone'.[111] Since the UK is in effective overall control and thus possessing jurisdiction over a part of Iranian territory, the ICCPR and ECHR are applicable. Both treaties require the UK to ensure or secure the human rights of all individuals within its jurisdiction, even against violations by other actors. A punishment of stoning for adultery undoubtedly qualifies at the very least as inhuman treatment under either of these treaties.

As for IHL, Article 43 of the Hague Regulations provides that '[t]he authority of the legitimate power having in fact passed into the hands of the occupant, the latter shall take all the measures in his power to restore, and ensure, as far as possible, public order and safety, while respecting, unless absolutely prevented, the laws in force in the country'. Article 64 of the Fourth Geneva Convention is even more pertinent for our example, stipulating that '[t]he penal laws of the occupied territory shall remain in force, with the exception that they may be repealed or suspended by the Occupying Power in cases where they constitute a threat to its security or an obstacle to the application of the present Convention'.

[110] Some scholars have resorted to a different, yet with respect ultimately unconvincing, approach to avoiding this norm conflict, by claiming that IHL like IHRL also imposes a necessity requirement for civilians taking a direct part in hostilities, and possibly even combatants—see N. Melzer, *Targeted Killing in International Law* (Oxford University Press, 2008), 228, 336 *et seq*, esp. at 397; Droege, above note 6, at 526. See also Abresch's review of Melzer's book, in (2009) 20 *EJIL* 449. Another step in this direction was made in the recently published ICRC Interpretative Guidance on the Notion of Direct Participation in Hostilities under International Humanitarian Law, available at <http://www.icrc.org/Web/eng/siteeng0.nsf/htmlall/direct-participation-report_res/$File/direct-participation-guidance-2009-icrc.pdf>, at 77 *et seq*. For commentary, see D. Akande, 'Clearing the Fog of War? The ICRC's Interpretive Guidance on Direct Participation in Hostilities', *EJIL: Talk!*, 4 June 2009, available at <http://www.ejiltalk.org/clearing-the-fog-of-war-the-icrcs-interpretive-guidance-on-direct-participation-in-hostilities/>.

[111] Art. 104, Penal Code of Iran, available at <http://www.wluml.org/english/newsfulltxt.shtml?cmd[157]=x-157-555118>.

The example that I have given is at least that of an apparent norm conflict. On one hand, IHRL commands the UK to take all possible measures to prevent the stoning of adulterers in the territory that it has occupied. On the other hand, because it considers occupation to be a temporary situation which requires deference to the displaced sovereign, IHL prohibits the UK from changing the laws of the occupied country, particularly its penal laws.[112] Can these two contradictory obligations somehow be reconciled?[113]

To do that, we would either have to read down the IHL obligation, or the IHRL obligation. When it comes to IHL, Article 43 of the Hague Regulations is a somewhat easier target—it prohibits the occupant from altering the domestic law of the occupied territory unless it is 'absolutely prevented' from maintaining it in force. We can therefore say that the UK's IHRL obligation to put a stop to any inhuman treatment does, in fact, 'absolutely prevent' it from respecting Iran's domestic law.[114] Article 64 of the Fourth Geneva Convention is less malleable—the domestic penal law must be kept in force, with only two possible exceptions. Does maintaining stoning for adultery constitute a threat to the security of the UK, the occupying power? Hardly. Is it an obstacle to the application of the Convention itself? Well, no, not really. There is no other obligation of the UK derived from the Convention that it could not fulfil while allowing the courts of the *ancien régime* to go about their brutal business. In reality, therefore, Article 64 can only be read down forcibly, if it is in effect rewritten to accommodate change in domestic laws which are incompatible with the occupant's (and perhaps even the displaced sovereign's) IHRL obligations.

If IHL cannot consistently and reasonably be read down to accommodate IHRL, how about IHRL? We could say something like this: the UK's positive obligation to secure or ensure respect for human rights in occupied territories even by non-state actors is one of due diligence. It requires the state only to do all that it reasonably can to prevent violations. Because in this particular instance doing so would require the UK to violate some of its other obligations under international law, the UK's positive obligation should be read as not requiring the UK to do so.

This is a textually and conceptually perfectly reasonable reading down of IHRL. But recall what I have said above—though a price must be paid if IHRL is to be applied jointly with IHL, that price must not be too steep. I personally—and I suspect most other international lawyers—would never read down IHRL to accommodate stoning for adultery, even if as a technical matter it may be perfectly appropriate to do so.

This, however, is simply because of a policy and value judgment that I have made on the issue, because I *care* more about human rights than about consistency in international law in this particular instance. It is the values enshrined in the IHRL

[112] See generally M. Sassòli, 'Legislation and Maintenance of Public Order and Civil Life by Occupying Powers', (2005) 16 *EJIL* 661.

[113] On transformative occupations generally, see A. Roberts, 'Transformative Military Occupation: Applying the Laws of War and Human Rights', (2006) 100 *AJIL* 580; G. Fox, *Humanitarian Occupation* (Cambridge University Press, 2008).

[114] See Sassòli, above note 112, at 676; Thienel, above note 40, at 126–7.

treaties that would make their reading down forced and inappropriate. Legally, then, there is an unresolvable norm conflict, with no norm having priority over the other. The UK would have a political choice to make as to which obligation to keep, and which to breach, and that is that. Yes, we would all like it to choose the ICCPR and the ECHR over the Fourth Geneva Convention in this particular instance, especially because it is likely that it would not suffer many ill consequences for its breach of the latter, but we would do so because of our own value judgment and a political calculation that this is how things should be, not because the law warrants this choice.

This example is particularly instructive because of how potentially far-reaching it could be. First, unlike the previous examples of targeting and detention, it does not deal merely with a state's negative obligations towards individuals, whose rights it must not violate through the actions of its own organs or agents, but with its positive obligation to secure or ensure respect for human rights even by other actors. Secondly, in my previous examples a state would not incur any responsibility for having to abide by a stricter IHRL standard as opposed to a looser IHL standard, though as a practical matter its liberty to act might be significantly curtailed. In this example, however, conflicting IHRL and IHL *obligations* meet head on, and one cannot be fulfilled without violating the other. Finally, the example again exposes the shifting boundary between interpretation and legislation, and the need to somehow maintain it. As a matter of policy, we do not want to *entirely* abandon the rule that the occupant must not alter the legal system of the occupied territory—by and large, this is a good rule[115]—we just want to carve out a human rights exception to that rule. Absent a legislative change, however, this can probably be done only by either violating or doing violence to the Fourth Geneva Convention.

6. Concluding Remarks

Improved enforcement and further humanization of IHL through the application of IHRL is the ultimate goal of this entire project—to bring states that created these treaties to the logical, legal, and moral conclusions that are mandated by their text, but even more so by their fundamental normative underpinnings, the values that inspire them. This imperative is of particular cogency in an extraterritorial context. As we have seen, to have any hope of advancing this goal we must be practical and be prepared to apply IHRL standards in a more flexible fashion. Yet we also need to drop the pretence, inspired by a false analogy with domestic public law, that states have in their infinite wisdom created two legal regimes which mesh together perfectly. They simply do not, either in the texts of the treaties, or in the values

[115] Generally on the evolution of the conservationist premise of belligerent occupation as a temporary condition during which the order imposed by the displaced sovereign should not be varied see N. Bhuta, 'The Antinomies of Transformative Occupation', (2005) 16 *EJIL* 721.

and policy considerations behind them.[116] Frequently, their relationship is that of conflict.

To address these norm conflicts, we have a variety of tools at our disposal. The most helpful are those of norm conflict avoidance, which are interpretative in nature. Among these tools, but not among the most useful, is *lex specialis*. To again take preventive detention as an example, if faced with a pure IHL/IHRL case arising from the occupation of Iraq, the European Court might in the end forcibly read down Article 5 ECHR as if setting an arbitrariness standard that could accommodate IHL like Article 9 ICCPR, but this would not be as a consequence of the *lex specialis* principle, but because it weighed the competing policy considerations so as to warrant such a result even in the absence of a derogation.[117] As is so often the case with legalese Latin, *lex specialis* is descriptively misleading, vague in meaning, and of little practical use in application. It should be discarded generally, and it should especially not be used to describe the relationship between IHL and IHRL as a whole.[118]

More often than not, IHL treaties and human rights treaties can be reconciled and interpreted harmoniously. But there will be instances, there *are* instances, where this quest for harmony will fail, when the two bodies of law cannot be reconciled, when all legitimate methods of norm conflict avoidance and resolution will be exhausted, and when ultimately a political choice will have to be made as to which of the conflicting norms should be given priority over the other.

That choice can rarely, if ever, be made by a court—sometimes it is only the legislator who produced the antinomy who can provide the remedy for it. The more forced the methods of avoidance, the less legitimate they are; the line between interpretation and legislation may often be a fine one, but it is nonetheless still there.[119] And when all methods of norm conflict avoidance and resolution fail, there is nothing shameful in admitting defeat and saying that in this or that particular instance IHL and IHRL cannot be reconciled. Just as a UK court can, after doing its best, ultimately say that a particular act of Parliament is incompatible with the Human Rights Act, so it can say that the internment of POWs and civilians during the occupation of Iraq was lawful under IHL, but unlawful under the ECHR.[120]

It is, of course, not an easy thing for a court to say, nor should it be. No wonder then that courts rarely, if ever, openly acknowledge the possibility of unresolvable conflict. To take *Al-Skeini* as the prime example, it is often far easier for courts to deny that IHRL applies at all, based on entirely arbitrary criteria (viz. that the ECHR applies to a person killed in the custody of UK troops in Basra, but not to

[116] See also Schabas, above note 9.

[117] See also Lindroos, above note 79, at 42.

[118] See also Prud'homme, above note 79.

[119] For a discussion on the frequently complex and ambiguous nature of this boundary, see A. Kavanagh, 'The Elusive Divide between Interpretation and Legislation under the Human Rights Act 1998', (2004) 24 *OJLS* 259.

[120] But see Ben-Naftali and Shany, above note 27, at 103.

persons killed by UK troops on patrol[121]), switching IHRL on when it becomes morally intolerable not to do so, and off when it becomes impracticable, rather than addressing hotly disputed substantive issues which may expose unresolvable norm conflicts. However, norm conflicts are in themselves not so undesirable that they must be avoided or resolved at any cost. They are immanent to the international legal system, due to its decentralized, non-hierarchical nature, and the consensual character of its law-making processes.[122] They occur as much in peacetime as in wartime, intraterritorially as extraterritorially—e.g. in *Soering* and *Matthews*. Indeed, exposing an unresolvable norm conflict may ultimately prove to be *more* productive then forcibly avoiding it, as nothing will give a better incentive to states to improve the normative framework within which we must operate. And, of course, a good place for states to start would be for them to use mechanisms which are already in place, such as derogations.

[121] See also Hampson, above note 21, at 570.
[122] See also Lindroos, above note 79, at 28.

General Conclusion

One purpose of this study was to dispel the numerous methodological misconceptions which abound both in the jurisprudence on the extraterritorial application of human rights treaties and in academic commentary. The greatest of these is the *Bankovic* fallacy that the notion of state jurisdiction in human rights treaties reflects that notion of jurisdiction in general international law which delimits the municipal legal orders of states. I hope to have shown that these two concepts are not the same, and that conflating them is a category error that would produce completely absurd results if actually applied consequentially (which it is not). This is not to say that the word 'jurisdiction' has a *special* meaning in human rights treaties. Rather, it has several *ordinary* meanings in general international law, and the interpreter has to establish which of these meanings the treaties in fact use. That meaning, I have argued, is one of state control over territory, and perhaps individuals.

Clarifying the conceptual and doctrinal confusion stemming mainly from the different meanings of the word 'jurisdiction' is not only its own reward—it is a necessary first step in properly interpreting the relevant treaties. This is how interpretation works—it begins from the text. But *interpretation* alone, in the narrower sense of establishing the semantic meaning of a legal text, is insufficient. Even if the semantic meaning of the word 'jurisdiction' was deciphered and the one that I favour adopted, we would then still have to *construct* the rules which translate this meaning to specific factual situations.

What is an 'area'? What counts as 'effective overall control'? To which obligations does this jurisdictional requirement apply? These questions can only be answered by looking at the object and purpose of human rights treaties, and the policy considerations which drive both courts and states in their decision-making. I have argued that the two considerations which matter most are universality and effectiveness, while the one that should matter the least (or not at all) is sovereignty or title over territory. The same problems of effectiveness that we have examined in an extraterritorial context can occur intraterritorially as well, ranging from an internal armed conflict to a total loss of state control over a portion of its territory. Hence, the truly vexing cases are not those which occur outside the territory which the state *owns*, but outside the territory which it *controls*.

This brings us to the choice between the several possible models of extraterritorial application. As we have seen, a purely spatial model of jurisdiction such as state control of an area would lead to numerous situations that would be excluded from the treaties' scope of application, with states being free to do as they will whenever they act in a territory that they do not themselves control. To an extent, these difficulties can be mitigated by applying the spatial model to ever smaller geographical areas, down to places or artificially constructed objects. This, however, can only be done up to a point, and this is where the personal model of jurisdiction as control over individuals steps in. Yet, in turn, there is no principled, non-arbitrary way of limiting this model and preventing its collapse, because the normative pull of universality is such that it is hard to rationalize why a state which is fully in control of the conduct of its own organs and agents should not have the obligation to refrain from violating the human rights of others when it is perfectly capable of complying with this obligation.

Therefore, the rule that I propose is clear and simple: the state obligation to respect human rights is not limited territorially; however, the obligation to secure or ensure human rights is limited to those areas that are under the state's effective overall control. This is, in my view, the only model of state jurisdiction and extraterritorial application that is stable in the long run. Not only does it have the benefit of clarity and predictability, it also provides the best balance among its competitors between the often conflicting demands of universality and effectiveness.

As I have endeavoured to explain, it is precisely the fear of courts—particularly of the European Court—that applying human rights treaties to extraterritorial situations would be impracticable and would unreasonably restrict the states' freedom to act that has led to the current state of the jurisprudence, where the issue of state jurisdiction and extraterritorial application is not *really* the purely preliminary, threshold question that it should be, but rather involves a furtive assessment of the merits. Almost all of confusions and contradictions in the case law are ultimately explicable by the policy tension between universality and effectiveness. Where that tension is best addressed, however, is on the merits of substantively very difficult and complex cases. But because courts are reluctant to address these merits, they fashion wholly artificial limitations on the territorial scope of human rights treaties (or for that matter, domestic constitutional protections of individual rights), only to carve out even more artificial exceptions from these limitations when it would be morally intolerable not to do so. To return again to the example of *Al-Skeini*, its result was not determined by the arcane intricacies of the concept of state jurisdiction, let alone by the validity of an analogy between a prison and an embassy, but by the moral assessment by the judges that it would be unacceptable for a person killed in captivity by British soldiers to be unprotected by the ECHR, coupled with the parallel fear that extending the ECHR to the other five applicants would simply go too far, requiring courts to micromanage the use of military force under Article 2 ECHR. We have seen many similar cases in municipal jurisprudence.

This fear on the part of judges that they are institutionally now out of their depth, and their fear of inflexibility in extraordinary situations, must be adequately addressed for the question of extraterritorial application to truly become preliminary and divorced from the merits. Addressing these fears is necessarily going to be a gradual, and often messy, process. Flexibility would need to be shown not only by judges, but also by applicants and human rights activists. For their part, governments should be prepared to actually use the tools provided to them by the treaties themselves to add needed flexibility—most importantly derogations. Likewise, for all the possibility of unresolvable norm conflict, when interpreting human rights treaties courts should in appropriate situations take into account with more confidence the rules of other branches of the law, such as IHL. However, we must always be mindful of the limits on our faculties; the integrity of human rights regimes must be protected, and bodies of law alien to them, such as IHL, can be imported only to an extent.

Let me conclude by saying that I certainly do not suffer from the naïve belief that my proposed model of extraterritorial application will be adopted any time soon. At least when it comes to Strasbourg, it requires a major departure from existing jurisprudence, above all from *Bankovic*. Not only are judges naturally risk-averse, and are loathe to overturn existing precedent. It is even harder for them do so while the ground has not been sufficiently prepared, as it were, by introducing adequate flexibility into the ECHR system that could accommodate its application to Afghanistan or Pakistan or what have you. In turn, it is harder still to introduce such flexibility without actually being able to substantively apply the ECHR in an extraterritorial setting. This vicious circle will be hard to break.

It is thus far more likely that the European Court will continue with its current approach for as long as it can, by grasping the rigid spatial model of *Bankovic* in one hand, and carving out exception after exception from it with the other, when not doing so would offend its notions of substantive justice. And, of course, the more this unhappy state of affairs persists, the more the whole approach is exposed as unprincipled and unworkable. Ironically, it is often the relatively *small* cases—your Stephenses, Medvedyevs, or Litvinenkos—which show this most clearly. I would also submit that the circle has already started to break with the European Court's engagement with politically controversial and legally and factually complex situations in Chechnya and Turkey, which it has managed to deal with reasonably well despite the difficulties. Similarly, despite all its problems, the application of the ECHR to Iraqi cases by both the Strasbourg and UK courts shows us that this can be done successfully and that human rights can have an impact, and, at that, without the sky falling.

It was not the purpose of this study to try to reconcile the conflicting strands of jurisprudence, particularly that of the Strasbourg court. This study is not a brief, nor am I counsel in an actual case, trying to persuade the judges of the wisdom and coherence of their prior case law, which my case supposedly fits perfectly. Rather, as I see it, the benefit—perhaps the sole benefit—of an academic examination is that it can be more honest than that. The current Strasbourg approach is so flawed that it cannot be fixed with a minor 'clarification' here or a 'distinguishing' there. What it needs is radical surgery.

Whether this will happen depends primarily on the level of pressure that the European Court, and to a lesser extent other courts and human rights bodies, are subjected to with respect to extraterritorial application. It is only if cases keep piling up, if advocacy groups maintain their efforts, and if tinkering with the current approach becomes manifestly insufficient to prevent intolerable results, that a radical change will truly become possible. This, I think, is a fundamental point. Again, the reason why we have so many cases on extraterritorial application today that we did not have decades ago is not globalization, nor is it because of some true novelty in state behaviour. States have always killed, detained, or otherwise mistreated people outside their territory (although perhaps not through the use of drones or radioactive poisons)—indeed, this was a major purpose of their existence. Rather, it is because we have through the preceding decades increasingly internalized human rights norms and universality as their foundation that litigation and activism based on the extraterritorial application of human rights treaties have become both possible and effective. It is only if these processes continue to bear fruit that universality can truly become unbound.

Bibliography

BOOKS AND MONOGRAPHS

Bascomb, N., *Hunting Eichmann: How a Band of Survivors and a Young Spy Agency Chased Down the World's Most Notorious Nazi* (HMH, 2009).

Bossyut, M., *Guide to the "travaux préparatoires" of the International Covenant on Civil and Political Rights* (Nijhoff, 1987).

Brownlie, I., *Principles of Public International Law* (Oxford University Press, 6th edn, 2003).

Burgers, J.H. and Danelius, H., *The United Nations Convention against Torture* (Nijhoff, 1988).

Cassese, A., *International Law* (Oxford University Press, 2nd edn, 2005).

Coomans, F. and Kamminga, M. (eds), *Extraterritorial Application of Human Rights Treaties* (Intersentia, 2004), hereinafter Coomans and Kamminga.

Crawford, J., *The Creation of States in International Law* (Oxford University Press, 2nd edn, 2006).

Denza, E., *Diplomatic Law* (Oxford University Press, 3rd edn, 2008).

Dinstein, Y., *The Conduct of Hostilities under the Law of International Armed Conflict* (Cambridge University Press, 2004).

—— *The International Law of Belligerent Occupation* (Cambridge University Press, 2009).

Donnelly, J., *Universal Human Rights in Theory and Practice* (Cornell University Press, 2nd edn, 2003).

Dworkin, R., *Is Democracy Possible Here?* (Princeton University Press, 2006).

Evans, M. (ed.), *International Law* (Oxford University Press, 2nd edn, 2006), hereinafter Evans.

Fox, G., *Humanitarian Occupation* (Cambridge University Press, 2008).

Fox, H., *The Law of State Immunity* (Oxford University Press, 2nd edn, 2008).

Gardiner, R., *Treaty Interpretation* (Oxford University Press, 2008).

Gibney, M. and Skogly, S., *Universal Human Rights and Extraterritorial Obligations* (University of Pennsylvania Press, 2010), hereinafter Gibney and Skogley.

Gondek, M., *The Reach of Human Rights in a Globalising World: Extraterritorial Application of Human Rights Treaties* (Intersentia, 2009).

Greenberg, K. et al. (eds), *The Torture Papers* (Cambridge University Press, 2005).

Hathaway, J., *The Rights of Refugees under International Law* (Cambridge University Press, 2005).

Henkin, L., *Age of Rights* (Columbia University Press, 1996).

Jennings, R. and A. Watts, A. (eds), *Oppenheim's International Law* (Longman, 9th edn, 1992).

Kamminga, M. and Scheinin, M. (eds), *The Impact of Human Rights Law on General International Law* (Oxford University Press, 2009).

Kershaw, I., *Hitler 1889–1936: Hubris* (Norton, 1999).

Koskenniemi, M., *From Apology to Utopia: The Structure of International Legal Argument* (Cambridge University Press, 2005).

Lauterpacht, H., *International Law and Human Rights* (Archon Books, 1968; earlier editions 1930 and 1950).

Lubell, N., *Extraterritorial Use of Force Against Non-State Actors* (Oxford University Press, 2010).

Meessen, K. (ed.), *Extraterritorial Jurisdiction in Theory and Practice* (Kluwer, 1996).

Melzer, N., *Targeted Killing in International Law* (Oxford University Press, 2008).

Meron, T., *The Humanization of International Law* (Nijhoff, 2006).

Mowbray, A., *The Development of Positive Obligations under the European Convention on Human Rights by the European Court of Human Rights* (Hart Publishing, 2004).

Nowak, M., *CCPR Commentary* (Engel, 2nd revised edition, 2005).

Pauwelyn, J., *Conflict of Norms in Public International Law: How WTO Law Relates to other Rules of International Law* (Cambridge University Press, 2003).

Provost, R., *International Human Rights and Humanitarian Law* (Cambridge University Press, 2002).

Raustiala, K., *Does the Constitution Follow the Flag? The Evolution of Territoriality in American Law* (Oxford University Press, 2009).

Robertson. A. H. (ed.), 3 *Collected Edition of the 'Travaux Préparatoires' of the European Convention on Human Rights* (Nijhoff, 1976).

Ryngaert, C., *Jurisdiction in International Law* (Oxford University Press, 2008).

Shaw, M., *International Law* (Cambridge University Press, 6th edn, 2008).

Simpson, B., *Human Rights and the End of Empire* (Oxford University Press, 2004).

Singer, N.J., 1A *Statutes and Statutory Construction* (West, 6th edn, 2002).

Skogly, S., *Beyond National Borders: States' Human Rights Obligations in International Cooperation* (Intersentia, 2006).

Solum, L., *Semantic Originalism* (2008), available at <http://ssrn.com/abstract=1120244>.

Spiro, P., *Beyond Citizenship: American Identity After Globalization* (Oxford University Press, 2008).

Thornberry, P., *International Law and the Rights of Minorities* (Oxford University Press, 1993).

van Dijk, P. et al, *Theory and Practice of the European Convention on Human Rights* (Intersentia, 4th edn, 2006).

Wilde, R., *International Territorial Administration: How Trusteeship and the Civilizing Mission Never Went Away* (Oxford University Press, 2008).

ARTICLES AND BOOK CHAPTERS

Abresch, W., 'A Human Rights Law of Internal Armed Conflict: The European Court of Human Rights in Chechnya', (2005) 16 *EJIL* 741.

Akande, D., 'Clearing the Fog of War? The ICRC's Interpretive Guidance on Direct Participation in Hostilities', *EJIL: Talk!,* 4 June 2009, available at <http://www.ejiltalk.org/clearing-the-fog-of-war-the-icrcs-interpretive-guidance-on-direct-participation-in-hostilities/>.

Akehurst, M., 'Jurisdiction in International Law', (1972–1973) 46 *BYBIL* 145.

—— 'The Hierarchy of the Sources of International Law', (1974–1975) 47 *BYBIL* 273.

Allott, P., 'The Concept of International Law', (1999) 10 *EJIL* 31.

Alston, P., Morgan-Foster, J., and Abresch, W., 'The Competence of the UN Human Rights Council and its Special Procedures in relation to Armed Conflicts: Extraterritorial Executions in the "War on Terror"', (2008) 19 *EJIL* 183.

Anderson, K., 'Predators Over Pakistan', *The Weekly Standard*, 8 March 2010, available at <http://ssrn.com/abstract=1561229>.

Aust, A., 'Treaties, Territorial Application', in *Max Planck Encyclopedia of Public International Law*, available at <http://www.mpepil.com>.

Bailin, A., 'Case preview—R (Smith) v Secretary of State for Defence—on appeal from [2009] EWCA Civ 441', *UKSCblog*, 13 March 2010, available at <http://ukscblog.com/case-preview-%E2%80%93-r-smith-v-secretary-of-state-for-defence-on-appeal-from-2009-ewca-civ-441>.

Bell, C., 'Capitulations', in *Max Planck Encyclopedia of Public International Law*, available at <http://www.mpepil.com>.

Ben-Naftali, O. and Shany, Y., 'Living in Denial: The Application of Human Rights Treaties in the Occupied Territories', (2003) 37 *Israel L Rev* 17.

Benvenisti, E., 'Occupation, Pacific', in *Max Planck Encyclopedia of Public International Law*, available at <http://www.mpepil.com>.

Bhuta, N., 'The Antinomies of Transformative Occupation', (2005) 16 *EJIL* 721.

—— 'Conflicting International Obligations and the Risk of Torture and Unfair Trial', (2009) 7 *JICJ* 1133.

Borgen, C., 'Resolving Treaty Conflicts', (2005) 37 *Geo. Wash. Int'l L. Rev.* 573.

Buergenthal, T., 'To Respect and to Ensure: State Obligations and Permissible Derogations', in L. Henkin (ed.), *The International Bill of Rights: The Covenant on Civil and Political Rights* (Columbia University Press, 1981).

Buyse, A., 'A Legal Minefield—The Territorial Scope of the European Convention', (2008) 1 *Inter-American and European Human Rights Journal* 269.

Cassese, A., 'The *Nicaragua* and *Tadić* Tests Revisited in Light of the ICJ Genocide Judgment in Bosnia', (2007) 18 *EJIL* 649.

Cassimatis, A., 'International Humanitarian Law, International Human Rights Law, and the Fragmentation of International Law', (2007) 56 *ICLQ* 623.

Cerna, C., 'Extraterritorial Application of the Human Rights Instruments of the Inter-American System', in Coomans and Kamminga 141.

Cerone, J., 'The Application of Regional Human Rights Law Beyond Regional Frontiers: The Inter-American Commission on Human Rights and US Activities in Iraq', *ASIL Insights*, 25 October 2005, available at <http://www.asil.org/insights051025.cfm>.

—— 'Jurisdiction and Power: The Intersection of Human Rights Law and the Law of Non-International Armed Conflict in an Extraterritorial Context', (2007) 40 *Israel L Rev* 72.

Chesney, R., 'Case Note: *Boumediene v. Bush*', (2008) 102 *AJIL* 848.

Cleveland, S., 'Embedded International Law and the Constitution Abroad', (2010) 110 *Colum. L. Rev.* 225.

Conderman, P., 'Status of Armed Forces on Foreign Territory Agreements (SOFA)', in *Max Planck Encyclopedia of Public International Law*, available at <http://www.mpepil.com>.

Coomans, F., 'Some Remarks on the Extraterritorial Application of the International Covenant on Economic, Social and Cultural Rights', in Coomans and Kamminga 183.

Costa, J.-P., 'L'Etat, le territoire et la Convention européenne des droits de l'homme', in M. Kohen (ed.), *Promoting Justice, Human Rights and Conflict Resolution through International Law: Liber Amicorum Lucius Caflisch* (Nijhoff, 2007) 179.

Craven, M., 'Legal Differentiation and the Concept of the Human Rights Treaty in International Law', (2000) 11 *EJIL* 489.

Crawford, J., 'Multilateral Rights and Obligations in International Law', (2006) 319 *Recueil des Cours* 325.

Cross, M. and Williams, S., 'Between the Devil and the Deep Blue Sea: Conflicted Thinking in the *Al-Saadoon* Affair', (2009) 58 *ICLQ* 689.

Czubik, P. and Szwedo, P., 'Consular Jurisdiction', in *Max Planck Encyclopedia of Public International Law*, available at <http://www.mpepil.com>.

Davis, H., 'Public Authorities as "Victims" under the Human Rights Act', (2005) 64 *CLJ* 315.

de Hoogh, A., 'Articles 4 and 8 of the 2001 ILC Articles on State Responsibility, the Tadić Case and Attribution of Acts of Bosnian Serb Authorities to the Federal Republic of Yugoslavia', (2001) 72 *BYBIL* 255.

De Schutter, O., 'Globalization and Jurisdiction: Lessons from the European Convention on Human Rights', (2006) 6 *Baltic Yearbook of International Law* 183.

Del Mar, K., 'The Requirement of "Belonging" under International Humanitarian Law', (2010) 21 *EJIL* 105.

Dennis, M., 'Application of Human Rights Treaties Extraterritorially in Times of Armed Conflict and Military Occupation', (2005) 99 *AJIL* 119.

—— and Surena, A., 'Application of the International Covenant on Civil and Political Rights in Times of Armed Conflict and Military Occupation: The Gap Between Legal Theory and State Practice', (2008) *EHRLR* 714.

Dimitrijevic, V., 'Customary Law as an Instrument for the Protection of Human Rights', (2006) *ISPI* 7 *Working Papers*, available at <http://www.ispionline.it/it/documents/wp_7_2006.pdf>.

Doswald-Beck, L., 'International humanitarian law and the Advisory Opinion of the International Court of Justice on the Legality of the Threat or Use of Nuclear Weapons', (1997) 37 *IRRC* 35.

—— 'The Right to Life in Armed Conflict: Does International Humanitarian Law Provide All the Answers?', (2006) 88 *IRRC* 881.

Droege, C., 'Elective Affinities? Human Rights and Humanitarian Law', (2008) 90 *IRRC* 501.

Endicott, T., '*Habeas Corpus* and Guantánamo Bay: A View from Abroad', *American Journal of Jurisprudence* (forthcoming), draft available on SSRN at <http://ssrn.com/abstract=982412>.

Fallon, R. and Meltzer, D., 'Habeas Corpus Jurisdiction, Substantive Rights, and the War on Terror,' (2007) 120 *Harv. L. Rev.* 2029.

Gillard, E.C., 'International Humanitarian Law and Extraterritorial State Conduct,' in Coomans and Kamminga 25.

Goldstone, R. and Hamilton, R., '*Bosnia v. Serbia*: Lessons from the Encounter of the International Court of Justice with the International Criminal Tribunal for the Former Yugoslavia', (2008) 21 *Leiden JIL* 95.

Goodwin-Gill, G., 'The Extra-Territorial Reach of Human Rights Obligations: A Brief Perspective on the Link to Jurisdiction', in L. Boisson de Chazournes and M. Kohen (eds), *International Law and the Quest for its Implementation/Le droit international et la quête de sa mise en œvre, Liber Amicorum Vera Gowlland-Debbas* (Brill/Martinus Nijhoff, 2010) 293.

Greer, S., 'Reflections of a Former President of the European Court of Human Rights', (2010) *EHRLR* 169.

Guilfoyle, D., 'Counter-Piracy Law Enforcement and Human Rights', (2010) 59 *ICLQ* 141.

—— 'ECHR Rights at Sea: Medvedyev and others v. France', *EJIL: Talk!*, 19 April 2010, available at <http://www.ejiltalk.org/echr-rights-at-sea-medvedyev-and-others-v-france/>.

Hampson, F., 'The Relationship between International Humanitarian Law and Human Rights Law from the Perspective of a Treaty Body', (2008) 90 *IRRC* 549.

Hannum, H., 'The Status of the Universal Declaration of Human Rights in National and International Law', (1995–1996) 25 *Georgia J. Int'l L.* 28.

Happold, M., 'Banković v. Belgium and the Territorial Scope of the European Convention on Human Rights', (2003) 3 *HRLR* 77.

Henkin, L., 'The Constitution as Compact and as Conscience: Individual Rights abroad and at Our Gates', (1985–1986) 27 *Wm. & Mary L. Rev.* 11.

Higgins, R., 'A Babel of Judicial Voices? Ruminations from the Bench', (2006) 55 *ICLQ* 791.

Jenkins, D., 'Habeas Corpus and Extraterritorial Jurisdiction after *Boumediene*: Towards a Doctrine of 'Effective Control' in the United States', (2009) 9 *HRLR* 306.

Kavanagh, A., 'The Elusive Divide between Interpretation and Legislation under the Human Rights Act 1998', (2004) 24 *OJLS* 259.

Keitner, C., 'Rights Beyond Borders', (2011) 36 *Yale J. Int'l L.* (forthcoming), draft available on SSRN at <http://ssrn.com/abstract=1480886>.

Kent, J.A., 'A Textual and Historical Case Against a Global Constitution', (2007) 95 *Geo. L. J.* 463.

King, H., 'The Extraterritorial Human Rights Obligations of States', (2009) 9 *HRLR* 521.

—— 'Unravelling the Extraterritorial Riddle: An Analysis of *R (Hassan) v. Secretary of State for Defence*', (2009) 7 *JICJ* 633.

Kolb, R., 'Occupation in Iraq since 2003 and the Powers of the UN Security Council', (2008) 90 *IRRC* 29.

Krieger, H., 'A Conflict of Norms: The Relationship between Humanitarian Law and Human Rights Law in the ICRC Customary Law Study', (2006) 11 *JCSL* 265.

Künnemann, R., 'Extraterritorial Application of the International Covenant on Economic, Social and Cultural Rights', in Coomans and Kamminga 201.

Lawson, R., 'Life after *Banković*: On the Extraterritorial Application of the European Convention on Human Rights', in Coomans and Kamminga 83.

—— 'Really out of Sight? Issues of Jurisdiction and Control in Situations of Armed Conflict under the ECHR', in A. Buyse (ed.), *Margins of Conflict: The ECHR and Transitions to and from Armed Conflict* (Intersentia, 2010) 57.

Leach, P., 'The British Military in Iraq—the Applicability of the *Espace Juridique* Doctrine under the European Convention on Human Rights', (2005) *PL* 448.

—— 'The Chechen Conflict: Analysing the Oversight of the European Court of Human Rights', (2008) *EHRLR* 732.

Lindroos, A., 'Addressing Norm Conflicts in a Fragmented Legal System: The Doctrine of *Lex Specialis*', (2005) 74 *Nord J Int'l L* 27.

Loucaides, L., 'Determining the Extra-territorial Effect of the European Convention: Facts, Jurisprudence and the *Banković* Case', (2006) 4 *EHRLR* 391.

Lowe, V., 'Jurisdiction', in Evans 335.

Lubell, N., 'Challenges in Applying Human Rights Law to Armed Conflict', (2005) 87 *IRRC* 737.

Mann, F., 'The Doctrine of International Jurisdiction Revisited After Twenty Years', (1984-III) 186 *RdC* 9.

McCarthy, C., 'Human Rights and the Laws of War under the American Convention on Human Rights', (2008) *EHRLR* 762.

McCrudden, C., 'Human Dignity and Judicial Interpretation of Human Rights', (2008) 19 *EJIL* 655.

McGoldrick, D., 'Extraterritorial Application of the International Covenant on Civil and Political Rights', in Coomans and Kamminga 41.

—— 'Human Rights and Humanitarian Law in the UK Courts', (2007) 40 *Israel L Rev* 527.

Meron, T., 'Extraterritoriality of Human Rights Treaties', (1995) 89 *AJIL* 78.

—— 'The Humanization of Humanitarian Law', (2000) 94 *AJIL* 243.

Milanovic, M., 'State Responsibility for Genocide', (2006) 17 *EJIL* 553.

—— 'Lessons for Human Rights and Humanitarian Law in the War on Terror: Comparing *Hamdan* and the Israeli *Targeted Killings* Case', (2007) 89 *IRRC* 373.

—— 'State Responsibility for Genocide: A Follow-Up', (2007) 18 *EJIL* 669.

—— 'From Compromise to Principle: Clarifying the Concept of State Jurisdiction in Human Rights Treaties', (2008) 8 *HRLR* 411.

—— 'Is Gaza Still Occupied by Israel?', *EJIL: Talk!*, 1 March 2009, available at <http://www.ejiltalk.org/is-gaza-still-occupied-by-israel/>.

—— 'Norm Conflict in International Law: Whither Human Rights?', (2009) 20 *Duke J. Comp. & Int'l L.* 69.

—— 'A Norm Conflict Perspective on the Relationship between International Humanitarian Law and Human Rights Law', (2009) 14 *JCSL* 459.

—— 'Territorial Application of the Convention and State Succession', in P. Gaeta, edn, *The UN Genocide Convention: A Commentary* (Oxford University Press, 2009) 473.

—— 'Al-Saadoon and Mufdhi Merits Judgment', *EJIL: Talk!*, 2 March 2010, available at <http://www.ejiltalk.org/al-saadoon-and-mufdhi-merits-judgment/>.

—— 'Grand Chamber Hearings and Preview of Al-Skeini and Al-Jedda', *EJIL: Talk!*, 9 June 2010, available at <http://www.ejiltalk.org/grand-chamber-hearings-and-preview-of-al-skeini-and-al-jedda/>.

—— 'More on Drones, Self-Defense, and the Alston Report on Targeted Killings', *EJIL: Talk!*, 5 June 2010, available at <http://www.ejiltalk.org/more-on-drones-self-defense-and-the-alston-report-on-targeted-killings/>.

—— 'What Exactly Internationalizes an Internal Armed Conflict?', *EJIL: Talk!*, 7 May 2010, available at <http://www.ejiltalk.org/what-exactly-internationalizes-an-internal-armed-conflict/>.

—— and Papic, T., 'As Bad As It Gets: The European Court of Human Rights' *Behrami and Saramati* Decision and General International Law', (2009) 58 *ICLQ* 267.

Miller, S., 'Revisiting Extraterritorial Jurisdiction: A Territorial Justification for Extraterritorial Jurisdiction under the European Convention', (2010) 20 *EJIL* 1223.

Modirzadeh, N., 'The Dark Sides of Convergence: A Pro-civilian Critique of the Extraterritorial Application of Human Rights Law in Armed Conflict', (2010) 86 *U.S. Naval War College International Law Studies (Blue Book) Series* 349.

Moor, L. and Simpson, A.W.B., 'Ghosts of Colonialism in the European Convention on Human Rights', (2006) 76 *BYBIL* 121.

Mujezinovic Larsen, K., 'Attribution of Conduct in Peace Operations: The 'Ultimate Authority and Control' Test', (2008) 19 *EJIL* 509.

—— '"Territorial Non-Application" of the European Convention on Human Rights', (2009) 78 *Nord J Int'l L* 73.

Neuman, G., 'Whose Constitution?', (1991) 100 *Yale L.J.* 909.

—— 'The Extraterritorial Constitution after *Boumediene v. Bush*', (2009) 82 *So. Cal. L. Rev.* 259.

—— 'Understanding Global Due Process', (2009) 23 *Geo. Immigr. L.J.* 365.

Nowak, M., 'Obligation of States to Prevent and Prohibit Torture in an Extraterritorial Perspective', in Gibney and Skogley 11.

O'Boyle, M., 'The European Convention on Human Rights and Extraterritorial Jurisdiction: A Comment on "Life After Banković"', in Coomans and Kamminga 125.

O'Keefe, R., 'Universal Jurisdiction: Clarifying the Basic Concept', (2004) 2 *JICJ* 735.

Oliver, D., 'The Frontiers of the State: Public Authorities and Public Functions under the Human Rights Act', (2000) *PL* 476.

Orakhelashvili, A., 'Restrictive Interpretation of Human Rights Treaties in the Recent Jurisprudence of the European Court of Human Rights', (2003) 14 *EJIL* 529.

Oxman, B., 'Jurisdiction of States' in *Max Planck Encyclopedia of Public International Law*, available at <http://www.mpepil.com>.

Pedersen, M., 'Territorial Jurisdiction in Article 1 of the European Convention on Human Rights', (2004) 73 *Nord J Int'l L* 279.

Pejic, J., 'Procedural Principles and Safeguards for Internment/Administrative Detention in Armed Conflict and Other Situations of Violence', (2005) 87 *IRRC* 375.

Piret, J.M., '*Boumediene v. Bush* and the Extraterritorial Reach of the US Constitution: A Step Towards Judicial Cosmopolitanism?', (2008) 4 *Utrecht Law Review* 81.

Prud'homme, N., '*Lex Specialis*: Oversimplifying a More Complex and Multifaceted Relationship?', (2007) 40 *Israel L Rev* 355.

Ress, G., 'Problems of extraterritorial human rights violations : the jurisdiction of the European Court of Human Rights : the Banković case', (2002) 12 *Italian Yearbook of International Law* 51.

Roberts, A., 'Transformative Military Occupation: Applying the Laws of War and Human Rights', (2006) 100 *AJIL* 580.

Rodley, N., 'The Extraterritorial Reach and Applicability in Armed Conflict of the International Covenant on Civil and Political Rights: a Rejoinder to Dennis and Surena', (2009) *EHRLR* 628.

Roosevelt, K., 'Detention and Interrogation in the Post-9/11 World', (2008) 42 *Suffolk U. L. Rev.* 1.

Ross, J., 'Jurisdictional Aspects of International Human Rights and Humanitarian Law in the War on Terror', in Coomans and Kamminga 9.

Roxstrom, E., Gibney, M., and Einarsen, T., 'The NATO Bombing Case (*Bankovic et al. v. Belgium et al.*) and the Limits of Western Human Rights Protection', (2005) 23 *B.U. Int'l L.J.* 55.

Ruth, A. and Trilsch, M., 'Banković v. Belgium (Admissibility)', (2003) 97 *AJIL* 168.

Sari, A., 'Jurisdiction and International Responsibility in Peace Support Operations: The Behrami and Saramati Cases', (2008) 8 *HRLR* 151.

Sassoli, M., 'Legislation and Maintenance of Public Order and Civil Life by Occupying Powers', (2005) 16 *EJIL* 661.

—— and Olson, L., 'The Relationship between International Humanitarian and Human Rights Law where it Matters: Admissible Killing and Internment of Fighters in Non-International Armed Conflicts', (2008) 90 *IRRC* 599.

Satterthwaite, M., 'Rendered Meaningless: Extraordinary Rendition and the Rule of Law', (2007) 75 *Geo. Wash. L. Rev.* 1333.

Schabas, W., '*Lex Specialis?* Belts and Suspenders? The Parallel Operation of Human Rights Law and the Law of Armed Conflict, and the Conundrum of *Jus ad Bellum*', (2007) 40 *Israel L Rev* 592.

Scheinin, M., 'Extraterritorial Effect of the International Covenant on Civil and Political Rights', in Coomans and Kamminga 73.

Schwelb, E., 'The Trieste Settlement and Human Rights', (1955) 49 *AJIL* 240.

Shany, Y., 'The Law Applicable to Non-Occupied Gaza: A Comment on *Bassiouni v. Prime Minister of Israel*', *Hebrew University International Law Research Paper No. 13-09*, 27 February 2009, available at <http://ssrn.com/abstract=1350307>.

Silverstone, B., 'R. (on the application of Smith) v Oxfordshire Assistant Deputy Coroner: Human Rights and the Armed Forces', (2009) 4 *EHRLR* 566.

Simma, B., 'International Human Rights and General International Law: A Comparative Analysis', (1995) 4 *Collected Courses of the Academy of European Law* 153.

Simpson, G., 'The Death of Baha Mousa', (2007) 8 *Melbourne JIL* 340.

Sloane, R., 'The Cost of Conflation: Preserving the Dualism of Jus Ad Bellum and Jus in Bello in the Contemporary Law of War', (2009) 34 *Yale J. Int'l L.* 47.

Spinedi, M., 'On the Non-Attribution of the Bosnian Serbs' Conduct to Serbia', (2007) 5 *JICJ* 829.

Stribopoulos, J., 'The Charter's Unstated Territorial Limits: *R. v. Hape*', *The Court*, 8 June 2007, available at <http://www.thecourt.ca/2007/06/08/the-charters-unstated-territorial-limits-r-v-hape/>.

Sudre, F., 'Les obligations positives dans la jurisprudence européenne des droits de l'homme', (1995) *Revue trimestrielle des droits de l'homme* 363.

Talmon, S., 'The Responsibility of Outside Powers for Acts of Secessionist Entities', (2009) 58 *ICLQ* 493.

Thienel, T., 'Cooperation in Iraq and the ECHR: An Awful Epilogue', *Invisible College*, 21 January 2009, available at <http://invisiblecollege.weblog.leidenuniv.nl/2009/01/21/cooperation-in-iraq-and-the-echr-an-awfu>.

—— 'Provisional Measures in the "*Case concerning Application of the International Convention on the Elimination of All Forms of Racial Discrimination" (Georgia v. Russian Federation)*', (2009) 1 *Goettingen JIL* 143.

—— 'The ECHR in Iraq', (2008) 6 *JICJ* 115.

—— 'Oral Argument in *Medvedyev v. France*', *Invisible College Blog*, 13 May 2008, available at <http://invisiblecollege.weblog.leidenuniv.nl/2008/05/13/oral-argument-in-medvedyev-v-france>.

—— 'The Georgian Conflict, Racial Discrimination and the ICJ: The Order on Provisional Measures of 15 October 2008', (2009) 9 *HRLR* 465.

Verdier, P.H., '*R. v. Hape*', (2008) 102 *AJIL* 143.

Verdirame, G., 'Human Rights in Wartime: A Framework for Analysis', (2008) *EHRLR* 689.

Verdross, A., '*Jus Dispositivum* and *Jus Cogens* in International Law', (1966) 60 *AJIL* 55.

Wilde, R., 'The "Legal Space" or "Espace Juridique" of the European Convention on Human Rights: Is It Relevant to Extraterritorial State Action?', (2005) 10 *EHRLR* 115.

—— 'Legal "Black Hole"? Extraterritorial State Action and International Treaty Law on Civil and Political Rights', (2005) 26 *Mich. J. Int'l L.* 739.

—— 'Triggering State Obligations Extraterritorially: The Spatial Test in Certain Human Rights Treaties', (2007) 40 *Israel L Rev* 503.

—— 'Complementing Occupation Law? Selective Judicial Treatment of the Suitability of Human Rights Norms', (2009) 42 *Israel L Rev* 80.

—— 'Compliance with human rights norms extraterritorially: "human rights imperialism"?', in L. Boisson de Chazournes and M. Kohen (eds), *International Law and the Quest for its Implementation/Le droit international et la quête de sa mise en œvre, Liber Amicorum Vera Gowlland-Debbas* (Brill/Martinus Nijhoff, 2010) 319.

Wouters, K, and Den Heijer, M., 'The *Marine I* Case: A Comment', (2009) 22 *International Journal of Refugee Law* 1.

Zimmermann, A., 'Extraterritorial Application of Human Rights Treaties—The Case of Israel and the Palestinian Territories Revisited', in I. Buffard, J. Crawford, A. Pellet, and S. Wittich (eds), *International Law Between Universalism and Fragmentation: Festschrift in Honour of Gerhard Hafner* (Brill, 2008) 747.

Index